Leszek Rutkowski

New Soft Computing Techniques for System Modelling, Pattern Classification and Image Processing

Springer-Verlag Berlin Heidelberg GmbH

Studies in Fuzziness and Soft Computing, Volume 143

Editor-in-chief
Prof. Janusz Kacprzyk
Systems Research Institute
Polish Academy of Sciences
ul. Newelska 6
01-447 Warsaw
Poland
E-mail: kacprzyk@ibspan.waw.pl

Further volumes of this series can be found on our homepage: springeronline.com

Vol. 124. X. Yu, J. Kacprzyk (Eds.)
Applied Decision Support with Soft Computing, 2003
ISBN 3-540-02491-3

Vol. 125. M. Inuiguchi, S. Hirano and S. Tsumoto (Eds.)
Rough Set Theory and Granular Computing, 2003
ISBN 3-540-00574-9

Vol. 126. J.-L. Verdegay (Ed.)
Fuzzy Sets Based Heuristics for Optimization, 2003
ISBN 3-540-00551-X

Vol 127. L. Reznik, V. Kreinovich (Eds.)
Soft Computing in Measurement and Information Acquisition, 2003
ISBN 3-540-00246-4

Vol 128. J. Casillas, O. Cordón, F. Herrera, L. Magdalena (Eds.)
Interpretability Issues in Fuzzy Modeling, 2003
ISBN 3-540-02932-X

Vol 129. J. Casillas, O. Cordón, F. Herrera, L. Magdalena (Eds.)
Accuracy Improvements in Linguistic Fuzzy Modeling, 2003
ISBN 3-540-02933-8

Vol 130. P.S. Nair
Uncertainty in Multi-Source Databases, 2003
ISBN 3-540-03242-8

Vol 131. J.N. Mordeson, D.S. Malik, N. Kuroki
Fuzzy Semigroups, 2003
ISBN 3-540-03243-6

Vol 132. Y. Xu, D. Ruan, K. Qin, J. Liu
Lattice-Valued Logic, 2003
ISBN 3-540-40175-X

Vol. 133. Z.-Q. Liu, J. Cai, R. Buse
Handwriting Recognition, 2003
ISBN 3-540-40177-6

Vol 134. V.A. Niskanen
Soft Computing Methods in Human Sciences, 2004
ISBN 3-540-00466-1

Vol. 135. J.J. Buckley
Fuzzy Probabilities and Fuzzy Sets for Web Planning, 2004
ISBN 3-540-00473-4

Vol. 136. L. Wang (Ed.)
Soft Computing in Communications, 2004
ISBN 3-540-40575-5

Vol. 137. V. Loia, M. Nikravesh, L.A. Zadeh (Eds.)
Fuzzy Logic and the Internet, 2004
ISBN 3-540-20180-7

Vol. 138. S. Sirmakessis (Ed.)
Text Mining and its Applications, 2004
ISBN 3-540-20238-2

Vol. 139. M. Nikravesh, B. Azvine, I. Yager, L.A. Zadeh (Eds.)
Enhancing the Power of the Internet, 2004
ISBN 3-540-20237-4

Vol. 140. A. Abraham, L.C. Jain, B.J. van der Zwaag (Eds.)
Innovations in Intelligent Systems, 2004
ISBN 3-540-20265-X

Vol. 141. G.C. Onwubolu, B.V. Babu
New Optimzation Techniques in Engineering, 2004
ISBN 3-540-20167-X

Vol. 142. M. Nikravesh, L.A. Zadeh, V. Korotkikh (Eds.)
Fuzzy Partial Differential Equations and Relational Equations, 2004
ISBN 3-540-20322-2

Leszek Rutkowski

New Soft Computing Techniques for System Modelling, Pattern Classification and Image Processing

Springer

Prof. Ph. D., D. Sc. Leszek Rutkowski
Department of Computer Engineering
Technical University of Czestochowa
Armii Krajowej 36
42-200 Czestochowa
Poland
E-mail: lrutko@kik.pcz.czest.pl

DOI 10.1007/978-3-540-40046-2

Library of Congress Cataloging-in-Publication-Data

Rutkowski, Leszek.
New soft computing techniques for system modelling, pattern classification and image processing / Leszek Rutkowski.
p. cm. -- (Studies in fuzziness and soft computing ; v. 143)
Includes bibliographical references and index.

1. Soft computing. 2. Neural networks (Computer science) I. Title. II. Series.
QA76.9.S63R89 2004
006.3--dc22

springeronline.com

Originally published by Springer-Verlag Berlin Heidelberg New York in 2004
MyCopy version of the original edition 2004

Cover design: E. Kirchner, Springer-Verlag, Heidelberg
Printed on acid free paper 62/3020/M - 5 4 3 2 1 0
www.springer.com/mycopy

Preface

Science has made great progress in the twentieth century, with the establishment of proper disciplines in the fields of physics, computer science, molecular biology, and many others. At the same time, there have also emerged many engineering ideas that are interdisciplinary in nature, beyond the realm of such orthodox disciplines. These include, for example, artificial intelligence, fuzzy logic, artificial neural networks, evolutional computation, data mining, and so on. In order to generate new technology that is truly human-friendly in the twenty-first century, integration of various methods beyond specific disciplines is required.

Soft computing is a key concept for the creation of such human-friendly technology in our modern information society. Professor Rutkowski is a pioneer in this field, having devoted himself for many years to publishing a large variety of original work. The present volume, based mostly on his own work, is a milestone in the development of soft computing, integrating various disciplines from the fields of information science and engineering. The book consists of three parts, the first of which is devoted to probabilistic neural networks. Neural excitation is stochastic, so it is natural to investigate the Bayesian properties of connectionist structures developed by Professor Rutkowski. This new approach has proven to be particularly useful for handling regression and classification problems in time-varying environments. Throughout this book, major themes are selected from theoretical subjects that are tightly connected with challenging applications.

The second part of the book explains the technique of vector quantization (VQ), which is a powerful tool for understanding the hidden structure in observed signals, and it plays a fundamental role in data-mining, signal processing and neural learning. A new, more general scope of VP is proposed for image compression. The third part analyzes various types of least square techniques from the point of view of fundamental learning properties for neural network design. Professor Rutkowski goes on to discuss hardware implementation using systolic technology.

The book has a clear motivation, and is sure to have a great impact on the field of soft computing. It is my pleasure to have benefited from the enthusiasm and energy with which Professor Rutkowski has devoted himself to this new emerging field.

August, 2003, Tokyo, Japan

Shun-ichi Amari
Director, RIKEN Brain Science Institute

Contents

1
Introduction

In the last decade we have been witnessing a re-orientation of traditional artificial intelligence methods towards soft computing techniques (see e.g. [10], [133], [319], [320]). This phenomena allows to solve difficult problems concerned with robotics, computer vision, speech recognition and machine translation. It was noted by Zadeh [319], [320] that soft computing techniques are characterized by tolerance for imprecision, uncertainty, and partial truth to achieve tractability, robustness and low solution cost. The main components of soft computing are (below we cite only pioneering works and excellent monographs):

a) fuzzy logic [129], [184], [207], [305] – [326] – a leading constituent of soft computing characterized by natural language description,

b) neural networks [11] – [23], [109], [110], [206], [269], [270], [331] – characterized by learning capabilities,

c) evolutionary computing [85], [96], [116] – characterized by global optimization properties,

d) rough sets [181] – [183], [247] – characterized by attribute reduction properties,

e) uncertain variables [39] – [43] – characterized by a remarkable capability to describe decision making problems,

f) probabilistic techniques [10], [79], [81], [175] – characterized by a rigorous framework for the representation of probabilistic knowledge and modelling of random phenomena.

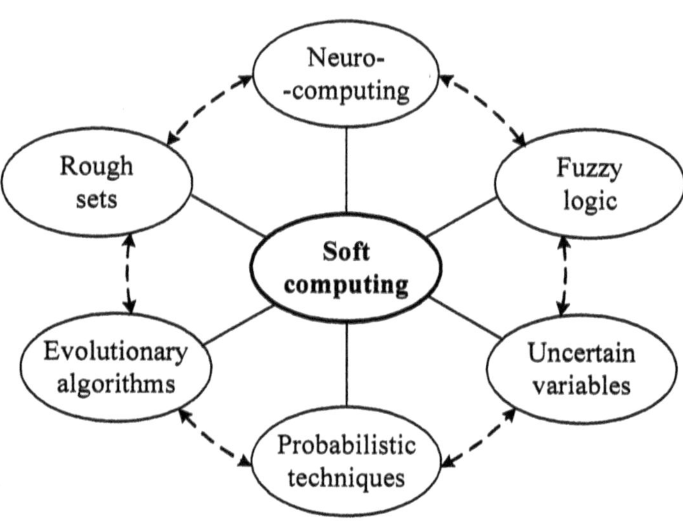

FIGURE 1.1. Soft computing techniques

In Fig. 1.1 we present the main components of soft computing. The constituent methodologies of soft computing are complementary and synergistic rather than competitive (see Zadeh [323]). Therefore, several intelligent combinations of soft computing techniques have been developed (see e.g. [10]). Among them the most popular are neuro-fuzzy, neuro-genetic and fuzzy-genetic combinations. Moreover, sometimes different soft computing techniques are based on similar or equivalent underlying mathematics. Typical examples include statistical classifiers which have a fuzzy interpretation (see [144]), probabilistic techniques which are translated into neural network structures [256] and neural networks with radial bases functions corresponding to a fuzzy system [133].

This book presents original concepts in selected soft computing methodologies and summarizes previous works of the author published among others in IEEE Transactions on Neural Networks, IEEE

Transactions on Signal Processing, IEEE Transactions on Systems, Man and Cybernetics, IEEE Transactions on Information Theory, IEEE Transactions on Automatic Control, IEEE Transactions on Pattern Analysis and Machine Intelligence, IEEE Transactions on Circuits and Systems and Proceedings of the IEEE.

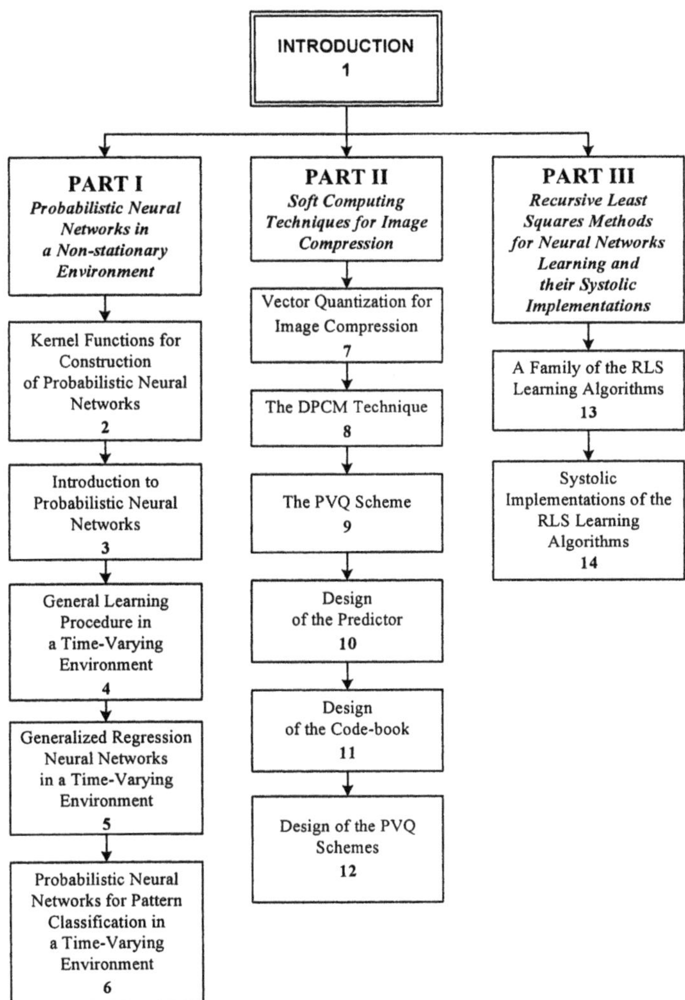

FIGURE 1.2. Contents of the book

The book is divided into three major parts (see Fig. 1.2):

a) Part I (Chapters 2 – 6) – Probabilistic Neural Networks in a Non-stationary Environment

b) Part II (Chapters 7 – 12) – Soft Computing Techniques for Image Compression

c) Part III Chapters 13 – 14) – Recursive Least Squares Methods for Neural Networks Learning and Their Systolic Implementations.

In Part I we present original results concerning probabilistic neural networks working in a time-varying environment. Although mathematics of probabilistic neural networks (PNN) have been developed in the sixties and seventies (see e.g. [45], [178]) and next rediscovered by Specht [256], [257], no one prior to the author's work had developed PNN working in a time-varying environment. The current state of knowledge as regards non-stationary processes is significantly poorer than that in the area of non-stationary signals. Many signals are treated as stationary only because in this way it is easier to analyse them; in fact, they are non-stationary. Non-stationary processes are undoubtedly more difficult to analyse and their diversity makes the application of universal tools impossible. On the other hand a lot of phenomena have a non-stationary character, e.g. the conventer-oxygen process of steelmaking, the change of catalyst properties in an oil refinery or in the process of carbon dioxide conversion.

In Part I we derive general regression neural networks (GRNN) that are able to follow changes of the best models described by time-varying regression functions. Moreover, we formulate the problem of pattern classification in a non-stationary environment as a prediction problem and we design a probabilistic neural network to classify patterns having time-varying probability distributions. Part I is organized into 6 chapters. In Chapter 2 we present kernel functions on which the construction of PNN will be based. Chapter 3 is a short introduction to PNN in a stationary environment. Moreover, in this chapter we extend the idea of the classical PNN to the recursive PNN with a gain $\frac{1}{n}$. In the following chapters we replace the gain $\frac{1}{n}$ by a more general a_n (like in stochastic approximation methods) in order to enhance the recursive PNN for tracking non-stationary signals.

Since the existing theories do not allow to analyze "enhanced" recursive PNN in a time-varying environment, in Chapter 4 we present appropriate theorems which are very useful in the next chapters. In Chapter 5 we present the GRNN working in a non-stationary environment. We formulate the theorem for the convergence of the GRNN in probability and with probability one to optimal characteristics. We design the GRNN tracking various non-stationarities. In Chapter 6 we describe the problem of pattern classification in a time-varying environment. Estimates of time-varying discriminant functions are presented and classification rules are proposed. It is shown that our PNN approach Bayes (time-varying) decision discriminant functions. Moreover, we investigate the speed of convergence of classification rules. The PNN based on the Parzen kernel and the orthogonal series kernel are discussed in detail. A specific case of non-stationarity of the "movable argument" type is also elaborated. The theorems and corollaries presented in Part I are proven in Appendices A, B and C.

Part II covers fundamental concepts of vector quantization (VQ) and image compression. It is well known that soft computing is of great importance for data compression. The applications include facsimile transmission, teleconferencing, HDTV, audio recording, image compression and image understanding. In Part II we present original research results concerning the predictive vector quantization (PVQ) which combines two techniques, i.e.

$$\text{PVQ} = \text{DPCM} + \text{VQ}$$

where VQ stands for Vector Quantization and DPCM stands for Differential Pulse Code Modulation. In Chapters 7 – 12 we show how to design the PVQ scheme and its major components: the code-book and the predictor. The code-book is based on a competitive neural network, whereas the predictor is presented in two versions: a) based on parametric models and statistical estimates of matrix coefficients, b) based on neural networks. For the first time in literature a detailed description of four PVQ design schemes is studied: the "open-loop" design, the "closed-loop" design, the modified "closed-loop" design and the neural design. The schemes are illustrated by a comprehensive simulation study.

In Part III we develop a family of neural network learning algorithms based on the recursive least square procedures adopted from the filters theory [265]. In Chapter 13 we present the classical RLS algorithm and its robust counterparts, called QQ^T– RLS and UD – RLS algorithms, which are less liable to round-up error accumulation within the classical RLS algorithm. The learning algorithms will be derived under the assumption that each neuron in every layer receives inputs from neurons in all previous layers (not only from the directly preceding layer). We will investigate and demonstrate increased computational abilities of the pyramid neural network structure. In Chapter 14 we present systolic architectures corresponding to the RLS learning algorithms developed in Chapter 13. A systolic array is a network of processors which rhythmically compute and pass data through the system. Since it operates like the pumping action of the human heart, the name "systolic" is commonly used. The systolic array can be directly implemented in the VLSI system. Therefore, systolic arrays proposed in Chapter 14 are of great potential use for design of neuro-computer architectures.

The author gratefully acknowledges the material quoted from his previous works published by IEEE, Springer-Verlag and Elsevier Science. For a complete listing of quoted articles the reader is referred to the "References". The author also gratefully acknowledges Jarosław Bilski, Robert Cierniak and Jacek Smoląg, his former Ph.D. students, for material being a part of our joint research and included in Part II and Part III of the book. I am also very grateful to Prof. Janusz Kacprzyk, the Editor of the book series "Studies in Fuzziness and Soft Computing", for his encouragement to publish the book.

Part I

Probabilistic Neural Networks in a Non-stationary Environment

2

Kernel Functions for Construction of Probabilistic Neural Networks

2.1 Introduction

All probabilistic neural networks studied in this book are based on a sequence $\{K_n\}$, $n = 1, 2, ...,$ of bivariate Borel – measurable functions (so-called general kernel functions) defined on $A \times A$, $A \subset R^p$, $p \geq 1$. The concept of general kernel functions stems from the theory of nonparametric density estimation and was first suggested in [87].

In the next sections we present examples of functions K_n on which the construction of probabilistic neural networks is based. We will use ideas of the two methods: Parzen's approach and orthogonal series.

2.2 Application of the Parzen kernel

Sequence K_n based on the Parzen kernel in the multi-dimensional version takes the following form:

$$K_n(x,u) = h_n^{-p} K\left(\frac{x-u}{h_n}\right) \tag{2.1}$$

where h_n is a certain sequence of numbers and K is an appropriately selected function. Precise assumptions concerning sequence h_n

and function K that ensure the convergence of probabilistic neural networks will be given in Chapters 3, 5 and 6. It is convenient to assume that function K can be presented in the form

$$K(x) = \prod_{i=1}^{p} H\left(x^{(i)}\right) \tag{2.2}$$

Then, sequence K_n is expressed by means of formula

$$K_n(x, u) = h_n^{-p} \prod_{i=1}^{p} H\left(\frac{x^{(i)} - u^{(i)}}{h_n}\right) \tag{2.3}$$

Examples of functions H and K are given in Table 2.1.

TABLE 2.1. Examples of functions H and K

Kernel type	Definition $H(u)$ $p = 1$	Definition $p > 1$
1) Uniform	$\frac{1}{2}$ if $\lvert u\rvert \le 1$ 0 if $\lvert u\rvert > 1$	2^{-p} if $\lvert u^{(i)}\rvert \le 1,\ i = 1, ..., p$ 0 otherwise
2) Triangular	$1 - \lvert u\rvert$ if $\lvert u\rvert \le 1$ 0 if $\lvert u\rvert > 1$	$\prod_{i=1}^{p}\left[1 - \lvert u^{(i)}\rvert\right]$ if $\lvert u^{(i)}\rvert \le 1$ 0 otherwise
3) Gaussian	$(2\pi)^{-\frac{1}{2}} e^{-\frac{1}{2}u^2}$	$(2\pi)^{-\frac{p}{2}} e^{-\frac{1}{2}\lVert u\rVert^2}$, $\lVert u\rVert^2 = u^T u$
4) Picard	$\frac{1}{2}e^{-\lvert u\rvert}$	$2^{-p}e^{-\lVert u\rVert}$, $\lVert u\rVert = \sum_{i=1}^{p}\lvert u^{(i)}\rvert$
5) Cauchy	$\pi^{-1}\left(1 + u^2\right)^{-1}$	$\pi^{-p}\prod_{i=1}^{p}\left(1 + \lvert u^{(i)}\rvert^2\right)^{-1}$
6) Féjér de Valée Pouissin	$(2\pi)^{-1}\left(\frac{\sin\frac{u}{2}}{\frac{u}{2}}\right)^2$	$(2\pi)^{-p}\prod_{i=1}^{p}\left(\frac{\sin\frac{u^{(i)}}{2}}{\frac{u^{(i)}}{2}}\right)^2$
7) Parabolic	$\frac{3}{4\sqrt{5}}\left(1 - \frac{u^2}{5}\right)$ if $\lvert u\rvert \le \sqrt{5}$ 0 if $\lvert u\rvert > \sqrt{5}$	$\left(\frac{3}{4\sqrt{5}}\right)^p \prod_{i=1}^{p}\left(1 - \frac{u^{(i)2}}{5}\right)$ if $\lvert u^{(i)}\rvert \le \sqrt{5},\ i = 1, ..., p$ 0 otherwise
8) Mexican hat	$H(u) = \frac{3}{2\sqrt{2\pi}}$ $\cdot\left(1 - \frac{u^2}{3}\right)e^{-\frac{1}{2}u^2}$	$\frac{3}{2\sqrt{2\pi}}\prod_{i=1}^{p}\left(1 - \frac{u^{2(i)}}{3}\right)e^{-\frac{1}{2}u^{2(i)}}$

2.3 Application of the orthogonal series

Let $g_j(.)$, $j = 0, 1, 2, ...,$ be a complete orthonormal system in $L_2(\Delta)$, $\Delta \in R$. Then, as it is known [173], the system composed of all possible products

$$\left\{ \begin{array}{c} \Psi_{j_1,...,j_p}\left(x^{(1)},...,x^{(p)}\right) = g_{j_1}\left(x^{(1)}\right)...g_{j_p}\left(x^{(p)}\right) \\ j_k = 0, 1, 2, ..., \ k = 1, ..., p \end{array} \right\} \tag{2.4}$$

is a complete orthonormal system in $L_2(A)$, where

$$A = \underbrace{\Delta \times ... \times \Delta}_{p-\text{times}}$$

It constitutes the basis for construction of the following sequence K_n:

$$K_n(x,u) = \sum_{j_1=0}^{q} \cdots \sum_{j_p=0}^{q} g_{j_1}\left(x^{(1)}\right) \tag{2.5}$$

$$\cdot g_{j_1}\left(u^{(1)}\right) \cdots g_{j_p}\left(x^{(p)}\right) g_{j_p}\left(u^{(p)}\right)$$

where q depends on the length of the learning sequence, i.e. $q = q(n)$. It can be given in a shortened form as

$$K_n(x,u) = \sum_{|\underline{j}| \le q} \Psi_{\underline{j}}(x)\, \Psi_{\underline{j}}(u) \tag{2.6}$$

where $\underline{j} = (j_1, ..., \ j_p)$ and $|\underline{j}| = \max_{1 \le k \le p}(j_k)$

Remark 2.1
In some applications, better asymptotic properties of probabilistic neural networks are obtained through a slight modification of formula (2.6):

$$K_n(x,u) = \sum_{|\underline{j}| \le q} \prod_{k=1}^{p} \left(1 - \frac{|j_k|}{q+1}\right) \Psi_{\underline{j}}(x)\, \Psi_{\underline{j}}(u) \tag{2.7}$$

In order to construct kernel (2.7), the idea of the so-called Cesaro averages [268], known in the theory of orthogonal series, was used. From (2.4) it follows that for the construction of an orthonormal function system of many variables it is enough to know an orthonormal function system of one variable.

Now, we will present 5 basic orthonormal systems of one variable.

a) Hermite orthonormal system

Hermite orthonormal system has the form

$$g_j(x) = \left(2^j j\,!\,\pi^{\frac{1}{2}}\right)^{-\frac{1}{2}e-\frac{x^2}{2}} H_j(x) \tag{2.8}$$

where

$$H_0(x) = 1,\ H_j(x) = (-1)^j e^{x^2} \frac{d_j}{dx^j} e^{-x^2},\ j = 1, 2, ... \tag{2.9}$$

are Hermite polynomials on a straight line, i.e. $\Delta = (-\infty, \infty)$. Functions g_j of this system are bounded as follows [268]

$$\max_x |g_j(x)| \le c_1\, j^{-\frac{1}{12}},\ j = 1, 2, ... \tag{2.10}$$

b) Laguerre orthonormal system

Laguerre orthonormal system has the form

$$g_j(x) = e^{-\frac{x}{2}} L_j(x) \tag{2.11}$$

where

$$L_0(x) = 1,\ L_j(x) = (j!)^{-1} e^x \frac{d_j}{dx^j}\left(x^j e^{-x}\right),\quad j = 1, 2, ... \tag{2.12}$$

are Laguerre polynomials, where $\Delta = [0, \infty)$. Functions g_j of this system are bounded as follows [268]

$$\max_x |g_j(x)| \le c_2\, j^{-\frac{1}{4}},\quad j = 1, 2, ... \tag{2.13}$$

c) Fourier orthonormal system has the form

$$\frac{1}{\sqrt{b-a}}, \sqrt{\frac{2}{b-a}} \cos 2\pi j \frac{x-a}{b-a}, \sqrt{\frac{2}{b-a}} \sin 2\pi j \frac{x-a}{b-a} \tag{2.14}$$

for $j = 1, 2, ...$ and $\Delta = [a, b] \subset R$. There is an obvious inequality

$$\max_x |g_j(x)| \le \text{const.} \tag{2.15}$$

d) Legendre orthonormal system has the form

$$g_j(x) = \left(j + \frac{1}{2}\right)^{\frac{1}{2}} P_j(x) \tag{2.16}$$

where

$$P_0(x) = 1, \quad P_j(x) = \left(2^j j!\right)^{-1} \frac{d^j}{dx^j}\left(x^2 - 1\right)^j \tag{2.17}$$

for $j = 1, 2, \ldots$ are Legendre polynomials, where $\Delta = [-1, 1]$. Functions g_j of this system are bounded as follows [237]

$$\max_x |g_j(x)| \le c_3\, j^{\frac{1}{2}}, \quad j = 1, 2, \ldots \tag{2.18}$$

e) Haar orthonormal system has the form [8]

$$g_m^k(x) = \begin{cases} 2^m & \text{for} \quad \frac{2k-2}{2^{m+1}} < x < \frac{2k-1}{2^{m+1}} \\ -2^m & \text{for} \quad \frac{2k-1}{2^{m+1}} < x < \frac{2k}{2^{m+1}} \\ 0 & \text{for} \quad \text{others } x \in [0, 1] \end{cases} \tag{2.19}$$

where $m = 0, 1, \ldots$ and $k = 1, \ldots, 2^m$. Haar functions can be numbered with one index, defining

$$g_j(x) = g_m^k(x) \tag{2.20}$$

where $j = 2^m + k - 1$ and $k = 1, \ldots, 2^m$.
From the construction of the system it follows that

$$\max_x |g_j(x)| \le c_4\, j^{\frac{1}{2}} \tag{2.21}$$

Remark 2.2
Inequalities (2.10), (2.13), (2.15), (2.18) and (2.21) can be expressed in a shortened form as

$$\max_x |g_j(x)| \le G_j \tag{2.22}$$

where $G_j = \text{const. } j^d$, and

$$d = \begin{cases} -\frac{1}{12} & \text{for the Hermite system} \\ -\frac{1}{4} & \text{for the Laguerre system} \\ 0 & \text{for the Fourier system} \\ \frac{1}{2} & \text{for the Legendre system} \\ \frac{1}{2} & \text{for the Haar system} \end{cases} \tag{2.23}$$

Let us notice that

$$\sum_{j=0}^{q} G_j = O\left(q^{d+1}\right) \tag{2.24}$$

and

$$\sum_{j=0}^{q} G_j^2 = 0\left(q^{2d+1}\right) \tag{2.25}$$

Formulas (2.24) and (2.25) will be helpful to read the conditions of the convergence of algorithms constructed in Chapters 5 and 6, based on the above orthonormal systems.

Remark 2.3
Hermite, Laguerre and Legendre polynomials satisfy the following recurrent dependencies [268]:
1)

$$H_{j+1}(x) = -2x\ H_j(x) - 2jH_{j-1}(x), \quad \text{for } j = 1, 2, ..., \tag{2.26}$$

where $H_0(x) = 1,\ H_1(x) = -2x$
2)

$$L_{j+1}(x) = \left((2j+1-x)L_j(x) - jL_{j-1}(x)\right) / \ / (j+1), \text{ for } j = 1, 2, ..., \tag{2.27}$$

where $L_0(x) = 1,\ L_1(x) = 1 - x$
3)

$$P_{j+1}(x) = \left((2j+1)xP_j(x) - jP_{j-1}(x)\right) / (j+1), \quad \text{for } j = 1, 2, ..., \tag{2.28}$$

where $P_0(x) = 1,\ P_1(x) = x$.

In connection with the above, functions of the Hermite, Laguerre and Legendre orthonormal systems can be generated in a recurrent way:

1) for the Hermite system

$$\left.\begin{aligned} g_0(x) &= \pi e^{\frac{-x^2}{2}} \\ g_1(x) &= -2^{\frac{1}{2}}\pi\, x e^{-\frac{x^2}{2}} = -2^{\frac{1}{2}} x g_0(x) \\ g_{j+1}(x) &= -\left(2/(j+1)\right)^{\frac{1}{2}} x g_j(x) \\ &\quad -\left(j/(j+1)\right)^{\frac{1}{2}} \cdot g_{j-1}(x) \\ &\text{for}\quad j = 1, 2, ..., \end{aligned}\right\} \qquad (2.29)$$

2) for the Laguerre system

$$\left.\begin{aligned} g_0(x) &= e^{-\frac{x}{2}} \\ g_1(x) &= e^{-\frac{x}{2}}(1-x) = g_0(x)(1-x) \\ g_{j+1}(x) &= \left((2j+1-x)\, g_j(x) - j g_{j-1}(x)\right)/(j+1) \\ &\text{for}\quad j = 1, 2, ..., \end{aligned}\right\} \qquad (2.30)$$

3) for the Legendre system

$$\left.\begin{aligned} g_0(x) &= (0,5)^{\frac{1}{2}} \\ g_1(x) &= (1,5)^{\frac{1}{2}}\, x \\ g_{j+1}(x) &= \left((2j+3)(2j+1)\right)^{\frac{1}{2}} x g_j(x) \\ &\quad -\left((2j+3)/(2j-1)\right)^{\frac{1}{2}} j g_{j-1}(x)/(j+1) \\ &\text{for}\quad j = 1, 2, ... \end{aligned}\right\} \qquad (2.31)$$

Moreover, for the mentioned above systems, true are Christoffel-Derboux's formulae (Sansone [237], Szegö [268]). They allow to express formula (2.5) in a simpler way:

$$\sum_{j=0}^{q} g_j(x)\, g_j(y) = C(q)\, \frac{g_{q+1}(x)\, g_q(y) - g_q(x)\, g_{q+1}(y)}{y - x} \tag{2.32}$$

where

$$C(q) = \begin{cases} \left(\frac{q+1}{2}\right)^{\frac{1}{2}} & \text{for the Hermite system} \\ q+1 & \text{for the Laguerre system} \\ -1 & \text{for the Legendre system} \end{cases} \tag{2.33}$$

In other words, to calculate K_n, it is enough to know the value of the q-th and $(q+1)$-th functions of an orthonormal system, whereas both functions are generated recurrently. This facilitates greatly the application of algorithms, simplifying numerical problems.

In some applications it is convenient to use multiple Fourier series. We present three different multiple Fourier series.

a) Expansions based on Dirichlet's kernel

It is well known [334] that the functions

$$\left\{ (2\pi)^{-p/2}\, e^{ikx},\ k = (k_1, ..., k_p),\ kx = \sum_{j=1}^{p} k_j x^{(j)} \right. \tag{2.34}$$

$$(k_j = 0,\ \pm 1,\ \pm 2, ...,\ j = 1, ..., p)$$

are orthonormal and complete over the p-dimensional cube

$$Q = \left\{ -\pi \le x^{(j)} \le \pi,\ j = 1, ..., p \right\}. \tag{2.35}$$

For any integrable function R defined on Q its multiple Fourier expansion takes the form

$$R(x) \sim \sum c_k e^{ikx}, \tag{2.36}$$

where

$$c_k = (2\pi)^{-p} \int_Q R(x)\, e^{-ikx} dx. \tag{2.37}$$

The partial sums of expansion (2.36) are given by

$$S_q(x) = \sum_{|k_j| \leq q} c_k^{ikx} = \pi^{-p} \int_Q R(t)\, D_q(x-t)\, dt, \tag{2.38}$$

where

$$D_q(x) = \prod_{j=1}^{p} D_q\left(x^j\right), \tag{2.39}$$

$$D_q(u) = \frac{1}{2} + \sum_{k=1}^{q} \cos ku = \frac{\sin\left(q+\frac{1}{2}\right)u}{2\sin\frac{u}{2}} \tag{2.40}$$

is Dirichlet's kernel of order q.

b) Expansion based on Fejer's kernel

The first arithmetic means of (2.36) are given by

$$\begin{aligned} \sigma_q(x) &= \frac{1}{(q+1)^p} \sum_{|k_j| \leq q} S_{k_1,\ldots,k_p}(x) \\ &= \pi^{-p} \int_Q R(t)\, F_q(x-t)\, dt, \end{aligned} \tag{2.41}$$

where F_q is called the multidimensional Fejer's kernel of order q (see, i.e., [173] or [334]):

$$F_q(x) = \prod_{j=1}^{p} \phi_q\left(x^{(j)}\right), \tag{2.42}$$

$$\phi_q(u) = \frac{1}{2(q+1)} \left(\frac{\sin\frac{1}{2}(q+1)u}{\sin\frac{1}{2}u} \right)^2. \tag{2.43}$$

The multidimensional Fejer's kernel of order q has the following properties:

$$|F_q(x)| \leq \text{const. } N^p, \tag{2.44}$$

$$\pi^{-p} \int_Q F_q(x)\, dx = 1. \tag{2.45}$$

c) Expansions based on de la Vallee Poussin's kernel

In [173] the following kernel was suggested:

$$V_q(x) = \prod_{j=1}^{p} P_q\left(x^{(j)}\right), \tag{2.46}$$

where

$$P_q(u) = \frac{\cos qu - \cos 2qu}{4q\left(\sin\frac{u}{2}\right)^2}. \tag{2.47}$$

Formula (2.46) is called the multidimensional de la Valle Poussin's kernel. It has the following properties:

$$\pi^{-p}\int_Q |V_q(x)|\,dx \leq \text{const.}, \tag{2.48}$$

$$\pi^{-p}\int_Q V_q(x)\,dx = 1. \tag{2.49}$$

We define

$$k_q(x) = \pi^{-p}\int_Q R(t)\,V_q(x-t)\,dt. \tag{2.50}$$

Remark 2.4

For any square integrable function R

$$\|S_q - R\|_{L_2} \to 0, \tag{2.51}$$

$$\|\sigma_q - R\|_{L_2} \to 0, \tag{2.52}$$

$$\|k_q - R\|_{L_2} \to 0, \tag{2.53}$$

as $q \to \infty$, see [173] and [334]. Moreover,

$$S_q(x) \to R(x) \quad \text{as} \quad q \to \infty \tag{2.54}$$

at almost all points $x \in Q$ if R is a square integrable (see [246]) and

$$\sigma_q(x) \to R(x) \quad \text{as} \quad q \to \infty \tag{2.55}$$

at almost all points $x \in Q$ if R is an integrable (see [334]).

Remark 2.5
It is easily seen that the general kernel K_n takes the form

$$K_n(x,u) = \pi^{-p} D_q(x-u) \tag{2.56}$$

$$K_n(x,u) = \pi^{-p} F_q(x-u) \tag{2.57}$$

$$K_n(x,u) = \pi^{-p} V_q(x-u) \tag{2.58}$$

where $q = q(n)$, for the Dirichlet, Fejer and de la Vallee Poussin multiple kernels, respectively.

2.4 Concluding remarks

In this chapter we presented Parzen kernels and orthogonal series kernels for constructing probabilistic neural networks presented in Chapters 3, 5 and 6. It should be noted that various kernels are a basic tool in other estimation problems, e.g. they are also used in the potential functions method [6], the radial bases method [50], [133], the wavelets method [55], [162] and the nearest neighbor techniques [57], [69], [156].

3

Introduction to Probabilistic Neural Networks

3.1 Introduction

Probabilistic neural networks (PNN), introduced by Specht [255] – [258], have their predecessors in the theory of statistical pattern classification. In the fifties and sixties problems of statistical pattern classification in the stationary case were accomplished by means of parametric methods, using the available apparatus of statistical mathematics (e.g. [35], [75], [89], [90], [293]). The knowledge of the probability density to an accuracy of unknown parameters was assumed and the parameters were estimated based on the learning sequence. Typical techniques included maximum likelihood and Bayesian approaches. Having observed tendencies present in literature within the last twenty years we should say that these methods have been almost completely replaced by the non-parametric approach (see e.g. [67], [70], [71], [79], [80], [81], [104], [105], [113], [114], [122], [175], [191], [195], [272], [296], [297]). In the non-parametric approach it is assumed that a functional form of probability densities is unknown. The latter are estimated by non-parametric estimators. For the construction of non-parametric estimators of the density function, three principal methods were used in literature (below only pioneering works on this subject are quoted):

a) method based on so-called kernels (functions selected in a special way), suggested by Rosenblatt [204], developed by Parzen [178], and generalized for the multi-dimensional case by Cacoullos [45].

b) method based on orthogonal series, presented by Čencov [58], developed by Schwartz [241], Kronmal and Tarter [141], Walter [289] and other authors.

c) method based on the concept of nearest neighbours developed in [57] and [156].

It is well known that these techniques are convergent in the probabilistic sense, e.g. in probability or with probability one. Moreover, pattern classification procedures derived from non-parametric estimates are convergent – when the length of the learning sequence increases – to Bayes' rules. Asymptotically optimal pattern classification rules were examined by several authors [62], [98], [101], [208], [209], [222], [236], [300]. The PNN studied in literature implement non-parametric estimation techniques in a parallel fashion. They are characterized by fast training and convergence to the Bayes optimal decision surface. For interesting applications of the PNN the reader is referred to [139], [167], [193], [203], [264]. The crucial problem in these applications is the choice of the smoothing parameter. Most techniques are based on vector quantization [44], [327], cluster analysis [258] or the genetic algorithm [160]. A short survey of other available methods is given in [127]. In this chapter we review probabilistic neural networks. Moreover, we extend that idea to the PNN with a general kernel. We also present generalized regression neural networks working in a stationary environment. Additionally, we introduce recursive probabilistic neural networks for density and regression estimation, and for pattern classification. The concept of recursive probabilistic neural networks will be very useful in the next chapters. Based on this concept we will derive in Chapters 5 and 6 probabilistic neural networks working in a time-varying environment.

3.2 Probabilistic neural networks for density estimation

a) Non-recursive procedures
Let $X_1, ..., X_n$ be a sequence of independent, identically distributed random variables taking values in $A \subset R^p$ and having a probability density function f. The general estimator of the probability density function f is given by the following formula

$$\widehat{f}_n(x) = \frac{1}{n} \sum_{i=1}^{n} K_n(x, X_i) \tag{3.1}$$

where K_n is a sequence of functions described in Chapter 2.
In Fig. 3.1 we show a PNN corresponding to formula (3.1).

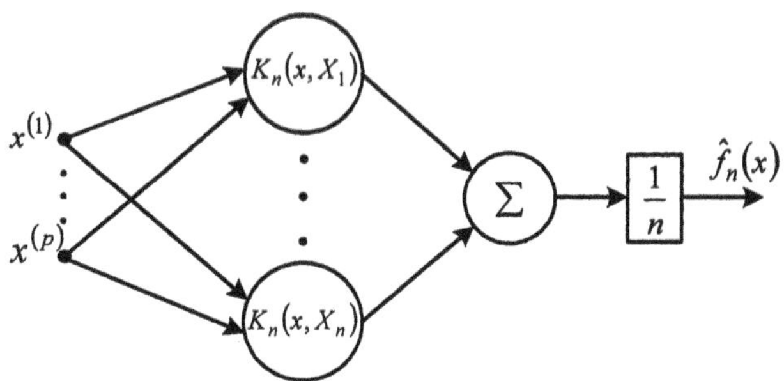

FIGURE 3.1. Probabilistic neural network for density estimation

A concrete form of estimator (3.1) depends on the kernel K_n. Below we present several examples of estimator (3.1) with various kernels K_n.

Example 3.1 (Rosenblatt [204])
In this example we give some heuristic motivations for estimator (3.1) with kernel (2.1). Let $p = 1$. From the definition of a probability density f we have

$$f(x) = \lim_{h \to 0} \frac{P\left(x - \frac{h}{2} < X < x + \frac{h}{2}\right)}{h} \tag{3.2}$$

The naive estimate of f is given by

$$\widehat{f}_n(x) = \frac{\text{number of observations falling in } (x-h,\ x+h)}{h} \tag{3.3}$$

Observe that estimator (3.3) is a histogram which mimics equality (3.2). The histogram is based on the observations "local" to x, where every point x is the center of a sampling interval $(x-h,\ x+h)$. The width h of the interval $(x-h,\ x+h)$ goes to zero as $n \rightarrow \infty$ and controls the amount by which the data are averaged to give estimate (3.3). To present the estimator more transparently, we define the kernel function

$$K(u) = \begin{cases} \frac{1}{2} & \text{if} \quad |u| \leq 1 \\ 0 & \text{otherwise} \end{cases} \tag{3.4}$$

The histogram estimator (3.3) expressed in terms of (3.4) takes the form

$$\widehat{f}_n(x) = \frac{1}{nh_n} \sum_{i=1}^{n} K\left(\frac{x - X_i}{h_n}\right) \tag{3.5}$$

where $h = h_n \xrightarrow{n} 0$.

Example 3.2 (Parzen [178])
Parzen investigated properties of estimator (3.5) with kernels satisfying the conditions

$$\sup |K(y)| < \infty$$

$$\int |K(y)| \,\mathrm{dy} < \infty$$

$$\lim_{y \longrightarrow \infty} |yK(y)| = 0$$

$$\int_R K(y) \,\mathrm{dy} = 1$$

Assuming that

$$h_n > 0,\ h_n \xrightarrow{n} 0 \ \text{ and } \ nh_n \xrightarrow{n} \infty$$

Parzen [178] showed that

$$E\left[\widehat{f}_n(x) - f(x)\right]^2 \xrightarrow{n} 0 \tag{3.6}$$

in continuity points x of f.

Example 3.3 (Cacullous [45])
The kernel estimator (3.5) was extended to the multivariate case by Cacoullos as follows

$$\widehat{f}_n(x) = \frac{1}{nh_n^p} \sum_{i=1}^{n} K\left(\frac{x - X_i}{h_n}\right) \tag{3.7}$$

$$h(n) > 0, \; h_n \xrightarrow{n} 0, \; nh_n^p \xrightarrow{n} \infty \tag{3.8}$$

and

$$\sup_x |K(x)| < \infty, \quad \int K(x)\, dx = 1,$$

$$\int |K(x)|\, dx < \infty, \quad \lim_{\|x\| \to \infty} \|x\|^p |K(x)| = 0$$

Under above conditions convergence (3.6) holds. Let us assume that function K is of the form

$$K(x) = (2\pi)^{-\frac{1}{2}p} e^{-\frac{1}{2}\|x\|^2}$$

where $\|x\|^2 = x^T x$. Then we can rewrite estimator (3.7) as follows:

$$\widehat{f}_n(x) = \frac{1}{(2\pi)^{\frac{p}{2}} nh_n^p} \sum_{i=1}^{n} \exp\left(-\frac{(x - X_i)^T (x - X_i)}{2h_n^2}\right) \tag{3.9}$$

Observe that

$$\begin{aligned} (x - X_i)^T (x - X_i) = -2 \Big(& x^{(1)} X_i^{(1)} \\ & + x^{(2)} X_i^{(2)} + \ldots + x^{(p)} X_i^{(p)}\Big) \\ & + \left(x^{(1)}\right)^2 + \left(x^{(2)}\right)^2 + \ldots + \left(x^{(p)}\right)^2 \\ & + \left(X_i^{(1)}\right)^2 + \left(X_i^{(2)}\right)^2 + \ldots + \left(X_i^{(p)}\right)^2 \end{aligned} \tag{3.10}$$

Now assuming normalization of the vectors x and X_i formula (3.9) simplifies to

$$\widehat{f}_n(x) = \frac{1}{(2\pi)^{\frac{p}{2}} nh_n^p} \sum_{i=1}^{n} \exp\left(-\frac{(1 - x^T X_i)}{h_n^2}\right) \tag{3.11}$$

Figure 3.2 shows a neural realization of algorithm (3.11). The proposed net has p inputs and two layers. The first layer consists of n neurons and each neuron has p weights. The output layer has a single neuron with the linear activation function. We should emphasize that the proposed network does not require a training procedure (optimal choosing of connection weights). The succeeding coordinates of the observation vectors X_i, $i = 1, ..., n$, play the role of the weights.

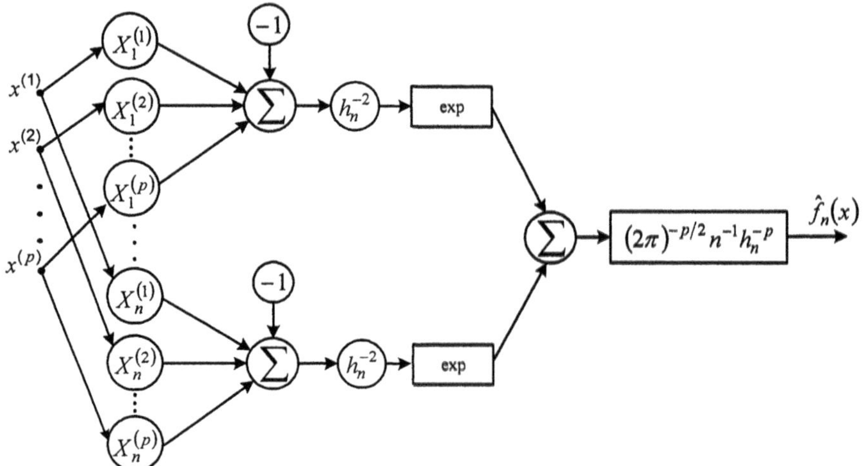

FIGURE 3.2. Probabilistic neural network for density estimation based on the Parzen kernel

Example 3.4 (Čencov [58])
We assume that f has the representation

$$f(x) \sim \sum_{j=0}^{\infty} a_j g_j(x), \tag{3.12}$$

where

$$a_j = E g_j(X) \tag{3.13}$$

and $g_j(\cdot)$, $j = 0, 1, 2, ...$, is a complete orthonormal system defined on A. As an estimator of density f we can take

$$\widehat{f}_n(x) = \sum_{j=0}^{q(n)} a_j g_j(x), \tag{3.14}$$

where

$$a_{jn} = \frac{1}{n} \sum_{i=1}^{n} g_j(X_i) \tag{3.15}$$

and $q(n)$ is a sequence of integers.

$$q(n) \to \infty$$

Density estimates (3.14) were introduced by Čencov [58] and studied by Schwartz [241], Kronmal and Tarter [141], and Walter [289] among others. For $q(n)$ satisfying

$$\frac{q(n)}{n} \xrightarrow{n} 0, \; q(n) \xrightarrow{n} \infty \tag{3.16}$$

estimator (3.14) is convergent in the mean integrated square error sense.

b) Recursive procedures

Let us modify estimator (3.1) as follows

$$\widehat{f}_n(x) = \frac{1}{n} \sum_{i=1}^{n} K_i(x, X_i) \tag{3.17}$$

Observe that estimator (3.17) is computationally equivalent to the recursive procedure

$$\begin{aligned} \widehat{f}_{n+1}(x) &= \widehat{f}_n(x) + \frac{1}{n+1}\left[K_{n+1}(x, X_{n+1}) - \widehat{f}_n(x)\right] \\ \widehat{f}_0(x) &= 0 \end{aligned} \tag{3.18}$$

or

$$\begin{aligned} \widehat{f}_{n+1}(x) &= \frac{n}{n+1}\widehat{f}_n(x) + \frac{1}{n+1}K_{n+1}(x, X_{n+1}) \\ \widehat{f}_0(x) &= 0 \end{aligned}$$

The PNN realizing procedure (3.18) is shown in Fig. 3.3.

A great advantage of the definition (3.18) over (3.1) is that $\widehat{f}_n$ can be computed by making use of the current observation X_n and the preceding estimator $\widehat{f}_{n-1}$. Thus the unknown probability density function is estimated sequentially. Estimator (3.18) with the Parzen kernel (see Fig. 3.4) was first introduced by Wolverton and Wagner [300] and independently by Yamato [301].

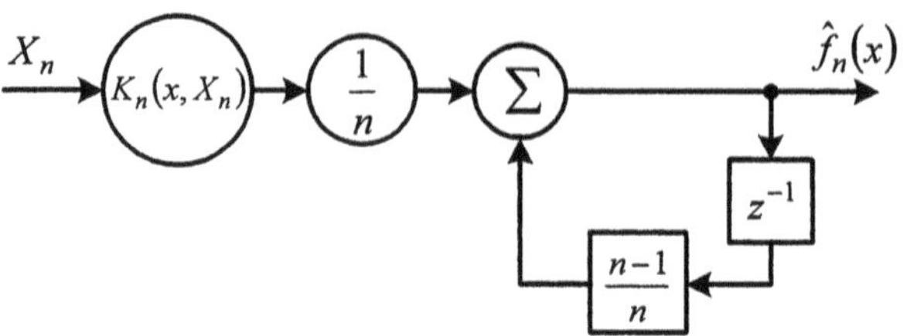

FIGURE 3.3. Recursive probabilistic neural network for density estimation

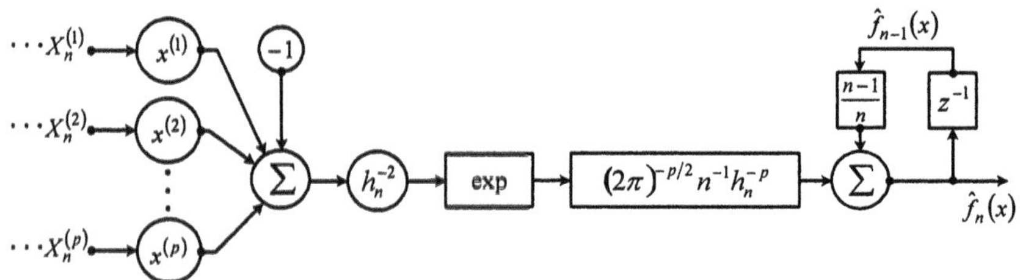

FIGURE 3.4. Recursive probabilistic neural network for estimation based on the Parzen kernel

Simulation 3.1

In the first simulation we apply the recursive PNN based on the Gaussian (the Parzen-type kernel) kernel for estimation of $N(0,1)$ distribution. In this case procedure (3.18) takes the form

$$\widehat{f}_{n+1} = \widehat{f}_n + \frac{1}{n+1}\left[\frac{1}{\sqrt{2\pi}} h_{n+1}^{-1} e^{-\frac{1}{2}\left(\frac{x-X_{n+1}}{h_{n+1}}\right)^2} - \widehat{f}_n(x)\right] \tag{3.19}$$

Let us assume that

$$h_n = kn^{-H}$$

In Fig. 3.5 we present the results for $n = 100$, $H = 0.4$ and varying k: a) $k = 0.5$, b) $k = 1$, c) $k = 2$.

Simulation 3.2

In this simulation we apply the recursive PNN given by (3.18) based on the triangular Parzen kernel (see Table 2.1) for estimation of uniform distribution. In Fig. 3.6 we show the results for $n = 1000$, $H = 0.4$ and varying k: a) $k = 0.5$, b) $k = 1$, c) $k = 2$.

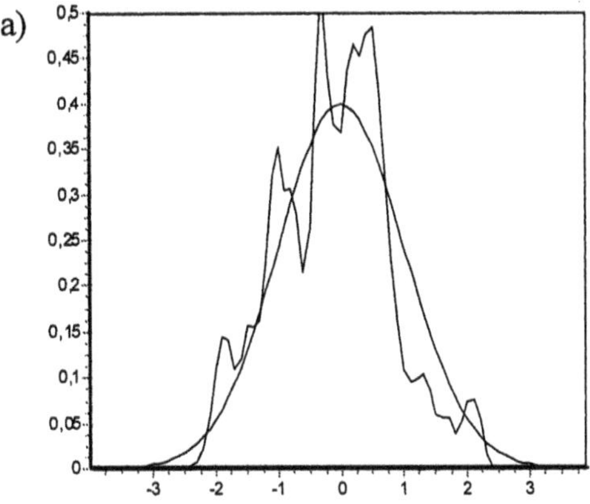

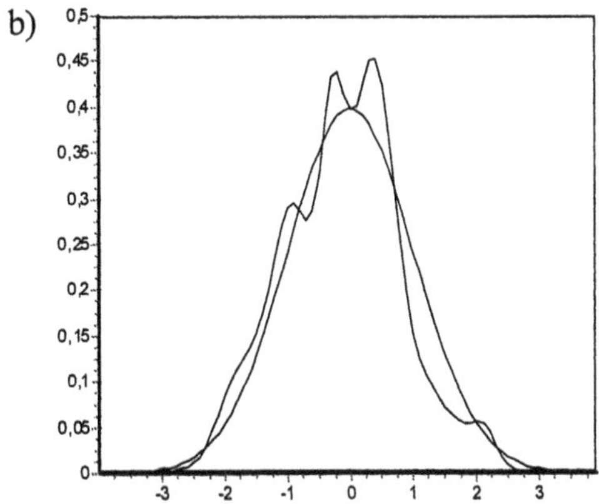

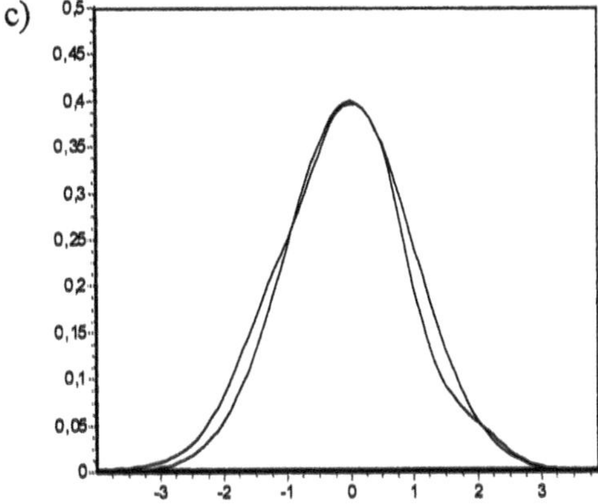

FIGURE 3.5. Estimation of stationary probability density – Simulation 3.1

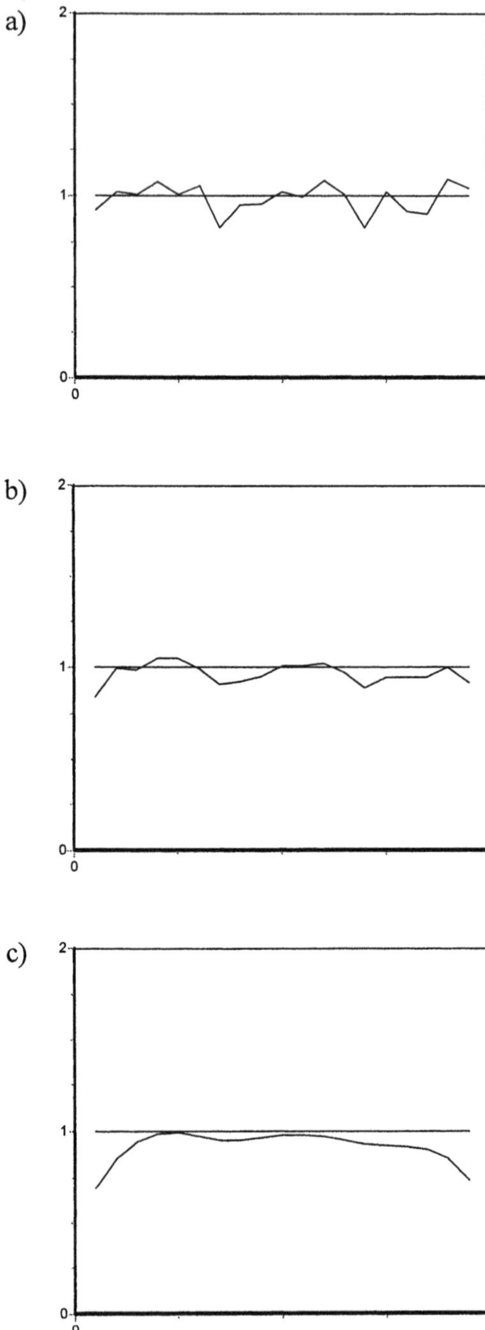

FIGURE 3.6. Estimation of stationary probability density – Simulation 3.2

Simulation 3.3
In this simulation we apply the recursive PNN given by (3.19) for estimation of $N(0,1)$ distribution. In Fig. 3.7 we present the results for $n = 100$, $k = 1$ and varying H: a) $H = 0.3$, b) $H = 0.5$, c) $H = 0.7$.

Simulation 3.4
In this simulation we apply the recursive PNN given by (3.19) for estimation of $N(0,1)$ distribution. In Fig. 3.8 we present the results for $k = 1$, $H = 0.3$ and varying n: a) $n = 10$, b) $n = 100$, c) $n = 200$.

Simulation 3.5
In this simulation we apply the recursive PNN given by (3.19) for estimation of $N(0,1)$ distribution. In Fig. 3.9 we present the results for $n = 100$, $H = 0.4$, $k = 1$ and varying Parzen kernels: a) uniform, b) triangular, c) Gaussian d) the Mexican hat.

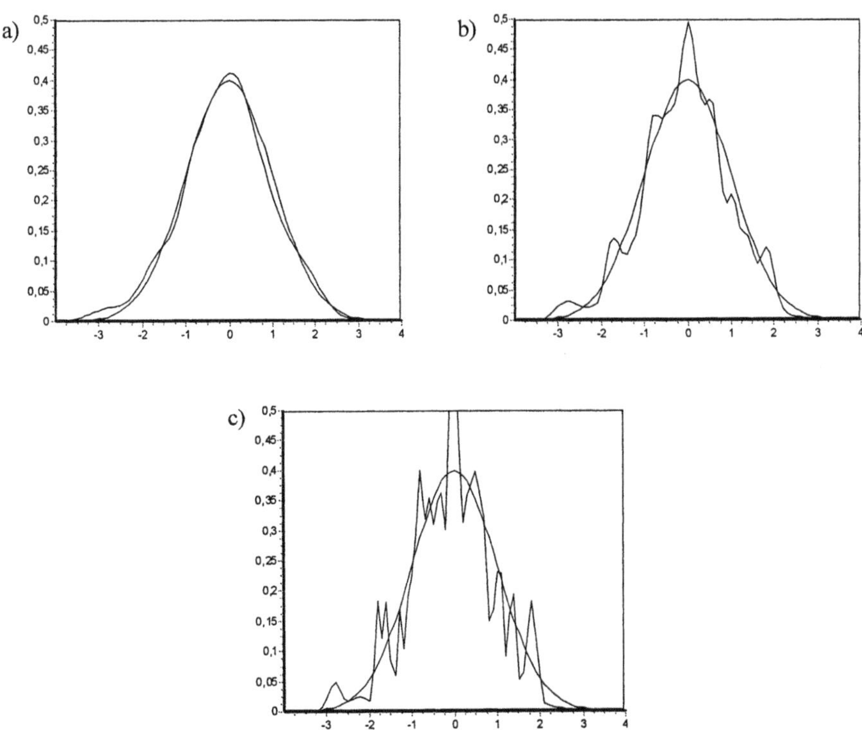

FIGURE 3.7. Estimation of stationary probability density – Simulation 3.3

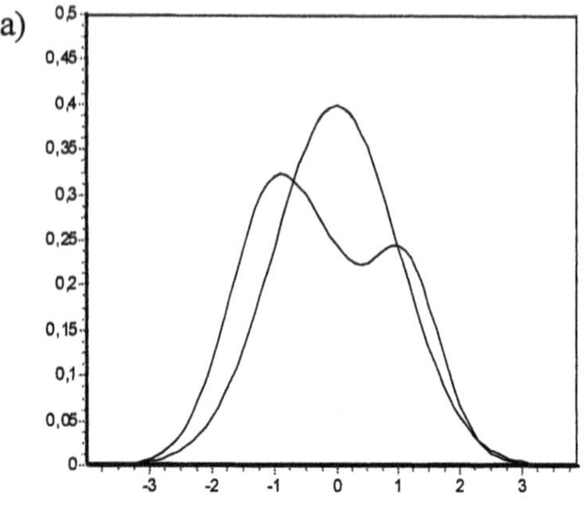

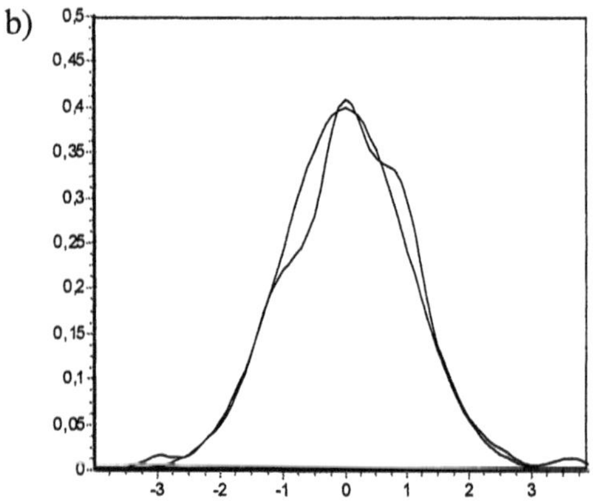

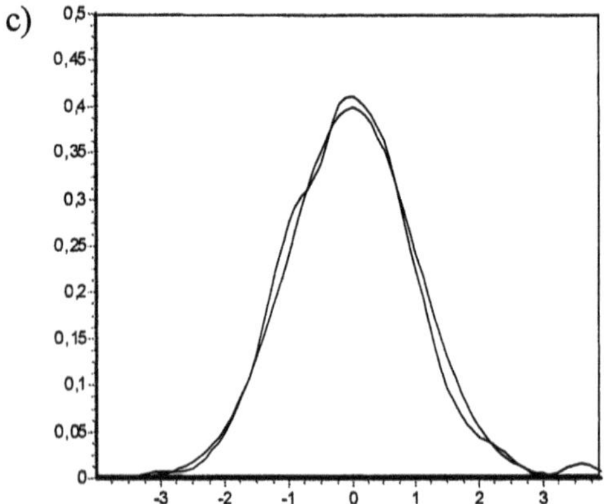

FIGURE 3.8. Estimation of stationary probability density – Simulation 3.4

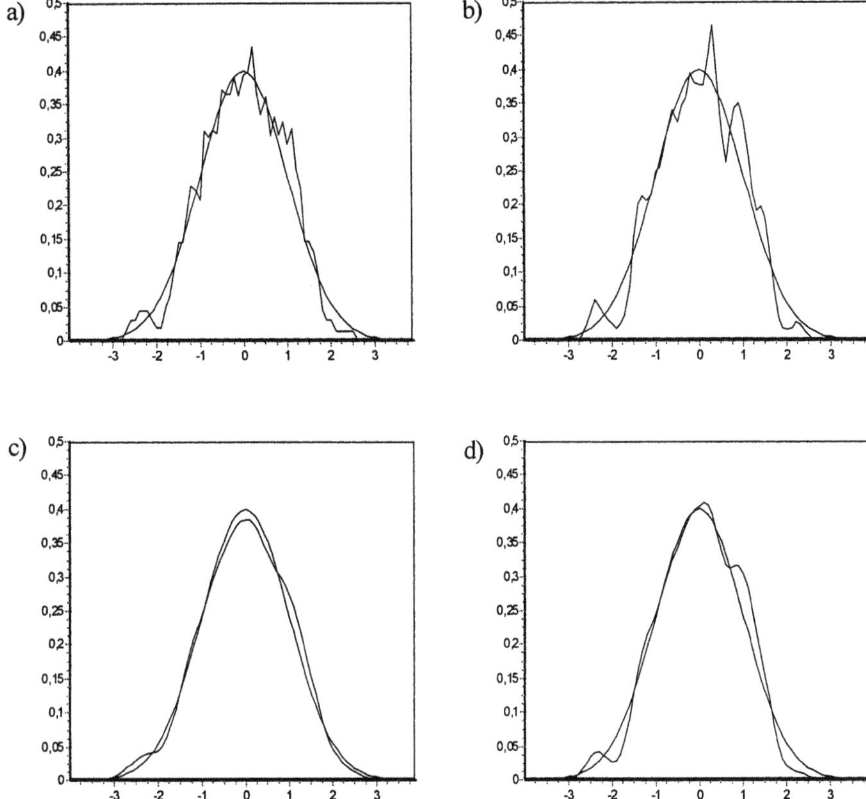

FIGURE 3.9. Estimation of stationary probability density – Simulation 3.5

The orthogonal series approach for recursive density estimation was developed by Rutkowski [208], [209]. In the next two examples we show recursion (3.18) with univariate and multivariate orthogonal series.

Example 3.5 (Rutkowski [208])
Let $p = 1$. For the kernel based on the orthogonal series (see Section 2.3) estimator (3.17) takes the form:

$$\widehat{f}_n(x) = \frac{1}{n} \sum_{i=1}^{n} \sum_{j=0}^{q(i)} g_j(X_i)\, g_j(x) \tag{3.20}$$

or

$$\widehat{f}_{n+1}(x) = \widehat{f}_n(x) + \tag{3.21}$$

$$+\frac{1}{n+1}\left[\sum_{j=0}^{q(n+1)} g_j\left(X_{n+1}\right) g_j\left(x\right) - \widehat{f}_n\left(x\right)\right]$$

where $\widehat{f}_0\left(x\right) = 0$. Rutkowski [208] showed that $E\left(\widehat{f}_n\left(x\right) - f_n\left(x\right)\right)^2$ $\xrightarrow{n} 0$ if

$$q\left(n\right) \xrightarrow{n} \infty, \quad \frac{1}{n^2}\sum_{i=1}^{n}\left(\sum_{j=0}^{q(i)} G_j^2\right)^2 \xrightarrow{n} 0, \tag{3.22}$$

Moreover, $\widehat{f}_n\left(x\right) \xrightarrow{n} f_n\left(x\right)$ with probability one if

$$q\left(n\right) \xrightarrow{n} \infty, \quad \sum_{n=1}^{\infty}\frac{1}{n^2}\left(\sum_{j=0}^{q(n)} G_j^2\right)^2 < \infty, \tag{3.23}$$

The weak and strong convergence holds at every point $x \in A$ at which

$$\sum_{j=0}^{q(n)} b_j g_j\left(x\right) \xrightarrow{n} f\left(x\right), \tag{3.24}$$

where

$$b_j = \int_A f\left(x\right) g_j\left(x\right) dx. \tag{3.25}$$

Simulation 3.6

In this simulation we apply the recursive PNN based on trygonometric orthogonal series (see formula 3.21) for estimation of $N(0,1)$ distribution. Let us assume that

$$q(n) = \left[kn^Q\right]$$

In Fig. 3.10 we present the results for $n = 100$, $k = 0.25$ and varying Q; a) $Q = 0.1$, b) $Q = 0.2$, c) $Q = 0.3$.

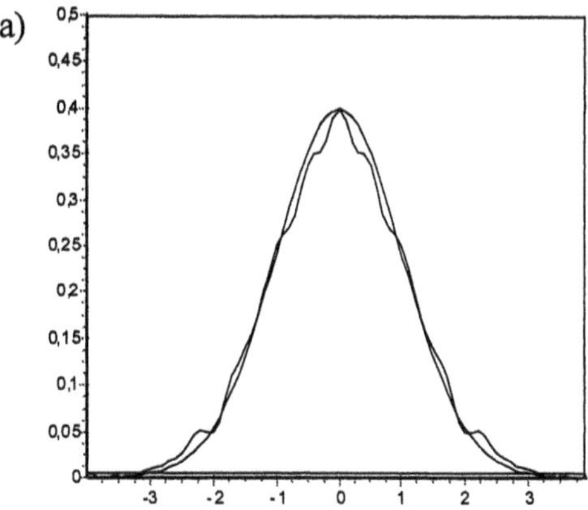

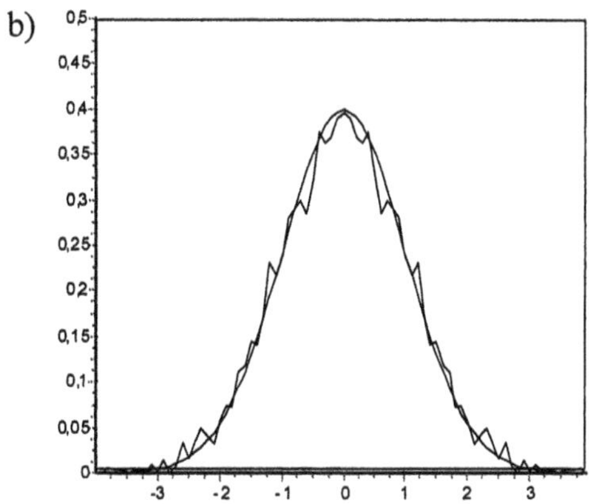

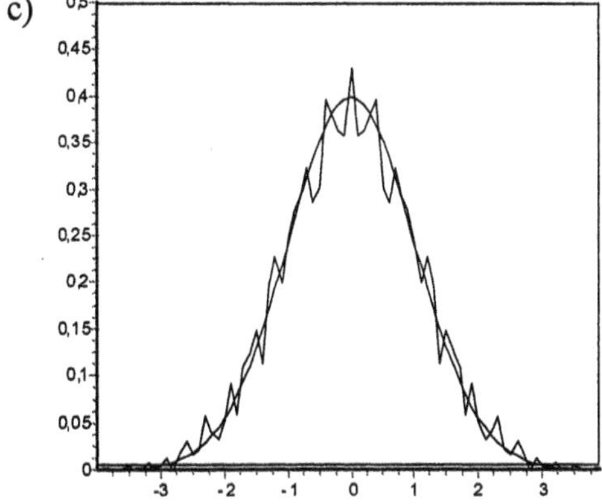

FIGURE 3.10. Estimation of stationary probability density – Simulation 3.6

Example 3.6 (Rutkowski [222])
Let $X_1, X_2, ...,$ be a sequence of independent observations of a random variable X having the Lebesgue density f and taking values in the p-dimensional cube $Q = [-\pi, \pi]^p$. The estimate f considered in this example is of the following form:

$$\widehat{f}_n(x) = \widehat{f}_{n-1}(x) + n^{-1}\left(\pi^{-p}F_q(x - X_n) - \widehat{f}_{n-1}(x)\right), \tag{3.26}$$

where $\widehat{f}_0(x) = 0$ and

$$F_q(x) = \prod_{i=1}^{p} \Phi_q\left(x^{(i)}\right), \tag{3.27}$$

$$\Phi_q(u) = \frac{1}{2(q+1)}\left(\frac{\sin\frac{1}{2}(q+1)u}{\sin\frac{1}{2}u}\right)^2. \tag{3.28}$$

The functions F_q are called multidimensional Fejer kernels of order q (see [173] and [334]). In algorithm (3.26) the number q depends on the number of observations n, i.e. $q = q(n)$.

Note that estimate (3.26) can be rewritten in the form

$$\widehat{f}_n(x) = n^{-1}\pi^{-p}\sum_{i=1}^{n} F_{q(i)}(x - X_i) \tag{3.29}$$

It is easy to verify that $\widehat{f}_n \geq 0$ and $\int_Q \widehat{f}_n(x)\,dx = 1$. Therefore, estimate (3.26) is itself the probability density function.
Procedure (3.29) is constructed in the spirit of sequential orthogonal series probability density estimates (Rutkowski [208], [209]).
If

$$n^{-2}\sum_{i=1}^{n} q^p(i) \xrightarrow{n} 0, \; q(n) \xrightarrow{n} \infty \tag{3.30}$$

then

$$E\left[\widehat{f}_n(x) - f(x)\right]^2 \xrightarrow{n} 0 \text{ a.e.} \tag{3.31}$$

If

$$\sum_{n=1}^{\infty} n^{-2}q^p(n) < \infty, \; q(n) \xrightarrow{n} \infty \tag{3.32}$$

then

$$\widehat{f}_n(x) \xrightarrow{n} f(x) \text{ a.e.} \tag{3.33}$$

with probability one. For details see Rutkowski [208], [209].

3.3 General regression neural networks in a stationary environment

Let (X, Y) be a pair of random variables. X takes values in a Borel set $A, A \subset R^p$, whereas Y takes values in R. Let f be the marginal Lebesgue density of X. Based on a sample $(X_1, Y_1), ..., (X_n, Y_n)$ of independent observations of (X, Y) we wish to estimate the regression function

$$\phi(x) = E[Y | X = x] \tag{3.34}$$

To estimate function (3.34) we propose the following formula

$$\widehat{\phi}_n(x) = \frac{\widehat{R}_n(x)}{\widehat{f}_n(x)} \tag{3.35}$$

where

$$\widehat{R}_n(x) = \frac{1}{n} \sum_{i=1}^{n} Y_i K_n(x, X_i) \tag{3.36}$$

and estimator $\widehat{f}_n$ is given by (3.1).

In Fig. 3.11 we show the neural-network implementation of estimator (3.35).

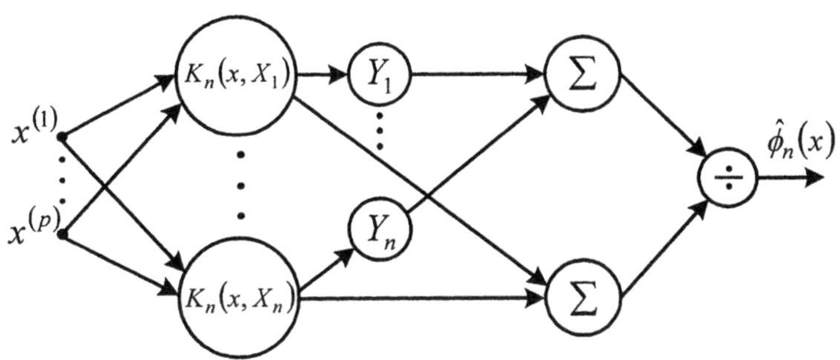

FIGURE 3.11. Scheme of generalized regression neural network

Example 3.7 (Nadaraya [168] and Watson [292])
Applying the Parzen kernel to estimator (3.35) we get

$$\widehat{\phi}_n(x) = \frac{\sum_{i=1}^{n} Y_i K\left(\frac{x - X_i}{h_n}\right)}{\sum_{i=1}^{n} K\left(\frac{x - X_i}{h_n}\right)} \tag{3.37}$$

Several results concerning the convergence of estimator (3.37) can be found in [62], [64], [65] and [68]. The neural network implementation of estimator (3.37) for $p \geq 1$ is depicted in Fig. 3.12.

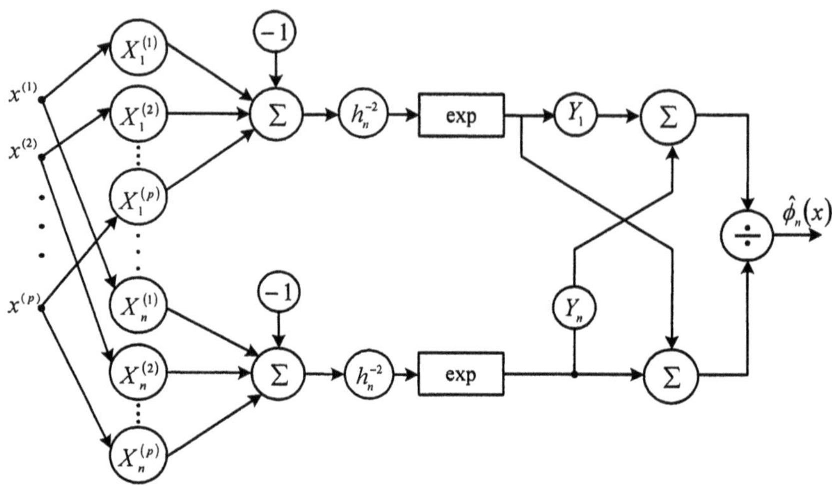

FIGURE 3.12. Generalized regression neural network based on the Parzen kernel

The recursive version of procedure (3.35) is given as follows

$$\widehat{\phi}_n(x) = \frac{\widehat{R}_n(x)}{\widehat{f}_n(x)} \tag{3.38}$$

where

$$\widehat{R}_n(x) = \frac{1}{n}\sum_{i=1}^{n} Y_i K_i(x, X_i) \tag{3.39}$$

or

$$\widehat{R}_{n+1}(x) = \widehat{R}_n(x) + \frac{1}{n+1}\left[Y_{n+1}K_{n+1}(x, X_{n+1}) - \widehat{R}_n(x)\right] \tag{3.40}$$

$\widehat{R}_0(x) = 0$ and estimator $\widehat{f}_n$ is given by (3.18). The block diagram of recursive formula (3.38) is depicted in Fig. 3.13 whereas Fig. 3.14 shows its Parzen – kernel version.

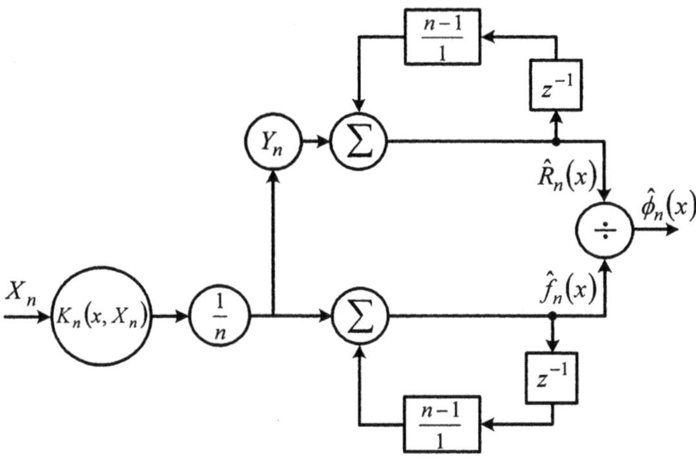

FIGURE 3.13. Recursive generalized regression neural network

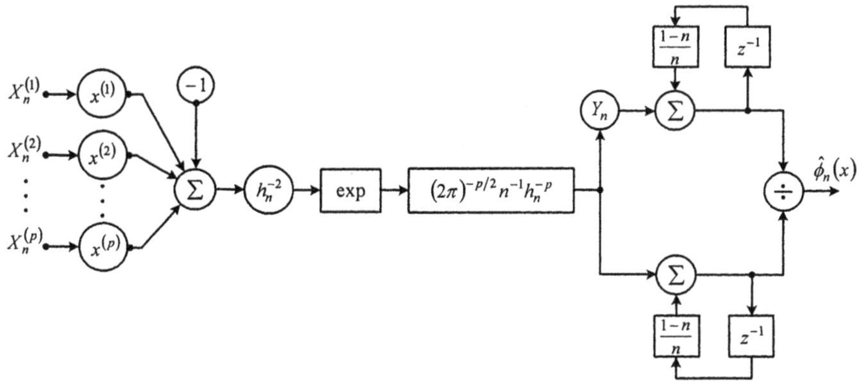

FIGURE 3.14. Recursive generalized regression neural network based on the Parzen kernel

Simulation 3.7

We consider the following stationary regression

$$\Phi_n^*(x_n) = 10x_n^2 + z_n$$

where x_n and z_n are realizations of $N(0,1)$ random variables.
The recursive GRNN given by (3.38), (3.40) and (3.18) and based on the Gaussian kernel has been applied with the following parameters: $H = 0.5$, $k = 1$.

The results are depicted in Fig. 3.15a which displays a comparison of a true regression and estimated by the GRNN for $n = 5000$. In Fig. 3.15b and Fig. 3.15c we show the convergence of the GRNN as the sample size grows large in points $x = 0.2$ and $x = 0.4$, respectively.

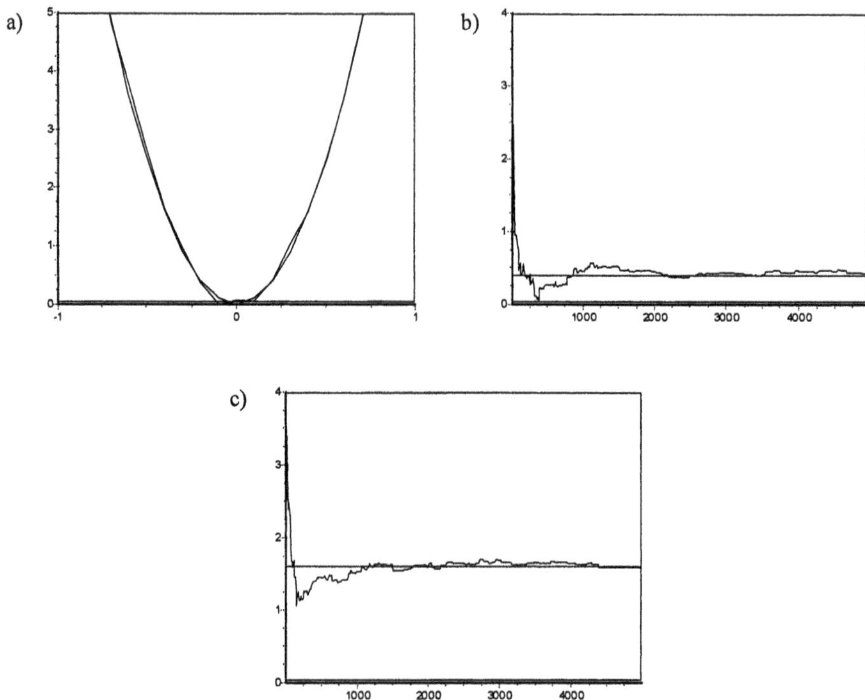

FIGURE 3.15. GRNN for estimation of a stationary regression function – Simulation 3.7

Simulation 3.8

We consider the following stationary regression

$$\Phi_n^*(x_n) = 10\cos(x_n) + z_n$$

where x_n and z_n are realizations of $N(0,1)$ random variables.
The GRNN based on the othogonal series has been applied with the following parameters:
$H = 0.5$, $k = 1$.

The results are depicted in Fig. 3.16a which displays a comparison of a true regression and estimated by the GRNN for $n = 5000$. In Fig.

3.16b and Fig. 3.16c we show the convergence of the GRNN as the sample size grows large in points $x = 0.2$ and $x = 0.4$, respectively.

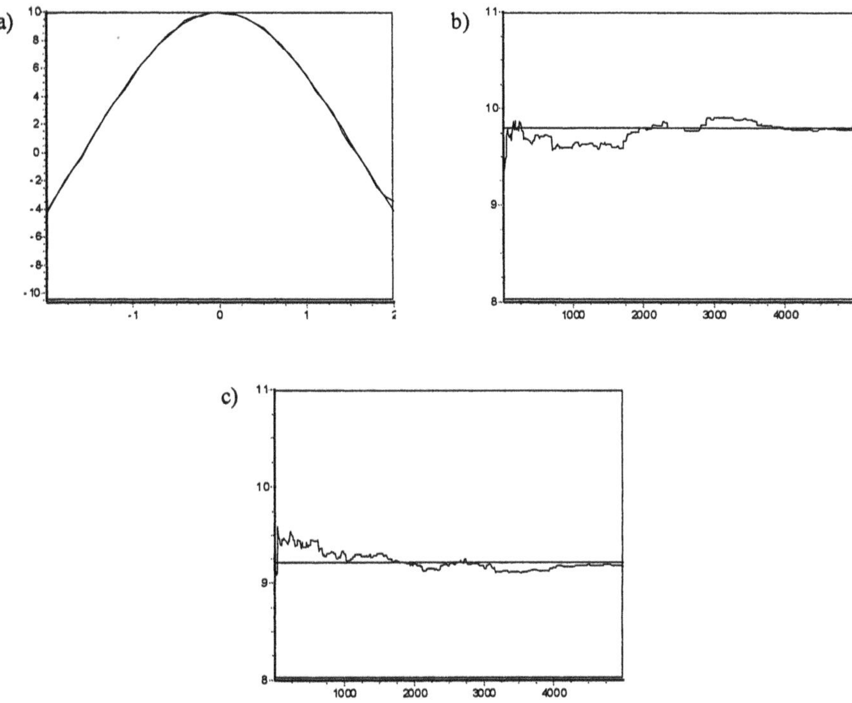

FIGURE 3.16. GRNN for estimation of a stationary regression function – Simulation 3.8

Remark 3.1
Observe that procedures (3.38), (3.40) and (3.18) can be alternatively expressed as follows:

$$\widehat{\phi}_n(x) = \frac{\widetilde{R}_n(x)}{\widetilde{f}_n(x)} \tag{3.41}$$

where

$$\widetilde{R}_{n+1}(x) = \widetilde{R}_n(x) + Y_n K_n(x, X_n) \tag{3.42}$$

and

$$\widetilde{f}_{n+1}(x) = \widetilde{f}_n(x) + K_n(x, X_n) \tag{3.43}$$

The block diagram of the appropriate PNN is shown in Fig. 3.17.

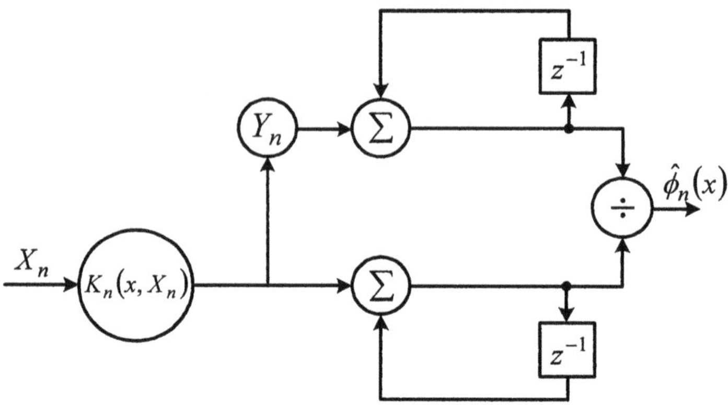

FIGURE 3.17. Simplified scheme of the recursive generalized regression neural network

Example 3.8 (Rutkowski [209])
In this example we present procedure (3.35) with the orthogonal series kernel. Let us define

$$R(x) = \phi(x) f(x) \tag{3.44}$$

We assume that functions R and f have representations

$$R(x) \sim \sum_{k=0}^{\infty} a_k g_k(x) \tag{3.45}$$

$$f(x) \sim \sum_{k=0}^{\infty} b_k g_k(x) \tag{3.46}$$

where

$$a_k = \int_A R(x) g_k(x) dx = E[Y g_k(X)] \tag{3.47}$$

$$b_k = \int_A f(x) g_k(x) dx = E[g_k(X)] \tag{3.48}$$

We propose a nonparametric estimate of regression $\phi(x)$ as follows:

$$\widehat{\phi}_n(x) = \widehat{R}_n(x) / \widehat{f}_n(x) \tag{3.49}$$

where

$$\widehat{R}_n(x) = \frac{1}{n} \sum_{i=1}^{n} \sum_{k=0}^{q(i)} Y_i g_k(X_i) g_k(x) \tag{3.50}$$

$$\widehat{f}_n(x) = \frac{1}{n}\sum_{i=1}^{n}\sum_{k=0}^{q(i)} g_k(X_i)\, g_k(x) \tag{3.51}$$

Observe that estimates (3.50) and (3.51) may be expressed as

$$\widehat{R}_{n+1}(x) = \widehat{R}_n(x) + \frac{1}{n+1} \tag{3.52}$$

$$\cdot \left[\sum_{k=0}^{q(n+1)} Y_{n+1} g_k(X_{n+1})\, g_k(x) - \widehat{R}_n(x)\right]$$

and

$$\widehat{f}_{n+1}(x) = \widehat{f}_n(x) + \frac{1}{n+1} \tag{3.53}$$

$$\cdot \left[\sum_{k=0}^{q(n+1)} g_k(X_{n+1})\, g_k(x) - \widehat{f}_n(x)\right]$$

Thus, the unknown regression function $\phi(x)$ is estimated recursively. It was shown (Rutkowski [209]) that if $EY^2 < \infty$, condition (2.22) is satisfied, and

$$q(n) \xrightarrow{n} \infty, \quad \frac{1}{n^2}\sum_{i=1}^{n}\left(\sum_{k=0}^{q(i)} G_k^2\right)^2 \xrightarrow{n} 0, \tag{3.54}$$

then

$$\widehat{\phi}_n(x) \xrightarrow{n} \phi(x) \quad \text{in probability}$$

at every point $x \in A$ at which series (3.45) and (3.46) converge to $R(x)$ and $f(x)$, respectively. Moreover, if $EY^2 < \infty$, condition (2.22) is satisfied, and

$$q(n) \xrightarrow{n} \infty, \quad \sum_{n=1}^{\infty}\frac{1}{n^2}\left(\sum_{k=0}^{q(n)} G_k^2\right)^2 < \infty \tag{3.55}$$

then

$$\widehat{\phi}_{nn}(x) \xrightarrow{n} \phi(x) \quad \text{with probability one}$$

at every point $x \in A$ at which series (3.45) and (3.46) converge to $R(x)$ and $f(x)$, respectively.

So far we have discussed the stochastic regression model. Now we assume that the inputs $x_1, ..., x_n$, are selected by an experimenter. Divide the cube $Q \in R^p$ into n mutually disjoint and totally exhaustive regions A_i, $i = 1, ..., n$. From each of these regions select and fix a point, so that we have $x_1, ..., x_n$, where $x_i \in A_i$, $i = 1, ..., n$. We require that

$$\max_{1 \le i \le n} \lambda (A_i) = O\left(n^{-1}\right), \tag{3.56}$$

where λ is the Lebesgue measure. As an estimator of $\phi(x)$ in the fixed design case we propose

$$\widehat{\phi}_n (x) = \sum_{i=1}^{n} Y_i \int_{A_i} K_n (x, u)\, du \tag{3.57}$$

Example 3.9 (Rutkowski [210], [212], [226], [227])
Let us partition interval $A = [0, 1]$ into n regions $A_1, ..., A_n$, where $A_i = [d_{i-1}, d_i]$, $d_0 = 0$, $d_n = 1$, and $\cup A_i = A$. Let the system be given by

$$Y_i = \phi(x_i) + Z_i, \ i = 1, ..., n \tag{3.58}$$

where input signals x_i are selected so that $x_i \in A_i$, y_i are the measured output signals, ϕ is a completely unknown function, and the errors Z_i are independent random variables with a zero mean and a finite variance, i.e.,

$$Ez_i = 0, \ Ez_i^2 = \sigma_i^2 \le \sigma^2, \quad i = 1, ..., n. \tag{3.59}$$

We expand the regression function $\phi(x) = E[y_i \,|x]$ in the orthogonal series

$$\phi(x) \sim \sum_{k=0}^{\infty} a_k g_k (x) \tag{3.60}$$

where

$$a_k = \int_A g_k(x)\, \phi(x)\, dx. \tag{3.61}$$

In Rutkowski [210] the following estimator of $\phi(x)$ was proposed

$$\widehat{\phi}_n (x) = \sum_{k=0}^{q(n)} g_k (x)\, \widehat{a}_{kn} \tag{3.62}$$

where

$$\widehat{a}_{kn} = \sum_{i=1}^{n} y_i \int_{A} g_k(x)\, dx \tag{3.63}$$

and $q(n)$ is a sequence of integers. It was shown (see Rutkowski [210]) that if $R(x)$ satisfies the Lipschitz condition, and

$$\max_{1 \le i \le n} |d_i - d_{i-1}| = 0\left(n^{-1}\right) \tag{3.64}$$

$$\frac{1}{n} \sum_{k=0}^{q(n)} G_k^2 \xrightarrow{n} 0, \quad q(n) \xrightarrow{n} \infty, \tag{3.65}$$

then

$$E\left[\widehat{R}_n(x) - R(x)\right]^2 \xrightarrow{n} 0$$

at every point $x \in A$ at which

$$\sum_{k=0}^{q(n)} a_k g_k(x) \xrightarrow{n} R(x) \tag{3.66}$$

Example 3.10 (Rutkowski [228])
The following three procedures for nonparametric fitting of an unknown function ϕ are based on kernels (2.39), (2.42), and (2.46) presented in Chapter 2:

$$\widehat{\phi}_n^{(1)}(x) = \Pi^{-p} \sum_{i=1}^{n} Y_i \int_{A_i} D_q(x-t)\, dt, \tag{3.67}$$

$$\widehat{\phi}_n^{(2)}(x) = \Pi^{-p} \sum_{i=1}^{n} Y_i \int_{A_i} F_q(x-t)\, dt, \tag{3.68}$$

$$\widehat{\phi}_n^{(3)}(x) = \Pi^{-p} \sum_{i=1}^{n} Y_i \int_{A_i} V_q(x-t)\, dt, \tag{3.69}$$

where number q depends on the number of observations n, i.e. $q = q(n)$. It will be convenient to extend domain of ϕ to the p-dimensional Euclidean space by periodicity. Let us define

$$\mathrm{MISE}^{(j)} = \int_{Q} E\left(\widehat{\phi}_n^{(j)}(x) - \phi(x)\right)^2 dx,\ j = 1, 2, 3 \tag{3.70}$$

as the mean integrated square error of (3.67) – (3.69). If $\phi \in L_2$, and

$$q(n) \xrightarrow{n} \infty, \quad \frac{q^p(n)}{n} \xrightarrow{n} 0, \tag{3.71}$$

$$q^p(n)\,\gamma^2(n) \xrightarrow{n} 0, \tag{3.72}$$

where

$$\gamma(n) = \max_{1 \le i \le n} \left\{ \sup_{x, x_i \in A_i} |R(x) - R(x_i)| \right\}, \tag{3.73}$$

then

$$MISE^{(j)} \longrightarrow 0, \; j = 1, 2, 3.$$

Proofs of the theorems are given in Rutkowski [228]. Suppose that $p = 1$, $Q = [-\pi, \pi]$, ϕ satisfies the Lipschitz condition

$$|\phi(x) - \phi(y)| \le C\,|x - y|$$

and

$$A_i = \left[-\pi + \frac{2\pi}{n} i, \quad -\pi + \frac{2\pi}{n}(i+1) \right]$$

for $i = 0, 1, ..., n-1$. In this case $\gamma(n) = O\left(n^{-1}\right)$ and conditions (3.71) and (3.72) reduce to

$$\frac{q(n)}{n} \xrightarrow{n} 0.$$

Example 3.11 (Gałkowski and Rutkowski [91], [92], [93])
Consider the p-dimensional space $Q_p = \{x \in [0,1]^p\}$. Let $n^{\frac{1}{p}} = N$ be an integer and $k = 1, ..., p$, $i_k = 1, ..., N$. Partition the unit interval $[0,1]$ on the k-th axis into N subsets Δx_{i_k}. Define

$$\Delta x_{i_1} \times \Delta x_{i_2} \times ... \times \Delta x_{i_p} = Q_{p, i_1 ... i_p} = Q_{p,i} \tag{3.74}$$

Let $Q_{p,i} \wedge Q_{p,j} = \varnothing$ for $i \ne j$ and $UQ_{p,i} = Q_p$. We assume that in model (3.58) inputs x_i, $i = [i_1, ..., i_p]$, are selected so that $x_i \in Q_{p,i}$ and propose the following algorithm:

$$\widehat{\phi}_n(x) = \sum_{i=1}^{N} Y_i \int_{Q_{p,i}} h_n^{-p} K\left(\frac{x-u}{h_n}\right) du \tag{3.75}$$

where $\mathbf{1} = [1, ..., 1]$ is $1 \times p$ vector, and

$$K(u) = \prod_{i=1}^{p} H(u_i), \; i = 1, ..., p$$

$$H(t) \geq 0, \quad \text{for} \quad t \in (-L, L), \quad L = \text{const.}$$

$$H(t) = 0, \quad \text{for} \quad t \notin (-L, L)$$

$$\int_{-L}^{L} H(t)\, dt = 1$$

$$\sup H(t) < \infty$$

We denote $|\Delta x_{i_k}|$ as the length of the interval Δx_{i_k}, $i_k = 1, ..., N$, $k = 1, ..., p$. If ϕ is continuous on $[0,1]^p$ and

$$EZ_r = 0 \quad EZ_r^2 = \sigma_r^2 \leq \text{const.}$$

$$\max_{1 \leq i_k \leq N} |\Delta x_{i_k}| = 0\left(n^{-\frac{1}{p}}\right), \; k = 1, ..., p \tag{3.76}$$

$$h_n \xrightarrow{n} 0, \quad n^{-1} h_n^{-p} \xrightarrow{n} 0 \tag{3.77}$$

then

$$E\left[\widehat{\phi}_n(x) - \phi(x)\right]^2 \xrightarrow{n} 0 \tag{3.78}$$

at every point $x \in (0,1)^p$.

If ϕ is continuous on $[0,1]^p$, and

$$EZ_r = 0, \; E|Z_r|^s \leq C_s < \infty, \; s > 2 \tag{3.79}$$

$$h_n \xrightarrow{n} 0, \; \sum_{n=1}^{\infty} n^{1-s} h_n^{-ps} < \infty \tag{3.80}$$

$$\sum_{n=1}^{\infty} \exp(-n h_n^p) < \infty \tag{3.81}$$

then

$$\widehat{\phi}_n(x) \xrightarrow{n} \phi(x) \tag{3.82}$$

with probability 1 at every point $x \in (0,1)^p$. A proof of convergence (3.78) and (3.82) is given in Gałkowski and Rutkowski [92].

3.4 Probabilistic neural networks for pattern classification in a stationary environment

Let $(X, Y), (X_1, Y_1), ..., (X_n, Y_n)$ be a sequence of i.i.d. pairs of random variables, Y takes values in the set of classes $S = \{1, ..., M\}$, whereas X takes values in $A \subset R^p$. The problem is to estimate Y from X and W_n, where $W_n = (X_1, Y_1), ..., (X_n, Y_n)$ is a learning sequence. Suppose that p_m and f_m, $m = 1, ..., M$ are the prior class probabilities and class conditional densities, respectively. We define a discriminant function of class j:

$$d_j(x) = p_j f_j(x) \tag{3.83}$$

Let $L(i, j)$ be the loss incurred in taking action $i \in S$ when the class is j. We assume $0 - 1$ loss function. For a decision function $\varphi : A \to S$ the expected loss is

$$R(\varphi) = \sum_{j=1}^{M} p_j \int_A L(\varphi(x), j) f_j(x)\, dx \tag{3.84}$$

A decision function φ^*which classifies every $x \in A$ as coming from any class m for which

$$p_m f_m(x) = \max_j p_j f_j(x) = \max_j d_j(x) \tag{3.85}$$

is a Bayes decision function and

$$R^* = R(\varphi^*) = \sum_{j=1}^{M} p_j \int_A L(\varphi^*(x), j) f_j(x)\, dx \tag{3.86}$$

is the minimal Bayes risk. The function $d_m(x)$ is called the Bayes discriminant function.

Let n_j be the number of observations from class j, $j = 1, ..., M$. We partition observations $X_1, ..., X_n$ into M subsequences

$$X_1^{(1)}, ..., X_{n_1}^{(1)}$$

$$X_1^{(2)}, ..., X_{n_2}^{(2)} \tag{3.87}$$

$$X_1^{(M)}, ..., X_{n_M}^{(M)}$$

As estimates of conditional densities f_j we apply estimator (3.1) in the form

$$\widehat{f}_{n_j}(x) = \frac{1}{n_j} \sum_{i=1}^{n_j} K_{n_j}\left(x, X_i^{(j)}\right) \tag{3.88}$$

The prior probabilities p_j are estimated by

$$\widehat{p}_j = \frac{n_j}{n} \tag{3.89}$$

Combining (3.83), (3.88) and (3.89) we get the following discriminant function estimate

$$\widehat{d}_{j,n}(x) = \frac{1}{n} \sum_{i=1}^{n_j} K_{n_j}\left(x, X_i^{(j)}\right) \tag{3.90}$$

and the corresponding classification procedure

$$\begin{aligned} \widehat{\varphi}_n(x) &= m \text{ if } \sum_{i=1}^{n_m} K_{n_m}\left(x, X_i^{(m)}\right) \geq \sum_{i=1}^{n_j} K_{n_j}\left(x, X_i^{(j)}\right) \\ \text{for } i &\neq m, \; i = 1, ..., M \end{aligned} \tag{3.91}$$

The probabilistic neural network realizing procedure (3.91) is shown in Fig. 3.18.

It was shown (see [61], [98], [236], [300]) that

$$R(\varphi_n) \xrightarrow{n} R^* \tag{3.92}$$

in probability (with pr. 1) if estimators (3.88) converge in probability (with pr. 1).

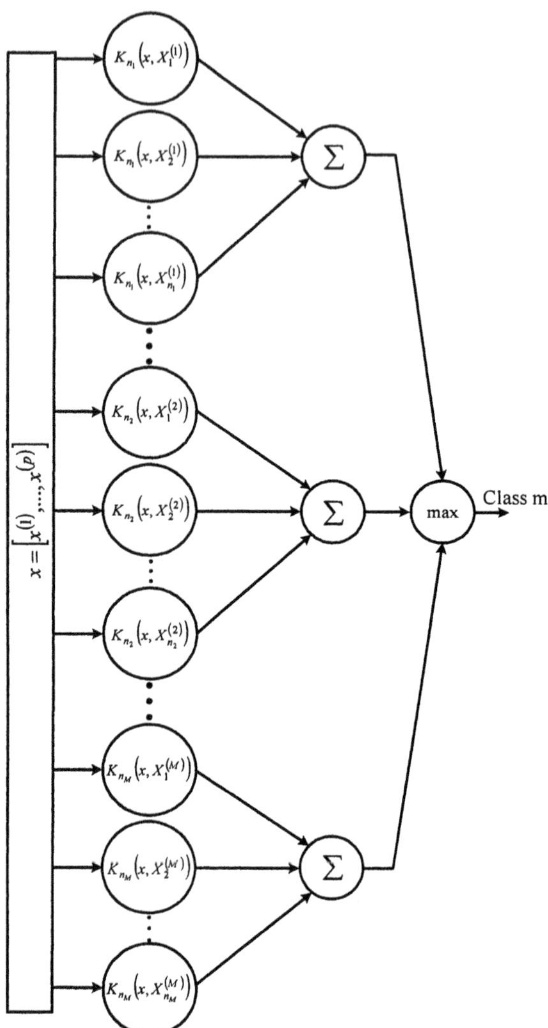

FIGURE 3.18. Probabilistic neural network for pattern classification

Example 3.12
For the Parzen kernel, procedure (3.91) classifies every $x \in A$ as coming from a class which maximizes

$$\frac{1}{h_{n_j}^p} \sum_{i=1}^{n_j} K\left(\frac{x - X_i^{(j)}}{h_{n_j}}\right)$$

for $j = 1, ..., M$.

We will now derive classification procedures from a general regression probabilistic neural network (3.35). Instead of partition (3.87) define

$$T_{ji} = \begin{cases} 1 & \text{if} \quad Y_i = j \\ 0 & \text{if} \quad Y_i \neq j \end{cases} \tag{3.93}$$

for $i = 1, ..., n$ and $j = 1, ..., M$.

Observe that discriminant functions d_j, $j = 1, ..., M$, can be presented in the form

$$d_j(x) = p_j f_j(x) = f(x) E\left[T_{ji} \left| X_i = x\right.\right] \tag{3.94}$$

where

$$f(x) = \sum_{j=1}^{M} p_j f_j(x) \tag{3.95}$$

Therefore, the estimates of discriminant functions derived from the regression model take the form

$$\widehat{d}_{j,n}(x) = \frac{1}{n} \sum_{i=1}^{n} T_{ji} K_n(x, X_i) \tag{3.96}$$

The classification procedure derived from estimator (3.96) takes the form

$$\widehat{\varphi}_n(x) = m \quad \text{if} \quad \sum_{i=1}^{n} T_{mi} K_n(x, X_i)$$

$$\geq \sum_{i=1}^{n} T_{ji} K_n(x, X_i) \tag{3.97}$$

$$\text{for} \quad i \neq m, \ i = 1, ..., M$$

Generalized regression neural network for pattern classification is presented in Fig. 3.19.

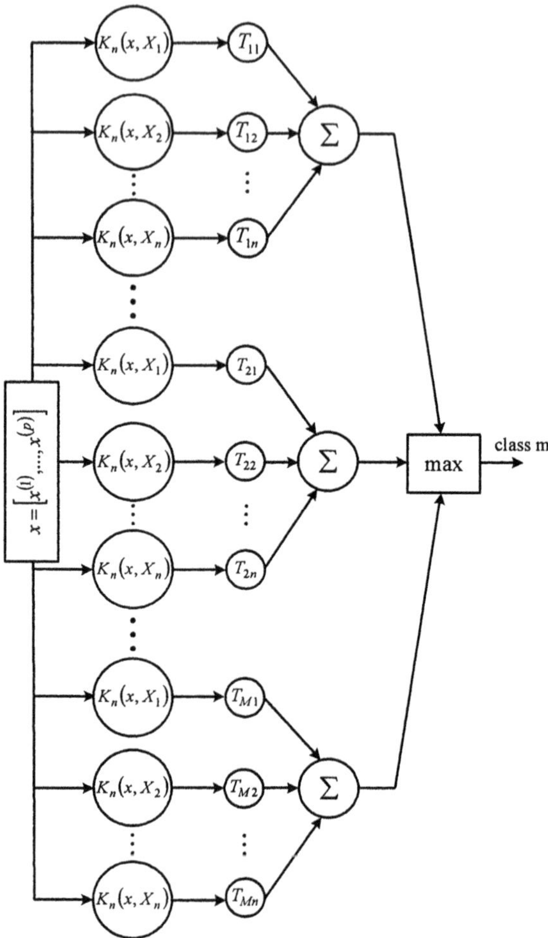

FIGURE 3.19. Generalized regression neural network for pattern classification

Example 3.13
For the Parzen kernel, procedure (3.97) classifies every $x \in A$ as coming from a class which maximizes

$$\sum_{i=1}^{n} T_{ji} K\left(\frac{x - X_i}{h_n}\right)$$

for $j = 1, \ldots, M$. The appropriate probabilistic neural network is shown in Fig. 3.20.

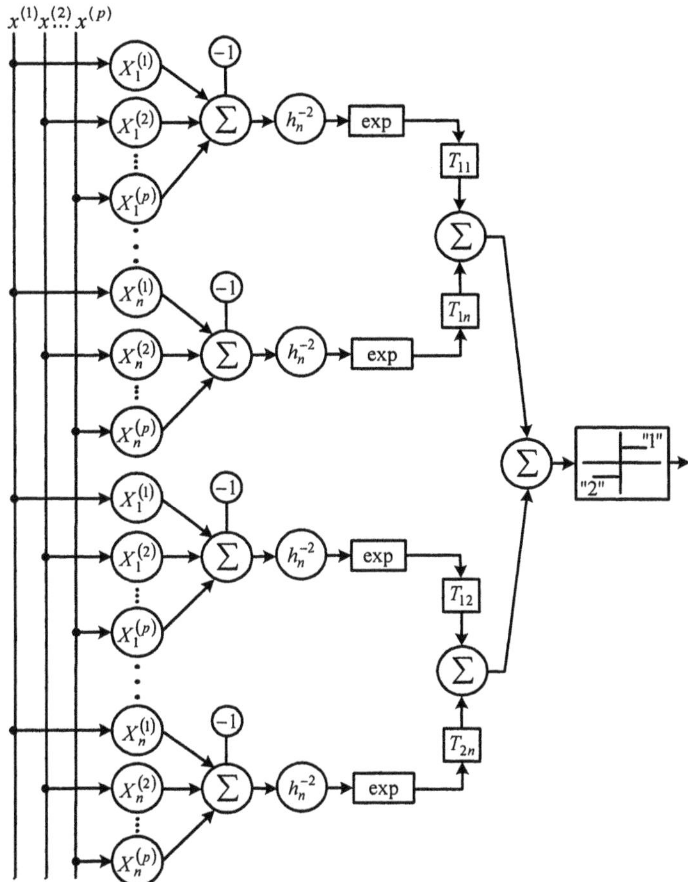

FIGURE 3.20. Generalized regression neural network based on the Parzen kernel for pattern classification ($M = 2$)

A recursive version of estimate (3.96) is given by

$$\widehat{d}_{j,n}(x) = \frac{1}{n}\sum_{i=1}^{n} T_{ji} K_i(x, X_i) \tag{3.98}$$

or alternatively in the form

$$\widehat{d}_{j,n+1}(x) = \widehat{d}_{j,n}(x) + \tag{3.99}$$

$$+\frac{1}{n+1}\left[T_{j,n+1} K_{n+1}(x, X_{n+1}) - \widehat{d}_{j,n}(x)\right]$$

The classification procedure becomes

$$\widehat{\varphi}_n(x) = m \quad \text{if} \quad \sum_{i=1}^{n} T_{mi} K_i(x, X_i) \geq \sum_{i=1}^{n} T_{ji} K_i(x, X_i) \tag{3.100}$$
$$\text{for} \quad i \neq m, \; i = 1, ..., M$$

The probabilistic neural network realizing procedure (3.100) is shown in Fig. 3.21 whereas its Parzen kernel version is depicted in Fig. 3.22.

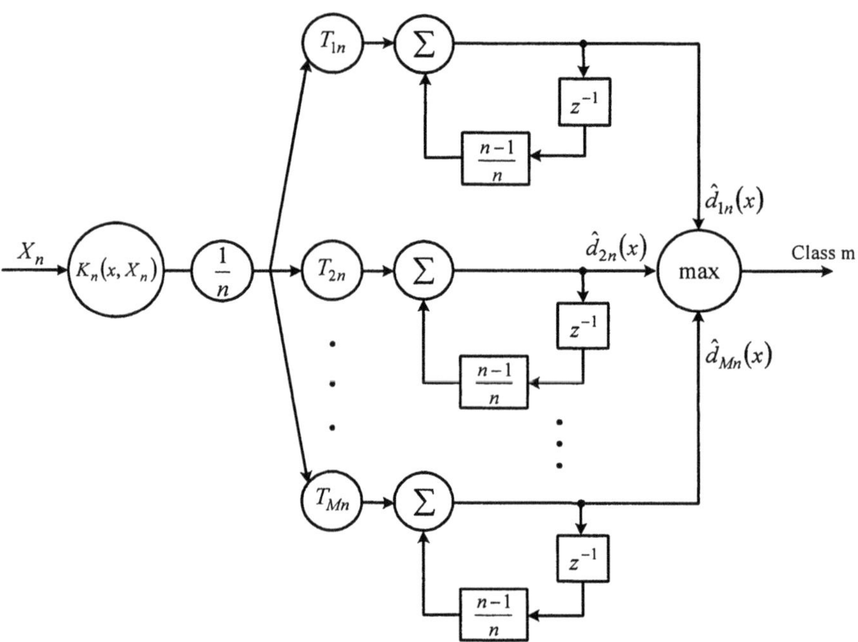

FIGURE 3.21. Recursive generalized regression neural network for pattern classification

The net in Fig. 3.22 consists of one neuron in the first layer having p inputs – coordinates of vector X_n, $n = 1, 2, \ldots$. Let us notice that the role of weights is played by the coordinates of vector x. The second layer consists of two neurons with the feedback typical for recurrent neural networks. When we classify L patterns, then the proposed structure should be copied L times. We shall obtain a neural network consisting of $3L$ neurons processing input observations in a parallel way.

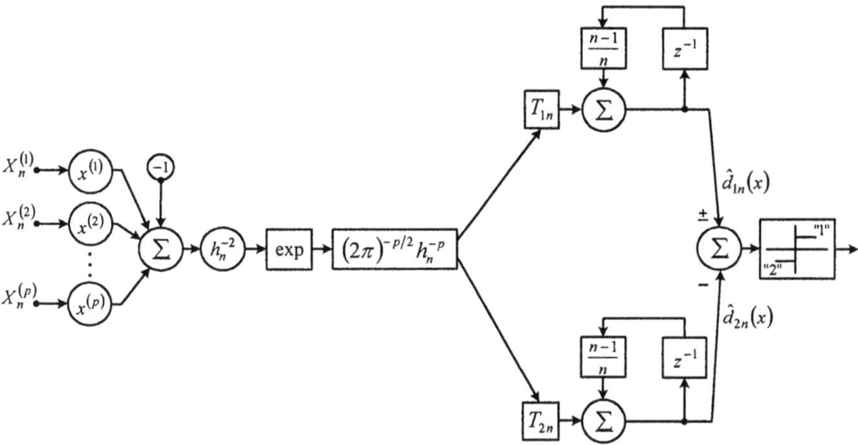

FIGURE 3.22. Recursive generalized regression neural network based on the Parzen kernel for pattern classification ($M = 2$)

Define

$$\widetilde{d}_{j,n}(x) = \sum_{i=1}^{n} T_{ji} K_i(x, X_i) \tag{3.101}$$

Simplified PNN realizing classification procedure (3.100) with function (3.101) is shown in Fig. 3.23 whereas its Parzen kernel version is depicted in Fig. 3.24.

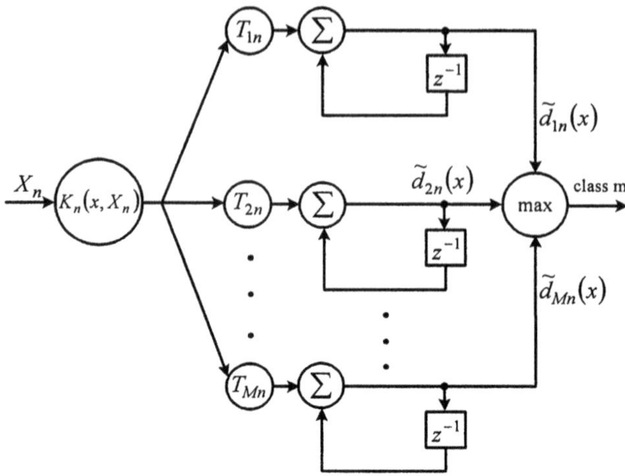

FIGURE 3.23. Simplified recursive generalized regression neural network for pattern classification

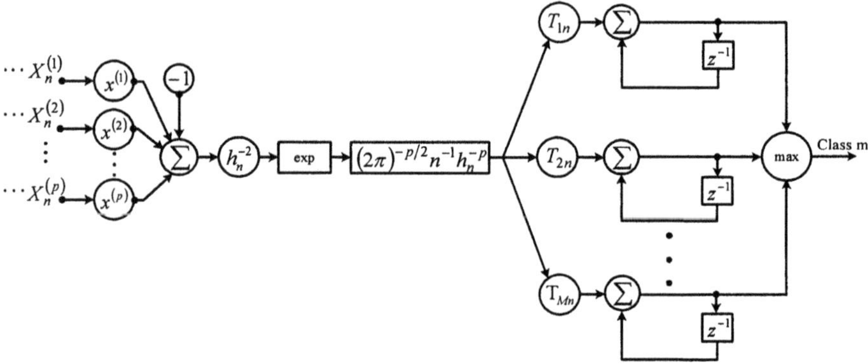

FIGURE 3.24. Simplified recursive generalized regression neural network based on the Parzen kernel for pattern classification

3.5 Concluding remarks

In this chapter we presented probabilistic neural networks based on the Parzen kernels and the orthogonal series. Other techniques for non-parametric estimation include maximum likelihood estimators and maximum penalized likelihood estimators [272], spline aproximation [283] – [288] and the nearest neighbor method [57], [66], [69], [156]. A very important problem is concerned with the choice of smoothing parameters. In literature several methods of data driven choices of smoothing parameters have been proposed [127]. Among them the most popular are: the reference method, the corss-validation method and the plug-in method.

4

General Learning Procedure in a Time-Varying Environment

4.1 Introduction

As we mentioned in Chapter 1, the problems of learning in non-stationary situations were rarely a subject of studies, even in the parametric case. Historically the first papers on this topic where occasionally published in the sixties and seventies. The proper tool for solving such a type of problems seemed to be the dynamic stochastic approximation technique [76], [77] as an extension of Robbins-Monro [201] and Kiefer-Wolfowitz [135] procedures for the non-stationary case. The traditional procedure of stochastic approximation was also used [290], [304] with a good effect for tracking the changing regression function root. The dynamic stochastic approximation procedure was also used for the construction of algorithms tracking time-varying moments of random variables with non-stationary probability distributions [53]. Some other results concerning learning in a time-varying environment are scattered in literature (see e.g. [111], [134], [202]).

In previous papers and books on learning in a time-varying environment a parametric approach was assumed [89], [277]. In practice, the degree of a priori knowledge is very low, which requires the developing of nonparametric learning methods. In this chapter, we will present a general learning procedure in the case when unknown probability distributions change with time. We will show that this procedure may be used for solving various identification, pattern classification and prediction problems. We will present and prove a set of theorems concerning procedure convergence, analyzing also the speed of convergence. It should be emphasized that in the last decade the problem of learning in a time varying environment has been almost intacted in the world literature. The results of this chapter will be a starting point in Chapter 5 to derive generalized regression neural networks in a time-varying environment, and in Chapter 6 to derive probabilistic neural networks for pattern classification in a time-varying environment.

4.2 Problem description

Let us consider a sequence $\{(X_n, Y_n)\}$, $n = 1, 2, ...$, of independent pairs of random variables, where
X_n – random variables having a probability density f_n taking values in the set $A \subset R^p$
Y_n – random variables taking values in the set $B \subset R$.
We assume that probability distributions of the above random variables are completely unknown.

Let us define the following function:

$$R_n(x) \stackrel{df}{=} f_n(x) E\left[Y_n \left| X_n = x\right.\right], \quad n = 1, 2, ... \tag{4.1}$$

From the assumption that the probability distributions are completely unknown, it follows that the sequence of functions (4.1) is also unknown. In this chapter, the goal of learning will be tracking the changing function R_n, $n = 1, 2, ...$.
Later we will show that, thanks to the fact that the learning goal was formulated in this way, various problems of identification, pattern classification and prediction can be solved.

4.3 Presentation of the general learning procedure

Now, the problem of construction – on the basis of the learning sequence $(X_1, Y_1), ..., (X_n, Y_n)$ – of a non-parametric procedure that would realize the learning goal specified in Section 4.2 arises. Thanks to proper processing of current information (observation of the learning sequence), this procedure should compensate for the lack of a priori information about probability distributions and relatively precisely track the changes of characteristics (4.1).

Let $\{K_n\}$, $n = 1, 2, ...$, be a sequence of bivariate Borel-measurable functions (so-called kernel functions) defined on $A \times A$ (examples of such functions were given in Chapter 2) and let $\{a_n\}$ be a sequence of numbers satisfying the following conditions:

$$a_n > 0, \; a_n \xrightarrow{n} 0, \; \sum_{n=1}^{\infty} a_n = \infty \tag{4.2}$$

In this chapter, we will consider a non-parametric learning procedure of the following type:

$$\widehat{R}_{n+1}(x) = \widehat{R}_n(x) + a_{n+1}\left[Y_{n+1}K_{n+1}(x, X_{n+1}) - \widehat{R}_n(x)\right] \tag{4.3}$$

$$n = 0, 1, 2, ... \quad \widehat{R}_0(x) = 0$$

A structural scheme of the system that realizes the learning algorithm (4.3) is shown in Fig. 4.1.

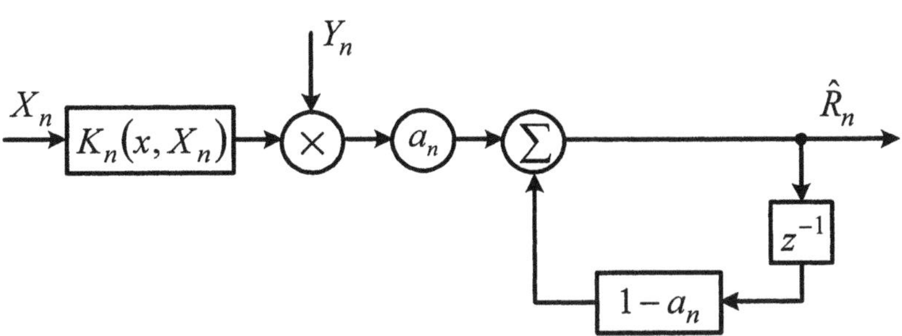

FIGURE 4.1. Block diagram of procedure (4.3)

Below, we will present various problems that can be solved with the use of procedure (4.3).

a) If $\{X_n\}$ is a sequence of random variables with identical probability distributions $f_n = f$, i.e.

$$R_n(x) = f(x)E\left[Y_n \,|X_n = x\right] \tag{4.4}$$

then procedure (4.3) can be used for the construction of identification algorithms of non-stationary plants, because function ϕ_n^* defined by

$$\phi_n^*(x) \stackrel{def}{=} R_n(x)/f(x) \tag{4.5}$$

is a characteristic of the best model (in moment n) of a non-stationary plant.

b) If $\{Y_n\}$ is a sequence of discrete random variables, then procedure (4.3) can be used for construction of empirical discrimination functions, because – as we will show in Chapter 6 – the expression

$$R_n(x) = f_n(x)E\left[Y_n \left| X_n = x\right.\right]$$

has an interpretation of the j-th discriminant function in moment n, i.e.

$$R_n(x) = p_{jn} f_{jn}(x) \tag{4.6}$$

for the fixed $j \in \{1, ..., M\}$ (p_{jn} is an a priori probability of occurrence of class j in moment n and f_{jn} is the probability density of this class). The problem of pattern classification in non-stationary situations will be discussed in detail in Chapter 6.

c) If $Y_n = 1$ (with probability 1) for $n = 1, 2, ...$, i.e.

$$R_n(x) = f_n(x) \tag{4.7}$$

then procedure (4.3) can be used for the tracking of the changing probability density function. This problem will be discussed in Chapter 6.

Remark 4.1

Algorithms that are similar in construction to procedure (4.3) were considered by other authors in the context of non-parametric estimation of the regression function (identification of static characteristics of stationary plants)

$$\phi(x) = E\left[Y_n \left| X_n = x\right.\right], \quad n = 1, 2, ..., \tag{4.8}$$

where $\{(X_n, Y_n)\}$ is a sequence of random variables with identical probability distributions:

a) Aizerman, Braverman and Rozonoer [6] used the following algorithm

$$\widehat{\phi}_{n+1}(x) = \widehat{\phi}_n(x) + \gamma_{n+1}(Y_{n+1} - \widehat{\phi}_n(X_{n+1}))\overline{K}\left(x, X_{n+1}\right) \tag{4.9}$$

where
$\overline{K}$ – potential function (appropriately selected function)
γ_n – sequence of numbers satisfying the following conditions:

$$\gamma_n > 0, \; \sum_{n=1}^{\infty} \gamma_n = \infty, \; \sum_{n=1}^{\infty} \gamma_n^2 < \infty. \tag{4.10}$$

b) Révész [197] ($A = [0,1], p = 1$) proposed algorithm

$$\widehat{\phi}_{n+1}(x) = \widehat{\phi}_n(x) + \frac{1}{n+1}\left(Y_{n+1} - \widehat{\phi}_n(x)\right) K_n(x, X_n), \tag{4.11}$$

where K_n is the sequence that was used for the construction of procedure (4.3).
The potential functions algorithm is closely connected with a stationary situation, whereas the Revesz's algorithm could be modified (by means of replacing sequence n^{-1} with a more general sequence of numbers) in an attempt to enhance it with tracking properties in a non-stationary situation. However, this algorithm is convergent, even in a relatively simple stationary case, under very complicated assumptions concerning function ϕ.

4.4 Convergence of general learning procedure

The learning process will be successful if the accomplishment of the learning goal is guaranteed. In other words, the convergence of procedure (4.3) in a specified sense is required. We will prove the convergence of procedure (4.3) both in the local (pointwise) and global (integral) sense. Moreover, we will investigate the speed of the convergence.

Remark 4.2
Throughout this chapter, assumption (4.2) concerning sequence $\{a_n\}$ remains valid. Moreover, herein we will use notation

$$r_n(x) = E\left[Y_n K_n(x, X_n)\right] \tag{4.12}$$

where $\{K_n\}$ is a sequence that occurs in procedure (4.3).

4.4.1 Local properties

The measure of quality of the learning process in a given point $x \in A$ can be

$$I_n(x) = \left|\widehat{R}_n(x) - R_n(x)\right| \tag{4.13}$$

Of course, sequence $I_n(x)$ in a given point $x \in A$ is a sequence of random variables. We will show that

$$EI_n^2(x) \xrightarrow{n} 0 \text{ and } I_n(x) \xrightarrow{n} 0 \text{ with pr. 1} \tag{4.14}$$

Theorem 4.1
If in a certain point x, the following conditions are satisfied

$$a_n \text{ var } [Y_n K_n(x, X_n)] \xrightarrow{n} 0 \tag{4.15}$$

$$a_n^{-1} |r_n(x) - R_n(x)| \xrightarrow{n} 0 \tag{4.16}$$

$$a_n^{-1} |R_{n+1}(x) - R_n(x)| \xrightarrow{n} 0 \tag{4.17}$$

then

$$EI_n^2(x) \xrightarrow{n} 0 \tag{4.18}$$

Theorem 4.2
If in a certain point x, the following conditions are satisfied

$$\sum_{n=1}^{\infty} a_n^2 \text{ var } [Y_n K_n(x, X_n)] < \infty \tag{4.19}$$

$$\sum_{n=1}^{\infty} a_n^{-1} (r_n(x) - R_n(x))^2 < \infty \tag{4.20}$$

$$\sum_{n=1}^{\infty} a_n^{-1} (R_{n+1}(x) - R_n(x))^2 < \infty \tag{4.21}$$

then

$$I_n(x) \xrightarrow{n} 0 \text{ with pr. 1} \tag{4.22}$$

4.4.2 Global properties

As the global measure of the learning process quality, we take

$$I_n = \int \left(\widehat{R}_n(x) - R_n(x)\right)^2 dx \tag{4.23}$$

We will show that

$$EI_n \xrightarrow{n} 0 \text{ and } I_n \xrightarrow{n} 0 \text{ with pr. } 1 \tag{4.24}$$

Theorem 4.3
If the following conditions are satisfied

$$a_n \int \text{var } [Y_n K_n(x, X_n)] \, dx \xrightarrow{n} 0 \tag{4.25}$$

$$a_n^{-2} \int (r_n(x) - R_n(x))^2 \, dx \xrightarrow{n} 0 \tag{4.26}$$

$$a_n^{-2} \int (R_{n+1}(x) - R_n(x))^2 \, dx \xrightarrow{n} 0 \tag{4.27}$$

then

$$EI_n \xrightarrow{n} 0$$

Theorem 4.4
If the following conditions are satisfied

$$\sum_{n=1}^{\infty} a_n^2 \int \text{var } [Y_n K_n(x, X_n)] \, dx < \infty \tag{4.28}$$

$$\sum_{n=1}^{\infty} a_n^{-1} \int (r_n(x) - R_n(x))^2 \, dx < \infty \tag{4.29}$$

$$\sum_{n=1}^{\infty} a_n^{-1} \int (R_{n+1}(x) - R_n(x))^2 \, dx < \infty \tag{4.30}$$

then

$$I_n \xrightarrow{n} 0 \text{ with pr. } 1 \tag{4.31}$$

Example 4.1
Conditions (4.17), (4.21), (4.27) and (4.30), concerning the way in

which functions R_n ($n = 1, 2, ...$) change, take a simpler form depending on the type of non-stationarity. For example, let us assume that

$$R_n(x) = \alpha_n R(x) \tag{4.32}$$

Then, conditions (4.17) and (4.21), as well as (4.27) and (4.30) if $R \in L_2$, can be written as

(i)

$$a_n^{-1} |\alpha_{n+1} - \alpha_n| \xrightarrow{n} 0 \tag{4.33}$$

(ii)

$$\sum_{n=1}^{\infty} a_n^{-1} (\alpha_{n+1} - \alpha_n)^2 < \infty \tag{4.34}$$

Let us assume that sequence a_n in procedure (4.3) is of the type

$$a_n = k/n^a, \quad k > 0, \quad 0 < a < 1 \tag{4.35}$$

and sequence α_n is of the type

$$\alpha_n = \text{ const. } n^t, \quad t > 0 \tag{4.36}$$

Now, the following question arises: how fast can sequence α_n diverge to infinity so that the conditions (4.33) and (4.34) could be met. It turns out that condition (4.33) is satisfied if

$$0 < t < 1 \tag{4.37}$$

whereas condition (4.34) is satisfied if

$$0 < t < \frac{1}{2} \tag{4.38}$$

The non-stationarity of the above type will be discussed in detail in Chapter 5 – in the context of modelling non-stationary plants, taking into account all conditions of Theorems 4.1 and 4.2.

Remark 4.3

As mentioned before, in works [76], [77], [252], [278], [279] and [290], their authors used the stochastic approximation procedure for tracking the changing root θ_n of the regression function assuming that

$$\gamma_n^{-1} |\theta_{n+1} - \theta_n| \xrightarrow{n} 0 \tag{4.39}$$

or

$$\gamma_n^{-1} \left|\theta_{n+1} - \nu_n\left(\theta_n\right)\right| \xrightarrow{n} 0 \tag{4.40}$$

where γ_n is a sequence characterizing the stochastic approximation procedure and ν_n is a sequence of known functions. Thus, conditions (4.17), (4.21), (4.27) and (4.30) – concerning the way in which function R_n $(n = 1, 2, ...)$ changes, correspond to conditions that were given in the abovementioned works. Conditions of this type seem to characterize various problems of learning in non-stationary situations, if appropriate procedures are expected to have tracking properties.

4.4.3 Speed of convergence

We will investigate which factòrs are responsible for the speed of convergence of procedure (4.3), both in the local and global sense. The results will be presented in the form of two theorems.

Theorem 4.5
If the following conditions are satisfied

$$\text{var}\ \left[Y_n K_n\left(x, X_n\right)\right] = 0\left(n^{A_1}\right), \quad A_1 > 0 \tag{4.41}$$

$$\left|R_{n+1}\left(x\right) - R_n\left(x\right)\right| = 0\left(n^{-B_1}\right), \quad B_1 > 0 \tag{4.42}$$

$$\left|R_n\left(x\right) - r_n\left(x\right)\right| = 0\left(n^{-C_1}\right), \quad C_1 > 0 \tag{4.43}$$

$$a_n = \frac{k}{n^a}, \quad k > 0, \quad 0 < a \leq 1 \tag{4.44}$$

then

$$EI_n^2\left(x\right) \leq l_1 n^{-2C_1} + l_2 n^{-r_1} \tag{4.45}$$

where l_1, l_2 – positive constants and

$$r_1 = \min\left\{a - A_1,\ 2\left(B_1 - a\right),\ 2\left(C_1 - a\right)\right\} \tag{4.46}$$

with $r_1 > 0$ for $0 < a < 1$ and $0 < r_1 < 1$ for $a = 1$.

Theorem 4.6
If the following conditions are satisfied

$$\int \text{var}\ \left[Y_n K_n\left(x, X_n\right)\right] dx = 0\left(n^{A_2}\right), \quad A_2 > 0 \tag{4.47}$$

$$\int \left(R_{n+1}(x) - R_n(x)\right)^2 dx = 0\left(n^{-B_2}\right), \quad B_2 > 0 \tag{4.48}$$

$$\int \left(R_n(x) - r_n(x)\right)^2 dx = 0\left(n^{-C_2}\right), \quad C_2 > 0 \tag{4.49}$$

$$a_n = \frac{k}{n^a}, \quad k > 0, \quad 0 < a \leq 1 \tag{4.50}$$

then

$$EI_n \leq l_3 n^{-C_2} + l_4 n^{-r_2} \tag{4.51}$$

where l_3 and l_4 are positive constants and $r_2 = \min[a - A_2, B_2 - 2a, C_2 - 2a]$ with $r_2 > 0$ for $0 < a < 1$ and $0 < r_2 < 1$ for $a = 1$.

4.5 Quasi-stationary environment

It is obvious that procedure (4.3) can be applied in a special situation, when the non-stationarity of probability distributions decays with the increase of the number of observations, i.e.

$$R_n(x) = f_n(x)\, E\left[Y_n \mid X_n = x\right] \xrightarrow{n} R(x) \tag{4.52}$$

This situation is called the quasi-stationary situation. Convergence of procedure (4.3) in this case is a consequence of general theorems given in Section 4.4. It turns out that the conditions presented in Section 4.4 can be weakened in a quasi-stationary situation. For this purpose, in procedure (4.3), sequence $a_n = n^{-1}$ should be taken. Then procedure (4.3) can be written as

$$\widehat{R}_n(x) = n^{-1} \sum_{i=1}^{n} Y_i K_i(x, X_i) \tag{4.53}$$

The last two theorems allow the examination of the speed of the convergence of algorithms presented in Chapters 5 and 6.

Theorem 4.7
(Pointwise convergence of procedure (4.53) in quasi-stationary situations). Let us assume that

$$R_n(x) \xrightarrow{n} R(x) \tag{4.54}$$

and

$$r_n(x) \xrightarrow{n} R(x) \tag{4.55}$$

a) If

$$n^{-2}\sum_{i=1}^{n}\text{var }\left[Y_iK_i\left(x,X_i\right)\right]\xrightarrow{n}0 \tag{4.56}$$

then

$$EI_n^2\left(x\right)\xrightarrow{n}0 \tag{4.57}$$

b) If

$$\sum_{n=1}^{\infty}n^{-2}\text{ var }\left[Y_nK_n\left(x,X_n\right)\right]<\infty \tag{4.58}$$

then

$$I_n\left(x\right)\xrightarrow{n}0\ \text{ with pr. 1} \tag{4.59}$$

Remark 4.4
It should be noted that when we use Theorem 4.1 to establish convergence (4.57) we have to assume conditions (4.16) and (4.17) given by

$$a_n^{-1}\left|r_n\left(x\right)-R\left(x\right)\right|\xrightarrow{n}0,\quad a_n^{-1}\left|R_n\left(x\right)-R\left(x\right)\right|\xrightarrow{n}0 \tag{4.60}$$

Since $a_n \xrightarrow{n} 0$, conditions (4.54) and (4.55) of Theorem 4.7 are much weaker than those of Theorem 4.1. The same conclusion refers to convergence (4.59).

4.6 Problem of prediction

It would be interesting to investigate if procedure (4.3) on the basis of learning set

$$\left(X_1,Y_1\right),...,\left(X_n,Y_n\right) \tag{4.61}$$

allows to predict

$$R_{n+k}\left(x\right)=f_{n+k}\left(x\right)E\left[Y_{n+k}\left|X_{n+k}=x\right.\right] \tag{4.62}$$

for $k\geq 1$. In the considered situation, performance measures (4.13) and (4.23) take the form

$$I_{n,k}\left(x\right)=\left|\widehat{R}_n\left(x\right)-R_{n+k}\left(x\right)\right| \tag{4.63}$$

and

$$I_{n,k}=\int\left(\widehat{R}_n\left(x\right)-R_{n+k}\left(x\right)\right)^2dx \tag{4.64}$$

The following result is a corollary from Theorems 4.1 – 4.4 and 4.7 and allows to predict R_{n+k}, $k \geq 1$, on the basis of a learning set of the length n.

Corollary 4.1
(i) If the assumptions of Theorem 4.1 (4.7a) are satisfied, then

$$EI_{n,k}^2(x) \xrightarrow{n} 0 \tag{4.65}$$

(ii) If the assumptions of Theorem 4.2 (4.7b) are satisfied, then

$$I_{n,k}(x) \xrightarrow{n} 0 \text{ with pr. } 1 \tag{4.66}$$

(iii) If the assumptions of Theorem 4.3 are satisfied, then

$$EI_{n,k} \xrightarrow{n} 0 \tag{4.67}$$

(iv) If the assumptions of Theorem 4.4 are satisfied, then

$$I_{n,k} \xrightarrow{n} 0 \text{ with pr. } 1 \tag{4.68}$$

The next corollary follows immediately from Theorems 4.5 and 4.6.

Corollary 4.2
(i)

$$EI_{n,k}^2(x) \leq l_1 n^{-2C_1} + l_2 n^{-r_1} + l'(k)\, n^{-2B_1} \tag{4.69}$$

(ii)

$$EI_{n,k} \leq l_3 n^{-C_2} + l_4 n^{-r_2} + l''(k)\, n^{-B_2} \tag{4.70}$$

Above, symbols $l'(k)$ and $l''(k)$ denote positive constants which depend on k, the rest of the symbols are identical as in Theorems 4.5 and 4.6. Of course, the more steps k in prediction, the bigger the value of the right side of expressions (i) and (ii) because $l'(k)$ and $l''(k)$ increase with bigger k (this is shown in the proof of the above corollary).

4.7 Concluding remarks

Thanks to the general formulation of the problem of non-parametric learning, particularly thanks to the presentation of a set of general theorems on convergence, the results obtained can be used for solving various problems. In all theorems some conditions were imposed on sequence K_n but the form of the sequence was not precisely specified. Of course, for each of the 7 Theorems 4.1 – 4.7, the two corresponding corollaries related to the Parzen's kernel or the orthogonal series method could be formed. Appropriate corollaries from Theorems 4.1 – 4.7 will be drawn in Chapters 5 and 6 that are devoted to particular applications of procedure (4.3). Note that the general formulation of Theorems 4.1 – 4.7 does not exclude the selection of other functions K_n (that could perhaps be more suitable from the viewpoint of practical applications). It is worth pointing out that procedure (4.3) can be modified in the spirit of the dynamic stochastic approximation algorithm [76], [77] and [279]. It would then take the following form

$$\begin{gathered}\widehat{R}_{n+1}(x) = \widehat{R}_n^{\Delta}(x) + a_{n+1} \\ \cdot\left(Y_{n+1}K_{n+1}(x, X_{n+1}) - \widehat{R}_n^{\Delta}(x)\right), \qquad (4.71) \\ \widehat{R}_n^{\Delta}(x) = W_n\left(\widehat{R}_n^{\Delta}(x)\right), \quad \widehat{R}_0(x) = 0, \quad n = 0, 1, 2, \ldots,\end{gathered}$$

where the sequence of function W_n is known. Assumptions (4.17), (4.21), (4.27) and (4.30) change accordingly. For example, condition (4.17) takes the form

$$a_n^{-1}\left|R_{n+1}(x) - W_n(R_n(x))\right| \xrightarrow{n} 0. \qquad (4.72)$$

In a parametric case, similar modifications of classic Robbins-Monro procedures allow a significant widening of the class of the non-stationary cases. Applying the classic (non-modified) procedure of stochastic approximation, we can track the changing root of the regression function $\theta_n = \text{const.}\, n^t$ if $0 < t < 1$ [76]. The procedures of the dynamic stochastic approximation additionally allow us to consider the situation where $t > 1$ [76], [79] and [279]. In Chapter 5, we will mention the possibility of the application of procedure (4.71) for modelling plants with multiplicative and additive non-stationarity. However, contrary to procedures of dynamic stochastic approximation, the above modification of algorithm (4.3) widens the class of non-stationary cases only to a limited extent.

5

Generalized Regression Neural Networks in a Time-Varying Environment

5.1 Introduction

The generalized regression neural network (GRNN) was introduced by Specht [257] to perform general (linear or non-linear) regressions. The GRNN was applied to solve a variety of problems [180] like prediction, control, plant process modelling or general mapping problems. Other stochastically – based neural networks, the so-called probabilistic neural networks, are used for classification as we have mentioned in Chapter 3. An interesting study presenting a bridge between non-parametric estimation and artificial neural networks is given in [303]. The concept of the GRNN is based on non-parametric regression estimation commonly used in statistics. The essence of non-parametric estimation is non-limiting to an assumed – usually in an arbitrary way – parametric class of models. Such approach was applied by many authors (see e.g. [3], [4], [25], [62], [64] – [66], [68], [69], [91] – [94], [102], [138], [158], [190], [192], [209] – [214], [218] – [221], [223], [226] – [229], [261] – [263]) who created non-parametric algorithms based on the Parzen method, the orthogonal series or the nearest neighbor methods. More precisely, in stationary regression analysis we consider a random vector (X, Y), where X is R^p – valued and Y is R – valued. The problem is to find a (measurable) function $\phi : R^p \to R$ such that the L_2 risk

$$E\left[\phi(X) - Y\right]^2$$

attains minimum. The solution is the regression function

$$\phi^*(x) = E\left[Y \mid X = x\right]$$

Non-parametric procedures and generalized regression neural networks approach the best solution $\phi^*(x)$, as the sample size grows large.

The non-parametric methods discussed above could be applied only in stationary situations – where probability distributions do not change with time. However, in many cases the assumption concerning stationarity may be false, because usually properties of various processes depend on time. It is possible to enumerate the following examples: (i) the production process in an oil refinery, where non-stationarity is a result of a change of catalyst properties, (ii) the process of carbon dioxide conversion, where non-stationarity is also a result of catalyst aging, (iii) the vibrations of the atmosphere around a starting space rocket are a non-stationary process, because the force that stimulates the rocket to start is a function of parameters that change quickly, such as the speed of the rocket and the distance from the earth's surface, (iv) the converter-oxygen process of steelmaking, when thermal conditions in the converter may change between melts. In literature, there are three best-known parametric methods for modelling non-stationary systems (see e.g. [38], [253]):

a) Movable models method; for modelling non-stationary systems, the classic method of minimum squares is used and the data set is constantly updated through the elimination of the oldest data and simultaneous feeding of the newest data. The period of time during which the data set is collected is called the observation horizon.

b) Method based on the criterion of the minimum weighted sum squares; the minimum squares method is also used, but the elimination of the oldest data is carried out through assigning decreasing weights in the criterion of the minimum weighted sum squares.

c) Method of dynamic stochastic approximation; characteristics of the non-stationary plants are approximated by a linear model having time-varying coefficients which are estimated by means of the dynamic stochastic approximation method.

An important problem in method a) is the optimization of the observation horizon and in method b), the selection of weight coefficients. Unfortunately, the solution of such problems depends on the possession of a relative number of a priori information, such as e.g. the character of non-stationarity, the variance of disturbances and the form of the input signal. Similarly, a disadvantage of method c) is a necessity to know the way in which the linear model coefficients change.

Methods a), b) and c) that were discussed above do not allow to track the changing characteristics of the best models described by time-varying regression functions. Such a property is possessed by the GRNN constructed in this chapter. In the non-stationary regression we consider a sequence of random variables $\{X_n, Y_n\}$, $n = 1, 2, ...,$ having time-varying cumulative probability density functions $f_n(x, y)$. The problem is to find a measurable function $\phi_n : R^p \to R$ such that the L_2 risk

$$E\left[\phi_n(X) - Y\right]^2 \tag{5.1}$$

attains minimum. The solution is the regression function

$$\phi_n^*(x) = E\left[Y_n \left|X_n = x\right.\right], \quad n = 1, 2, ..., \tag{5.2}$$

changing with time.

In this chapter we propose a new class of generalized regression neural networks working in a non-stationary environment. The general regression neural networks studied in the next sections are able to follow changes of the best model i.e. time-varying regression functions given by (5.2). The novelty of this chapter is summarized as follows:

1. We present the adaptive GRNN tracking time-varying regression functions.

2. We prove the convergence of the GRNN based on general learning theorems presented in Chapter 4.

3. We design in detail special GRNN based on the Parzen and the orthogonal series kernels. In each case we precise conditions ensuring the convergence of the GRNN to the best models given by (5.2).

4. We investigate the speed of the convergence of the GRNN and compare the performance of specific structures based on the Parzen kernel and the orthogonal series kernel.

5. We study various non-stationarities (multiplicative, additive, "scale change", "movable argument") and design in each case the GRNN based on the Parzen kernel and the orthogonal series kernel.

As we have mentioned, the current state of knowledge as regards non-stationary processes is significantly poorer than in the case of stationary signals. In many applications signals are treated as stationary only because in this way it is easier to analyse them; in fact, they are non-stationary. Non-stationary processes are undoubtedly more difficult to analyse and their diversity makes application of universal tools impossible. In this context our results seem to be a significant contribution to the development of new techniques in the area of non-stationary signals. More specifically, our proposition advances the current state of knowledge in the following fields: a) stochastic – based neural networks, b) non-parametric regression estimation, c) modelling of time-varying plants. It should be emphasized that the methodology proposed in this chapter allows to solve problems that earlier could have been treated as "impossible to solve". For the illustration of the capability of our GRNN we may consider an application to modelling of non-stationary plants described by

$$Y_n = \phi_n^* (X_n) + Z_n$$

where ϕ_n^* is given by (5.2). Suppose that

(i)

$$\phi_n^* (x) = \alpha_n \phi (x)$$

(ii)

$$\phi_n^* (x) = \phi (x) + \beta_n$$

(iii)

$$\phi_n^* (x) = \phi (x \omega_n)$$

(iv)

$$\phi_n^*(x) = \phi(x - \lambda_n)$$

In the next sections, based on the learning sequence $(X_1, Y_1), (X_2, Y_2), \ldots$, we design the GRNN that allow to track $\phi_n^*(x)$ in cases (i) – (iv) despite the fact that we do not known the function ϕ and sequences $\alpha_n, \beta_n, \omega_n$ or λ_n.

5.2 Problem description and presentation of the GRNN

The problem of non-parametric regression boils down to finding an adaptive algorithm that could follow the changes of optimal characteristics expressed by formula (5.2). This algorithm should be constructed on the basis of a learning sequence, i.e. observations of the following random variables

$$(X_1, Y_1), (X_2, Y_2), \ldots$$

We assume that pairs of the above random variables are independent. In points x, where $f(x) \neq 0$, the characteristics of the best model (5.2) can be expressed as

$$\phi_n^*(x) = R_n(x)/f(x), \quad n = 1, 2, \ldots,$$

where $R_n(x) = \phi_n^*(x) f(x)$ corresponds to (4.1) if $f_n = f$. Because of this, the adaptive algorithm that is able to follow changes of unknown characteristics of the best model ϕ_n^*, will be constructed on the basis of a general procedure (4.3). The algorithm has the form

$$\widehat{\phi}_n(x) = \widehat{R}_n(x)/\widehat{f}_n(x) \tag{5.3}$$

where $\widehat{R}_n$ is expressed by means of formula (4.3) which is repeated here for the reader's convenience

$$\widehat{R}_{n+1}(x) = \widehat{R}_n(x) + a_{n+1}\left[Y_{n+1}K_{n+1}(x, X_{n+1}) - \widehat{R}_n(x)\right]$$

for $n = 0, 1, 2, \ldots$, $\widehat{R}_0(x) = 0$ and $\widehat{f}_n$ is a recurrent estimator of the density of the input signal f (see Section 3.2):

$$\widehat{f}_{n+1}(x) = \widehat{f}_n(x) + \frac{1}{n+1}\left(K_{n+1}(x, X_{n+1}) - \widehat{f}_n(x)\right) \tag{5.4}$$

Comparing (5.4) and (4.3) we realize that algorithm (5.4) is a special case of the general procedure (4.3); if $a_n = n^{-1}$and $Y_n = 1$. In Fig. 5.1 we show the block diagram of the GRNN applied to modelling non-stationary plants. It is understandable that sequences K_n that are in the numerator and denominator of expression (5.3) can be of a different type. If sequences K_n are of the same type (e.g. based on the Parzen kernel), they generally should meet different conditions. In the diagram (Fig. 5.1) of the system that realizes algorithm (5.3) sequence K_n' present in the numerator of (5.3) was differentiated from sequence K_n'' present in the denominator of that expression. In a situation where there is no doubt, corresponding indexes in sequences h_n' and h_n'' as well as $q^{'}(n)$ and $q''(n)$ will be omitted.

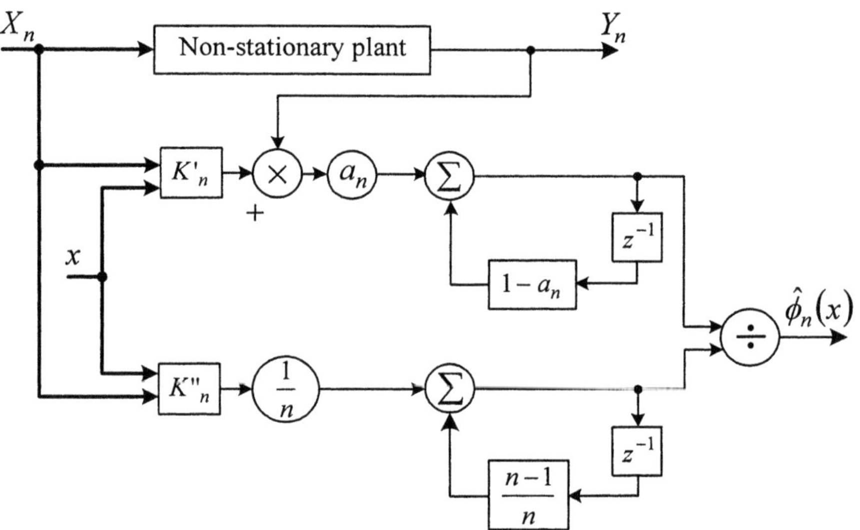

FIGURE 5.1. Block diagram of the GRNN applied to modelling of non-stationary plant

5.3 Convergence of the GRNN in a time-varying environment

The theorem presented below describes general conditions ensuring the convergence of algorithm (5.3).

Theorem 5.1
(pointwise convergence of algorithm (5.3) in probability and with pr.1). Let us assume that the following conditions are satisfied:
(i) Condition A:

$$\left|\widehat{R}_n(x) - R_n(x)\right| \xrightarrow{n} 0 \text{ in prob. (with prob. 1)} \tag{5.5}$$

(ii) Condition B:

$$\widehat{f}_n(x) \xrightarrow{n} f(x) > 0 \text{ in prob. (with prob. 1)} \tag{5.6}$$

(iii) Condition C:

$$|\phi_n^*(x)| \left|\widehat{f}_n(x) - f(x)\right| \xrightarrow{n} 0 \text{ in prob. (with prob. 1)} \tag{5.7}$$

Then, for algorithm (5.3) we have

$$\left|\widehat{\phi}_n(x) - \phi_n^*(x)\right| \xrightarrow{n} 0 \text{ in prob. (with prob. 1)} \tag{5.8}$$

Let us point out that condition A is satisfied when conclusions of Theorems 4.1 and 4.2 are true. Condition B reflects the requirement of the convergence of the estimator of the density function (expressed by formula (5.4)) and condition C imposes certain assumptions on the speed of this convergence. Of course, when ϕ_n^* is a bounded sequence, condition C boils down to condition B.

Now we will consider two methods of construction of algorithm (5.3). We will present procedures based on the Parzen kernel and on the orthogonal series method. In both cases, we will present assumptions that guarantee satisfaction of conditions A, B and C and, as a result, convergence (5.8). In this chapter, we use the following symbols

$$m_n' = \sup_x \left\{ \left(\text{var } [Y_n \,|X_n = x] + \phi_n^{*2}(x) \right) f(x) \right\} \tag{5.9}$$

and

$$m_n'' = E\left[Y_n - \phi_n^*(X_n)\right]^2 + \int \phi_n^{*2}(x) f(x)\, dx \tag{5.10}$$

In Sections 5.5 – 5.7, we will discuss in detail plants described by equation

$$Y_n = \phi_n^*(X_n) + Z_n$$

where

$$EZ_n = 0,\ EZ_n^2 = \sigma_z^2$$

In such a situation, expressions (5.9) and (5.10) take the form

$$m_n' = \sup_x \left\{ \left(\sigma_z^2 + \phi_n^{*2}(x) \right) f(x) \right\} \tag{5.11}$$

and

$$m_n'' = \sigma_z^2 + \int \phi_n^{*2}(x) f(x)\, dx \tag{5.12}$$

5.3.1 The GRNN based on Parzen kernels

The structural scheme of the system that realizes algorithm (5.3) on the basis of the Parzen kernel is depicted in Fig. 5.2. In order to differentiate sequences h_n and functions K present in the numerator and denominator of expression (5.3), symbols h_n' and h_n'' as well as K' and K'' are used. Condition A will be connected with the selection of sequence h_n' and conditions B and C with the selection of sequence h_n''.

Now, we will present assumptions that guarantee satisfaction of conditions A, B and C of Theorem 5.1.

a) Condition A

As we remember (Section 2.2), kernel K can be expressed in the following way:

$$K(x) = \prod_{i=1}^{p} H\left(x^{(i)}\right) \tag{5.13}$$

Let us assume that

$$\sup_{v \in R} |H(v)| < \infty \tag{5.14}$$

$$\int_R H(v)\, \mathrm{dv} = 1 \tag{5.15}$$

$$\int_R H(v)\, v^j \mathrm{dv} = 0, \quad j = 1, ..., r-1 \tag{5.16}$$

$$\int_R \left| H(v)\, v^k \right| \mathrm{dv} < \infty,\ k = 1, ..., r \tag{5.17}$$

For $r = 2$, the above conditions are satisfied by most functions H presented in Table 2.1. For $r = 4$, the conditions are met by the function

$$H(v) = \frac{3}{2\sqrt{2\pi}}\left(1 - \frac{v^2}{3}\right)e^{-\frac{1}{2}v^2}$$

also given in Table 2.1.

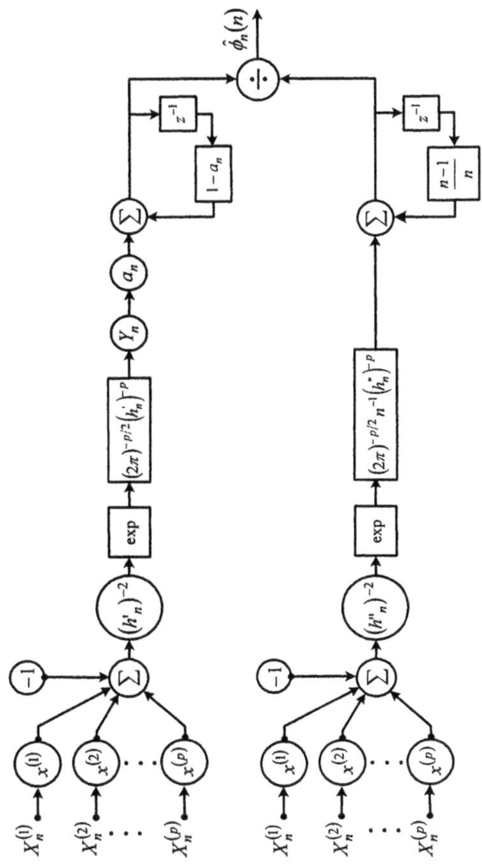

FIGURE 5.2. Block diagram of the GRNN based on the Parzen kernel

Let us introduce the following symbol:

$$D_n^{\underline{i}} = \sup_x \left| \frac{\delta^r}{\delta x^{(i_1)} \ldots \delta x^{(i_r)}} R_n(x) \right| \tag{5.18}$$

where

$$\underline{i} = (i_1, \ldots, i_r), \; i_k = 1, \ldots, p, \; k = 1, \ldots, r$$

We will associate parameter r (Sections 5.5 – 5.7) with smooth properties of function R_n ($n = 1, 2, ...$). The following corollaries from Theorems 4.1 and 4.2 guarantee satisfaction of condition A:

Corollary 5.1
Let us assume that function K satisfies conditions (5.13) – (5.17), $h_n \xrightarrow{n} 0$ and one of the following assumptions holds:

$$a_n h_n^{-p} m_n' \xrightarrow{n} 0 \tag{5.19}$$

or

$$a_n h_n^{-2p} m_n'' \xrightarrow{n} 0 \tag{5.20}$$

If function ϕ_n^* changes with time in such a way that

$$a_n^{-1} \left| \phi_{n+1}^* (x) - \phi_n^* (x) \right| \xrightarrow{n} 0 \tag{5.21}$$

and

$$a_n^{-1} h_n^r D_n^{\frac{i}{}} \xrightarrow{n} 0 \tag{5.22}$$

then

$$E\left[\widehat{R}_n (x) - R_n (x)\right]^2 \xrightarrow{n} 0$$

Corollary 5.2
Let us assume that function K satisfies conditions (5.13) – (5.17), $h_n \xrightarrow{n} 0$ and one of the following assumptions holds:

$$\sum_{n=1}^{\infty} a_n^2 h_n^{-p} m_n' < \infty \tag{5.23}$$

or

$$\sum_{n=1}^{\infty} a_n^2 h_n^{-2p} m_n'' < \infty \tag{5.24}$$

If function ϕ_n^* changes with time in such a way that

$$\sum_{n=1}^{\infty} a_n^{-1} \left(\phi_{n+1}^* (x) - \phi_n^* (x) \right)^2 < \infty \tag{5.25}$$

and

$$\sum_{n=1}^{\infty} a_n^{-1} h_n^{2r} \left(D_n^{\frac{i}{}} \right)^2 < \infty \tag{5.26}$$

then

$$\left|\widehat{R}_n(x) - R_n(x)\right| \xrightarrow{n} 0 \text{ with pr. } 1$$

It is worth mentioning that assumptions (5.19) and (5.23) are weaker than the alternative (5.20) and (5.24) as far as the selection of sequence h_n is concerned. However, while designing the system that realizes algorithm (5.3) one should also take into account information that may be possessed about functions ϕ_n^* and f, because the mentioned assumptions depend also on m_n' and m_n'' (formulas (5.11) and (5.12)). Assumptions (5.22) and (5.26) concern certain conditions of smooth properties of function R_n. As we will see later (Sections 5.5 – 5.7), they take a more concrete form for plants with various types of non-stationarity.

b) Condition B

A recurrent estimator (5.4) of density function with Parzen's kernel was suggested by Wolwerton and Wagner [300] and investigated in detail by Yamato [301], Davies [59], Devroy [63] and Greblicki and Krzyżak [99]. The convergence of this estimator could be obtained by the use of general Theorems 4.1 and 4.2. However, one should remember that these theorems concern a non-stationary situation, so the conditions obtained would be quite strong. Therefore, results concerning the convergence of estimator (5.4) will be taken from references [63] and [99] that were mentioned above. Let us assume that $h_n \xrightarrow{n} 0$ and kernel K satisfies conditions

$$K(x) \geq 0, \quad \int K(x)\,dx = 1, \qquad \sup K(x) < \infty \tag{5.27}$$

Devroye [63] showed that

$$n^{-2} \sum_{i=1}^{n} h_i^{-p} \xrightarrow{n} 0 \tag{5.28}$$

implies

$$\widehat{f}_n(x) \xrightarrow{n} f(x) \tag{5.29}$$

in probability, and

$$\sum_{n=1}^{\infty} n^{-2} h_n^{-p} < \infty \tag{5.30}$$

implies convergence (5.29) with pr.1, whereas both these convergences occur in the following points x:
(i) in every point of continuity of f if

$$\|x\|^p K(x) \xrightarrow{n} 0 \quad \text{when} \quad \|x\| \xrightarrow{n} 0 \tag{5.31}$$

(ii) in Lebesque points of function f if f is bounded,
(iii) in Lebesque points of function f if $K(x)$ satisfies the condition

$$\begin{aligned} K(x) &\neq 0 \text{ when } x \in B \subset R^p, \\ \mu(B) &< \infty, \ K(x) = 0 \text{ when } x \in R^p - B \end{aligned} \tag{5.32}$$

It is worth reminding that Lebesque points are points of continuity of the function and almost all points x. Examples of the functions K satisfying conditions (5.31) and (5.32) are given in Table 2.1. The speed of the convergence of procedure (5.4) can be evaluated by means of expression (see [99])

$$E\left[\widehat{f}_n(x) - f(x)\right]^2 \leq c_1 n^{-2} \sum_{i=1}^{n} h_i^{-p} + c_2 n^{-2} \left(\sum_{i=1}^{n} h_i^2\right)^2 \tag{5.33}$$

if density function f has continuous partial derivatives up to the 3rd order.

c) Condition C

As we have already mentioned, this condition imposes certain assumptions on the speed of convergence of estimator (5.4). Let us assume that density function f has continuous partial derivatives up to the 3rd order. With the use of reasoning similar to that in [63] and using results of [99] it is possible to show that convergence (5.7) in version "in probability" is implied by

$$|\phi_n^*(x)| \, n^{-1} \sum_{i=1}^{n} h_i^2 \xrightarrow{n} 0 \tag{5.34}$$

and

$$\phi_n^{*2}(x) \, n^{-2} \sum_{i=1}^{n} h_i^{-p} \xrightarrow{n} 0 \tag{5.35}$$

whereas convergence (5.7) with pr. 1 is implied by

$$\sum_{n=1}^{\infty} \phi_n^{*2}(x) \, n^{-2} h_n^{-p} < \infty \tag{5.36}$$

Analyzing the above assumptions, we may raise the following problem: how fast can ϕ_n^* grow to infinity (if ϕ_n^* an unbounded sequence) so that condition C could be satisfied at all? Let e.g. in a certain point x

$$|\phi_n^*(x)| = 0(n^\alpha), \ \alpha > 0$$

Of course, conditions (3.34) and (3.35) are now satisfied when

$$0 < \alpha < 1$$

while conditions (3.34) and (3.36) are satisfied when

$$0 < \alpha < \frac{1}{2}$$

In other words, algorithm (5.3) has tracking properties if function ϕ_n^* does not grow to infinity too fast. Other limitations as regards sequence ϕ_n^* result from assumptions formulated in Corollaries 5.1 and 5.2. This problem will be discussed in detail in the next sections, considering plants with various types of non-stationarity.

5.3.2 The GRNN based on the orthogonal series

The structural scheme of the system that realizes algorithm (5.3) on the basis of the orthogonal series method is shown in Fig. 5.3. In order to differentiate between sequences $q(n)$ that appear in the numerator and denominator of expression (5.3), symbols $q'(n)$ and $q''(n)$ were used.

Like in Section 5.3.1 we will specify assumptions ensuring satisfaction of conditions A, B and C of Theorem 5.1. Condition A will be connected with the selection of sequence $q'(n)$ and conditions B and C will be connected with the selection of sequence $q''(n)$.

a) Condition A

Let us denote

$$S_n(x) = \left| \sum_{|\underline{j}| \leq q} b_{\underline{j}n} \Psi_{\underline{j}}(x) - R_n(x) \right| \tag{5.37}$$

where

$$b_{\underline{j}n} = \int R_n(x) \Psi_{\underline{j}}(x)\, dx$$

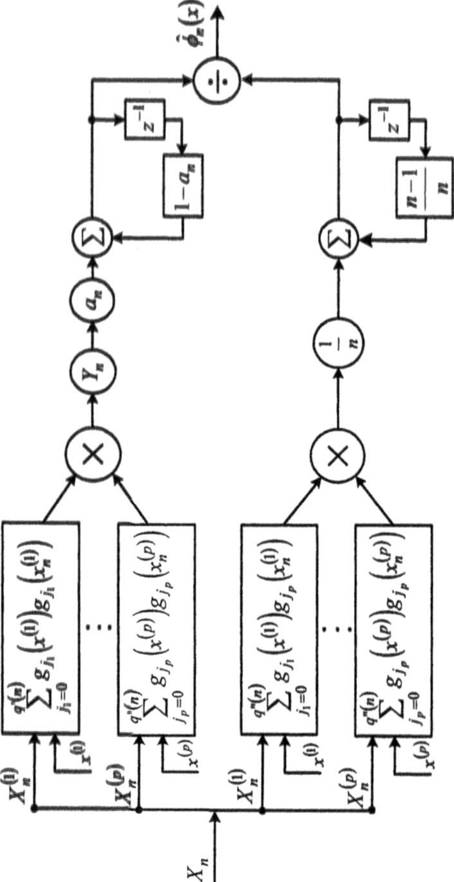

FIGURE 5.3. Block diagram of the GRNN based on the orthogonal series kernel

The following two corollaries from general Theorems 4.1 and 4.2 result in the meeting of condition A.

Corollary 5.3
Let us assume that $q(n) \xrightarrow{n} \infty$, condition (5.21) holds and one of the following two conditions is satisfied:

$$a_n \left(\sum_{j=0}^{q(n)} G_j^2 \right)^p m_n' \xrightarrow{n} 0 \tag{5.38}$$

or

$$a_n \left(\sum_{j=0}^{q(n)} G_j^2 \right)^{2p} m_n'' \xrightarrow{n} 0 \tag{5.39}$$

If

$$a_n^{-1} S_n(x) \xrightarrow{n} 0 \tag{5.40}$$

then

$$E\left[\widehat{R}_n(x) - R_n(x) \right]^2 \xrightarrow{n} 0$$

Corollary 5.4
Let us assume that $q(n) \xrightarrow{n} \infty$, assumption (5.25) holds and one of the following two conditions is satisfied:

$$\sum_{n=1}^{\infty} a_n^2 \left(\sum_{j=0}^{q(n)} G_j^2 \right)^{p} m_n' < \infty \tag{5.41}$$

or

$$\sum_{n=1}^{\infty} a_n^2 \left(\sum_{j=0}^{q(n)} G_j^2 \right)^{2p} m_n'' < \infty \tag{5.42}$$

If

$$\sum_{n=1}^{\infty} a_n^{-1} S_n^2(x) < \infty \tag{5.43}$$

then

$$\left| \widehat{R}_n(x) - R_n(x) \right| \xrightarrow{n} 0 \text{ with pr. } 1$$

The problem of convergence of (5.40) and (5.43) is not a standard problem of the orthogonal series theory because the expanded function R_n changes with the increase of n. Even if n is fixed, the problem of convergence of multidimensional series

$$\sum_{\underline{j}} b_{\underline{j}n} \Psi_{\underline{j}}(x) \rightarrow R_n(x) \tag{5.44}$$

is not trivial. It was investigated in more detail for the Fourier series (e.g. Sjölin [246]) but it is less known for the multidimensional Hermite series. For these two series, we will specify conditions (5.40) and (5.43). It is possible to show (the proof can be found in the Appendix) that these conditions take the following form

$$a_n^{-1} \left\| t_n^l \right\|_{L_2} q^{-ps}(n) \xrightarrow{n} 0 \tag{5.45}$$

$$\sum_{n=1}^{\infty} a_n^{-1} \left\| t_n^l \right\|_{L_2}^2 q^{-2ps}(n) < \infty \tag{5.46}$$

where

$$t_n^l(x, R_n) = \begin{cases} \prod_{k=1}^{p} \left(x^{(k)} - \partial / \partial x^{(k)} \right)^l R_n\left(x^{(1)}, ..., x^{(p)}\right) \\ \text{for the Hermite series} \\ \prod_{k=1}^{p} \partial^l / \partial x^l (k)\, R_n\left(x^{(1)}, ..., x^{(p)}\right) \\ \text{for the Fourier series} \end{cases}$$

and

$$s = \begin{cases} \frac{l}{2} - \frac{5}{12} & \text{for the Hermite series} \\ l - \frac{1}{2} & \text{for the Fourier series} \end{cases}$$

Parameter l we will be associated with smooth properties of function R_n, $(n = 1, 2, ...)$.

The above mentioned conditions were derived under the following assumptions:
i) Hermite series: $t_n^l \in L_2(R^p)$, $l \geq 1$ and condition (5.44) holds
ii) Fourier series: $t_n^l \in L_2(A)$, $l \geq 1$, $A = [-\pi, \pi]^p$, condition (5.44) holds, it is assumed that function t_n^l and its partial derivatives up to the order of $p-1$ are equal 0 on the boundary of A.

Conditions (5.45) and (5.46) are connected with the assessment of the "tail" of series (5.44); it follows from (5.37). Unfortunately, the assessment of the "tail" of an orthogonal series requires making rather complicated assumptions concerning functions expanded into this series. However, assumptions of this type are typical in all works devoted to the orthogonal series theory (see e.g. monograph [237]).

b) Condition B

A recurrent estimator (5.4) of a density function constructed on the basis of orthogonal series was proposed by Rutkowski in [208] and [209]. Let us assume that $q(n) \xrightarrow{n} \infty$. Now, the condition

$$n^{-2} \sum_{i=1}^{n} \left(\sum_{j=0}^{q(i)} G_j^2 \right)^{2p} \xrightarrow{n} 0 \tag{5.47}$$

implies a weak convergence (5.6) and

$$\sum_{n=1}^{\infty} n^{-2} \left(\sum_{j=0}^{q(n)} G_j^2 \right)^{2p} < \infty \tag{5.48}$$

implies a strong convergence (5.6) at every point x where

$$\sum_{|\underline{j}| \leq q} \Psi_{\underline{j}}(x)\, d_j \to f(x) \tag{5.49}$$

where $d_j = \int \Psi_{\underline{j}}(x)\, f(x)dx$. If (5.49) is true for almost every x, then a weak and a strong convergence of procedure (5.4) is also true almost everywhere. Convergence (5.49) depends on the orthonormal series used and on the properties of function f. In the one-dimensional case, $p = 1$, the following results are known:

- for the Fourier, Legendre, Laguerre and Hermite series various conditions imposed on function f, ensuring point and uniform convergence (5.49), were given by Sansone [237],
- for the Haar series, (5.49) is true in almost all points x for any function $f \in L_1$(Alexits [8]),
- for the Fourier series, (5.49) is true in almost all points x for any function $f \in L_2$ (Carleson [46]); a similar result may be obtained for the Legendre, Laguerre and Hermite series using theorems of equivalent convergence (Szegö [268]),
- for the Fourier series with Fejer's kernel (2.7), convergence (5.49) is true in almost all points x for any function $f \in L_1$ (Sansone [237]); this result may be extended for Laguerre and Hermite series with the help of the above mentioned theorems on equivalent convergence.

Unfortunately, in the multidimensional case, the conditions for convergence (5.49) are known only for the Fourier series:

- for any function $f \in L_2$ (5.49) is true in almost all points x (Sjölin [246]).
- for the multidimensional Fourier series with Fejer's kernel (2.7), convergence (5.49) is true uniformly if f is a continuous function (Nikolski [173]).

It is easily seen that

$$E\left[\widehat{f}_n(x) - f(x)\right]^2 \leq c_1 n^{-2} \sum_{i=1}^{n} q^{2pw}(i)$$

$$+c_2 n^{-2} \left(\sum_{i=1}^{n} q^{-ps}(i) \right)^2 \tag{5.50}$$

where

$$s = \begin{cases} \frac{l}{2} - \frac{5}{12} & \text{for theHermite series} \\ l - \frac{1}{2} & \text{for the Fourier series} \end{cases}$$

$$w = \begin{cases} \frac{5}{6} & \text{for the Hermite series} \\ 1 & \text{for the Fourier series} \end{cases}$$

c) Condition C

Concrete assumptions imposed on sequence $q(n)$ that appear in the denominator of (5.3) that guarantee the satisfaction of condition C can be derived with the use of reasoning similar to that in Rutkowski [209]. Now, we obtain a weak convergence (5.7) if

$$n^{-2} \phi_n^{2*}(x) \sum_{i=1}^{n} \left(\sum_{j=0}^{q(i)} G_j^2 \right)^{2p} \xrightarrow{n} 0 \tag{5.51}$$

and a strong convergence (5.7) if

$$\sum_{n=1}^{\infty} n^{-2} \phi_n^{2*}(x) \left(\sum_{j=0}^{q(i)} G_j^2 \right)^{2p} < \infty \tag{5.52}$$

In both cases we should also assume that

$$|\phi_n^*(x)|\, n^{-1} \left| \sum_{i=1}^{n} \left(\sum_{|\underline{j}| \leq q(i)} \Psi_{\underline{j}}(x) d_{\underline{j}} - f(x) \right) \right| \xrightarrow{n} 0 \tag{5.53}$$

Condition (5.53) can take a concrete form depending on the chosen orthogonal system and assumptions imposed on function f. Assuming that the orthogonal expansion of function f is convergent in point

x (in almost every point x), i.e.

$$\sum_{\underline{j}} \Psi_{\underline{j}}(x) d_{\underline{j}} = f(x)$$

we will use the Hermite and Fourier orthogonal systems. We define

$$t^l(x; f) = \begin{cases} \prod_{k=1}^{p} \left(x^{(k)} - \partial/\partial x^{(k)}\right)^l f(x) \\ \text{for the Hermite series, } l \geq 1 \\ \\ \left(\prod_{k=1}^{p} \partial/\partial x^{(k)}\right)^l f(x) \\ \text{for the Fourier series} \end{cases}$$

Let us assume that $t^l \in L_2$. It can be shown that condition (5.53) takes the form

$$|\phi_n^*(x)| \, n^{-1} \sum_{i=1}^{n} [q(i)]^{-ps} \xrightarrow{n} 0 \tag{5.54}$$

where

$$s = \begin{cases} \frac{l}{2} - \frac{5}{12} & \text{for the Hermite series} \\ l - \frac{1}{2} & \text{for the Fourier series} \end{cases}$$

In other words, the satisfaction of condition C depends on the smooth properties of an unknown function f. From (5.51) and (5.52) it follows that $|\phi_n^*|$ cannot grow to infinity too fast. If, e.g. $|\phi_n^*(x)| = 0(n^\alpha)$, $\alpha > 0$, then parameter α should be contained within the same bounds as in the case of the use of the algorithm based on the Parzen kernel (Section 5.3.1c).

5.4 Speed of convergence

The problem of investigating the speed of the convergence of procedure (5.3) which is a quotient of two algorithms is a relatively complex one. The following theorem allows us to assess the speed of the convergence of procedure (5.3) on the basis of the knowledge of the speed of the convergence of procedures (4.3) and (5.4):

Theorem 5.2
For any $\varepsilon > 0$, the following inequality holds

$$P\left(\left|\widehat{\phi}_n(x) - \phi_n^*(x)\right| > \varepsilon\right)$$
$$\leq \left(\frac{\varepsilon + 2}{\varepsilon\, f(x)}\right)^2 \left(E\left[\widehat{R}_n(x) - R_n(x)\right]^2 \right. \tag{5.55}$$
$$\left. + \left(\phi_n^{*2}(x) + 1\right) E\left[\widehat{f}_n(x) - f(x)\right]^2\right)$$

Using the above inequality, we will later assess the speed of convergence of procedure (5.3) used for modelling plants with particular types of non-stationarity.

In [99], in the context of stationary problems, the authors also considered non-parametric procedures (based on the Parzen kernel) that are a quotient of two algorithms. Moreover, they carried out an optimization of the speed of convergence. However, they assumed that $h_n' = h_n''$ which is justified in a stationary case. In a non-stationary case, sequences h_n' and h_n'' as well as $q'(n)$ and $q''(n)$ which are present in the numerator and denominator of expression (5.3) should usually satisfy different conditions, which makes their optimal selection difficult (e.g. in the sense of minimizing the right side of expression (5.54)). The matter is further complicated by the necessity of selection of sequence a_n (in a stationary case, $a_n = n^{-1}$) and by the influence of non-stationarity in the expression (5.55) that evaluates the speed of convergence of the procedure (5.3). That is why sequences a_n, $q(n)$ and $h(n)$ will not be the subject of optimization. We will be satisfied with the fact that the conditions given in this chapter allow us to design a system that realizes algorithm (5.3). It should be noted that a proper selection of sequences a_n, h_n' and h_n'' (or a_n, $q'(n)$, $q''(n)$) implies possession of tracking properties by this algorithm, which is by no means an easy task in a non-stationary situation. The expression (5.55), as was already mentioned, will be used several times in the next sections for analysis of the influence of various factors on the speed of convergence of procedure (5.3). In the next sections we will present an application of the GRNN to modelling non-stationary plants that are under the influence of additive disturbances with a zero mean and a finite variance, i.e.

$$Y_n = \phi_n^*(X_n) + Z_n,\ n = 1, 2, \ldots \tag{5.56}$$

5.5 Modelling of systems with multiplicative non-stationarity

Let us consider a plant described by equation (5.56) assuming that

$$\phi_n^*(x) = \alpha_n \phi(x), \tag{5.57}$$

where: α_n – unknown sequence of numbers, ϕ – unknown function. The plant with the above characteristic is shown in Fig. 5.4:

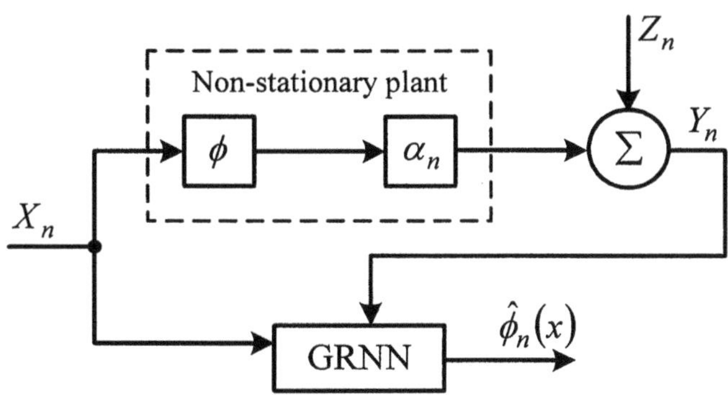

FIGURE 5.4. GRNN for modelling a plant with multiplicative non-stationarity

In Tables 5.1a, 5.1b, 5.2a and 5.2b, based on the results of Sections 5.3.1 and 5.3.2, we present the conditions implying convergence of algorithm (5.3) used for identification of a plant with multiplicative non-stationarity. Tables 5.1a and 5.1b give proper conditions for the algorithm based on the Parzen kernel, whereas Tables 5.2a and 5.2b give similar conditions for the algorithm based on the orthogonal series method. In order to specify these conditions more precisely, in Tables 5.2a and 5.2b, two specific multidimensional orthogonal series were considered: the Fourier series and the Hermite series. We should notice that now

$$R_n(x) = \alpha_n f(x) \phi(x)$$

Because of the "separability" of the non-stationary factor α_n assumptions (5.22) and (5.26) (as well as (5.40) and (5.43)), connected with smooth properties of function R_n, $n = 1, 2, ...$, reduce to assumptions concerning smooth properties of function $f\phi$, which simplifies significantly the convergence conditions described in Section 5.3.

TABLE 5.1a. Conditions for weak convergence of the GRNN based on the Parzen kernel – multiplicative non-stationarity

Condition	$\left\|\widehat{\phi}_n(x) - \varphi_n^*(x)\right\| \xrightarrow{n} 0$ in probability
(5.19)	$a_n h_n^{-p}\left(\alpha_n^2+1\right) \xrightarrow{n} 0$
(5.20)	$a_n h_n^{-2p}\left(\alpha_n^2+1\right) \xrightarrow{n} 0$
(5.21)	$a_n^{-1}\left\|\alpha_{n+1}-\alpha_n\right\| \xrightarrow{n} 0$
(5.22)	$a_n^{-1} h_n^r \left\|\alpha_n\right\| \xrightarrow{n} 0$
(5.34)	$\left\|\alpha_n\right\| n^{-1}\sum_{i=1}^n h_i^2 \xrightarrow{n} 0$
(5.28), (5.35)	$\left(\alpha_n^2+1\right) n^{-2}\sum_{i=1}^n h_i^{-p} \xrightarrow{n} 0$

TABLE 5.1b. Conditions for strong convergence of the GRNN based on the Parzen kernel – multiplicative non-stationarity

Condition	$\left\|\widehat{\phi}_n(x) - \varphi_n^*(x)\right\| \xrightarrow{n} 0$ with pr. 1
(5.23)	$\sum_{n=1}^\infty a_n^2 h_n^{-p}\left(\alpha_n^2+1\right) < \infty$
(5.24)	$\sum_{n=1}^\infty a_n^2 h_n^{-2p}\left(\alpha_n^2+1\right) < \infty$
(5.25)	$\sum_{n=1}^\infty a_n^{-1}\left(\alpha_{n+1}-\alpha_n\right)^2 < \infty$
(5.26)	$\sum_{n=1}^\infty a_n^{-1} h_n^{2r} \alpha_n^2 < \infty$
(5.34)	$\left\|\alpha_n\right\| n^{-1}\sum_{i=1}^n h_i^2 \xrightarrow{n} 0$
(5.30), (5.36)	$\sum_{n=1}^\infty \left(\alpha_n^2+1\right) n^{-2} h_n^{-p} < \infty$

Conditions in Tables 5.1a and 5.1b hold under the following assumptions:

(i) $\sup f(x) < \infty$, $\sup |f(x)\phi(x)| < \infty$

(ii) $\int \phi^2(x) f(x)\, dx < \infty$

(iii) $\sup \left| \frac{\partial^r}{\partial x^{(i_1)} \partial x^{(i_r)}} f(x) \phi(x) \right| < \infty$, $i_k = 1, ..., p$, $k = 1, ..., r$

(iv) f has continuous partial derivates up to the third order

(v) condition (5.34) should be neglected if α_n is a bounded sequence.

TABLE 5.2a. Conditions for weak convergence of the GRNN based on the orthogonal series method – multiplicative non-stationarity

Condition	$\left\|\widehat{\phi}_n(x) - \varphi_n^*(x)\right\| \xrightarrow{n} 0$ in probability
(5.21)	$a_n^{-1} \|\alpha_{n+1} - \alpha_n\| \xrightarrow{n} 0$
(5.38)	$a_n q^{(2d+1)p}(n) \left(\alpha_n^2 + 1\right) \xrightarrow{n} 0$
(5.39)	$a_n q^{(2d+1)2p}(n) \left(\alpha_n^2 + 1\right) \xrightarrow{n} 0$
(5.40)	$a_n^{-1} \|\alpha_n\| q^{-ps}(n) \xrightarrow{n} 0$
(5.47), (5.51)	$\left(\alpha_n^2 + 1\right) n^{-2} \sum_{i=1}^{n} q^{(2d+1)2p}(n) \xrightarrow{n} 0$
(5.54)	$\|\alpha_n\| n^{-1} \sum_{i=1}^{n} q^{-ps}(i) \xrightarrow{n} 0$

Conditions in Tables 5.2a and 5.2b hold under the following assumptions:

(i) $d = \begin{cases} 0 & \text{for the Fourier series} \\ -\frac{1}{12} & \text{for the Hermite series} \end{cases}$

(ii) $s = \begin{cases} l - \frac{l}{2} & \text{for the Fourier series} \\ \frac{l}{2} - \frac{5}{12} & \text{for the Hermite series} \end{cases}$

(iii) $\sup f(x) < \infty$, $\sup |f(x)\phi(x)| < \infty$

(iv) $\int \phi^2(x) f(x)\, dx < \infty$

(v) $\prod_{k=1}^{p} \left(x^{(k)} - \frac{\partial}{\partial x}^{(k)} \right)^l f(x) \phi(x) \in L_2$

(vi) $\prod_{k=1}^{p} \left(x^{(k)} - \frac{\partial}{\partial x}^{(k)} \right)^l f(x) \in L_2$

(vii) condition (5.54) should be neglected if α_n is a bounded sequence.

TABLE 5.2b. Conditions for strong convergence of the GRNN based on the orthogonal series method – multiplicative non-stationarity

Condition	$\left\| \widehat{\phi}_n(x) - \varphi_n^*(x) \right\| \xrightarrow{n} 0$ with pr. 1
(5.25)	$\sum_{n=1}^{\infty} a_n^{-1} (\alpha_{n+1} - \alpha_n)^2 < \infty$
(5.41)	$\sum_{n=1}^{\infty} a_n^2 q^{(2d+1)p}(n) \left(\alpha_n^2 + 1 \right) < \infty$
(5.42)	$\sum_{n=1}^{\infty} a_n^2 q^{(2d+1)2p}(n) \left(\alpha_n^2 + 1 \right) < \infty$
(5.43)	$\sum_{n=1}^{\infty} a_n^{-1} \alpha_n^2 q^{-2ps}(n) < \infty$
(5.48), (5.52)	$\sum_{n=1}^{\infty} \left(\alpha_n^2 + 1 \right) n^{-2} q^{(2d+1)2p}(n) < \infty$
(5.54)	$\lvert\alpha_n\rvert n^{-1} \sum_{i=1}^{n} q^{-ps}(i) \xrightarrow{n} 0$

Remark 5.1

Conditions (5.19) – (5.22) and (5.23) – (5.26) presented in Tables 5.1a and 5.1b concern the selection of sequence h_n'. Conditions (5.28), (5.34), (5.35) and (5.36) concern the selection of sequence h_n'' (according to symbols in Fig. 5.2). In a similar way, conditions (5.21), (5.38), (5.39) and (5.40) presented in Table 5.2a, and conditions (5.25), (5.41), (5.42) and (5.43) presented in Table 5.2b, concern the selection of sequence $q'(n)$. Conditions (5.47), (5.48), (5.51), (5.52) and (5.54) concern the selection of sequence $q''(n)$ (according to symbols in Fig. 5.3).

It is obvious that sequence α_n cannot change in a completely arbitrary way so that algorithm (5.3) could possess the tracking property. Nevertheless we will show that the class of considered sequences α_n is quite wide. From the point of view of maintaining tracking properties by algorithm (5.3), situations when regression (5.57) is for a certain x divergent to infinity or does not have a finite limit, seem to be particularly difficult. Such cases are illustrated by the following examples of sequences α_n:

a)

$$\alpha_n = c_0 + c_1 n^{t_1} + c_2 n^{t_2} + \ldots + c_k n^{t_k}$$

where c_0, c_1,..., c_k – real numbers and $t_j > 0$, $j = 1, \ldots, k$.

b)

$$\alpha_n = c_1 n^t + c_2 \log n + c_3$$

where c_1, c_2, c_3 are real numbers, $t > 0$

c)

$$\alpha_n = c_1 \sin A_n + c_2 \cos B_n + c_3$$

where $A_n = k_1 n^{t_1}$, $B_n = k_2 n^{t_2}$, c_1, c_2, c_3, k_1, k_2 are real numbers, t_1, $t_2 > 0$.

d)

$$\alpha_n = c_1 n^{t_1} \sin A_n + c_2$$

where $A_n = k_n{}^{-t_2}$, c_1, c_2, k are real numbers, t_1, $t_2 > 0$ and $t_1 > t_2$.

e)

$$\alpha_n = c_1 n^{t_1} \sin A_n + c_2 n^{t_2} \cos B_n + c_3$$

where $A_n = k_1 n^{\tau_1}$, $B_n = k_2 n^{\tau_2}$, c_1, c_2, c_3, k_1, k_2 are real numbers, t_1, t_2, τ_1, $\tau_2 > 0$.

f)

$$\alpha_n = c_1 n^{t_1} + c_2 \sin A_n + c_3 \cos B_n + c_4$$

where $A_n = k_1 n^{\tau_1}$, $B_n = k_2 n^{\tau_2}$, c_1, c_2, c_3, c_4, k_1, k_2 are any real numbers, t_1, t_2, τ_1, $\tau_2 > 0$.

Fig. 5.5. depicts how the above sequences α_n change with time. Let us now choose in algorithm (5.3) the following parameters:

$$h'_n = k'_1 n^{-H'}, \; h''_n = k''_1 n^{-H''}, \; H', H'' > 0, \; k'_1, k''_1 > 0$$

for the algorithm based on the Parzen kernel and

$$q'(n) = \left[k_2' n^{Q'}\right], \; q''(n) = \left[k_2'' n^{Q''}\right], \; Q', Q'' > 0, \; k_2', k_2'' > 0$$

for the algorithm based on the orthogonal series method ($[a]$ stands for the integer part of a). In both cases we take

$$a_n = k/n^a, \; 0 < a \leq 1, \; k > 0$$

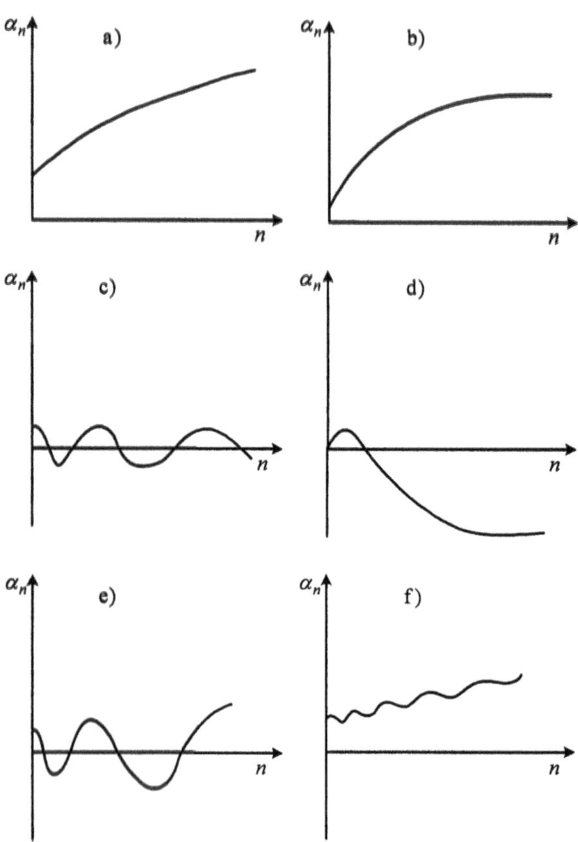

FIGURE 5.5. Changes with time of sequences α_n from examples a) - f)

Analysing all the conditions given in Tables 5.1a, 5.1b, 5.2a and 5.2b, it is possible to specify precisely within what limits the constants t, t_j, τ_j present in examples a) – f) should be contained so that algorithm (5.3) could possess tracking properties. The results are shown in Tables 5.3a and 5.3b.

It is worth emphasizing that for the designing a system that would realize algorithm (5.3), i.e. for a proper selection of sequences h_n, $q(n)$ and a_n, it is not necessary to precisely know sequences α_n that were specified in examples a) – e) but only to know the information contained in Tables 5.3a and 5.3b. For example, in order to track changes of the system described by

$$y_n = \left(c_1 n^t + c_2 \log n + c_3\right) \phi\left(x_n\right) + z_n$$

where t is an unknown parameter and ϕ is an unknown function, it is possible to use algorithm (5.3) if $0 < t < \frac{1}{3}$ for week convergence and $0 < t < \frac{1}{6}$ for strong convergence.

TABLE 5.3a. Conditions imposed on constants t, t_j, τ_j from examples a) – f); weak convergence

Example	$\left\lvert\widehat{\phi}_n(x) - \phi_n^*(x)\right\rvert \xrightarrow{n} 0$ in probability
a	$0 < t_j < \frac{1}{3},\ j = 1, ..., k$
b	$0 < t < \frac{1}{3}$
c	$0 < t_j < 1,\quad j = 1, 2,$
d	$0 < t_1 - t_2 < \frac{1}{3}$
e	$0 < t_j < \frac{1}{3},\quad j = 1, 2,$ $0 < t_j + \tau_j < 1,\quad j = 1, 2,$
f	$0 < t_1 < \frac{1}{3}$ $0 < \tau_j < 1,\quad j = 1, 2,$

We will now investigate the speed of convergence of algorithm (5.3). For this purpose one should:
1) use dependence (5.55),
2) determine constants A_1, B_1 and C_1 that are present in assumptions of Theorem 4.5 and then, with use of this theorem, evaluate

the speed of convergence of

$$E\left[\widehat{R}_n(x) - R_n(x)\right]^2 \xrightarrow{n} 0$$

3) on the basis of inequalities (5.33) or (5.50), evaluate the speed of convergence of

$$E\left[\widehat{f}_n(x) - f(x)\right]^2 \xrightarrow{n} 0$$

TABLE 5.3b. Conditions imposed on constants t, t_j, τ_j from examples a) – f); strong convergence

Example	$\left\|\widehat{\phi}_n(x) - \phi_n^*(x)\right\| \xrightarrow{n} 0$ with pr. 1
a	$0 < t_j < \frac{1}{6}, \ j = 1, ..., k$
b	$0 < t < \frac{1}{6}$
c	$0 < t_j < \frac{1}{2}, \quad j = 1, 2,$
d	$0 < t_1 - t_2 < \frac{1}{6}$
e	$0 < t_j < \frac{1}{6}, \quad j = 1, 2,$ $0 < t_j + \tau_j < \frac{1}{2}, \quad j = 1, 2,$
f	$0 < t_1 < \frac{1}{6}$ $0 < \tau_j < \frac{1}{2}, \quad j = 1, 2,$

Example 5.1

Assuming that

$$\alpha_n = \text{const.}\ n^t, \ t > 0$$

we will evaluate the speed of convergence of algorithm (5.3) based on the Parzen kernel and the orthogonal series method. We will assume that sequences h_n, $q(n)$ and a_n are of a power type.

a) Speed of convergence of algorithm (5.3) based on the Parzen's kernel.

In this case we have

$$A_1 = 2t + R'p$$
$$B_1 = 1 - t$$
$$C_1 = rH' - t,$$

where parameter r is connected with smooth properties of function ϕf. Omitting some simple calculations, we obtain

$$P\left(\left|\widehat{\phi}_n(x) - \phi_n^*(x)\right| > \varepsilon\right) \leq \left(\frac{\varepsilon + 2}{\varepsilon\, f(x)}\right)^2 n^{2t}\left(c_1 n^{-2rH'} + c_2 n^{-r_1} + c_3 n^{-4H''} + c_4 n^{-(1-H''p)}\right) \tag{5.58}$$

where

$$r_1 = \min\left[a - H'p,\ 2(1-a),\ 2\left(rH' - a\right)\right]$$

An optimal selection of parameters H', H'' and a, minimizing the right side of expression (5.58) seems to be a complicated problem. However, this expression may be used for designing a system that realizes algorithm (5.3) in such a way that it could possess tracking properties. Analysing expression (5.58) we realize that parameters H' and H'' should satisfy the following conditions:

$$\frac{t+a}{r} < H' < \frac{a-2t}{p}, \quad a < 1 - t$$

$$\frac{t}{2} < H'' < \frac{1-2t}{p}$$

We should point out that the maximum value of t, with which the algorithm has tracking properties, depends on parameter r specifying smooth properties of function ϕf. From the above inequalities it follows that if

$$t \in \left(0, \frac{1}{3} - \frac{p}{3r}\right), \quad r > p$$

then algorithm (5.3) is convergent. Let us notice that an increase of dimension p results in a decrease of the above range and an increase of smooth properties results in an increase of this range.

b) The speed of convergence of algorithm (5.3) based on the orthogonal series method.

Referring to symbols from Theorem 4.6, we obtain

$$A_2 = 2t + Q'(2d+1)p$$

$$B_2 = 1 - t$$

$$C_2 = pQ's - t$$

where
$d = -\frac{1}{12}, \quad s = \frac{l}{2} - \frac{5}{12}$ for the Hermite series
$d = 0, \quad s = l - \frac{1}{2}$ for the Fourier series,
and parameter l is connected with smooth properties of function ϕf.
The speed of convergence of procedure (5.3) can be now expressed in the following way

$$P\left(\left|\widehat{\phi}_n(x) - \phi_n^*(x)\right| > \varepsilon\right)$$
$$\leq \left(\frac{\varepsilon+2}{\varepsilon\, f(x)}\right)^2 n^{2t}\left(c_1 n^{-2pQ's} + c_2 n^{-r_1}\right. \tag{5.59}$$
$$\left. + c_3 n^{2pQ''(2d+1)-1} + c_4 n^{-2Q''ps}\right)$$

where

$$r_1 = \min\left[a - Q'(2d+1)p, \; 2(1-a), \; 2(pQ's - a)\right]$$

Analysing the above inequality it is possible to say that algorithm (5.3) has tracking properties if

$$t \in \left(0, \frac{1}{3} - \frac{1}{3s}\right)$$

with use of the Fourier system and

$$t \in \left(0, \frac{1}{3} - \frac{5}{18}\frac{1}{s}\right)$$

with use of the Hermite system.

Of special interest is the fact that the maximum value of t with which the algorithm still has tracking properties does not depend on the dimension p.

In Simulations 5.1 – 5.3 we investigate the GRNN applied to track time-varying regressions:

$$\Phi_n^*(x_n) = 10x_n^2 n^{0.1} + z_n$$

Simulation 5.1
We assume, that x_n and z_n are realizations of $N(0,1)$ random variables. The GRNN based on the Gaussian kernel and given by (5.3) has been applied with the following parameters:
$a = 0.8,\ H' = H'' = 0.5,\ k_1' = k_2'' = 1.$

The results are depicted in Fig. 5.6a, Fig. 5.6b and Fig. 5.6c. Figure 5.6a displays a comparison of a true regression and estimated by the GRNN for $n = 1000$. Figures 5.6b and 5.6c show the tracking of the non-stationary regression with a changing n in points $x = 0.2$ and $x = 0.4$, respectively.

Simulation 5.2
We assume, that x_n and z_n are realizations of uniformly distributed random variables on $[0,1]$ and $[-\frac{1}{2}, \frac{1}{2}]$, respectively. The GRNN based on the Parzen triangular kernel has been applied with the following parameters:
$a = 0.8,\ H' = H'' = 0.5,\ k_1' = k_2'' = 1.$

The results are depicted in Fig. 5.7a, Fig. 5.7b and Fig. 5.7c. In Fig. 5.7a we display a comparison of a true regression and estimated by the GRNN for $n = 1000$. Figure 5.7b and Fig. 5.7c show the tracking of the non-stationary regression with a changing n in points $x = 0.2$ and $x = 0.4$, respectively.

Simulation 5.3
We assume, that x_n and z_n are realizations of $N(0,1)$ random variables. The GRNN based on the trigonometric orthogonal series has been applied with the following parameters:
$a = 0.8,\ Q' = 0.55,\ Q'' = 0.4.$

The results are depicted in Fig. 5.8a, Fig. 5.8b and Fig. 5.8c. In Figure 5.8a we display a comparison of a true regression and estimated by the GRNN for $n = 1000$. Figure 5.8b and Fig. 5.8c show the tracking of the non-stationary regression with a changing n in points $x = 0.2$ and $x = 0.4$, respectively.

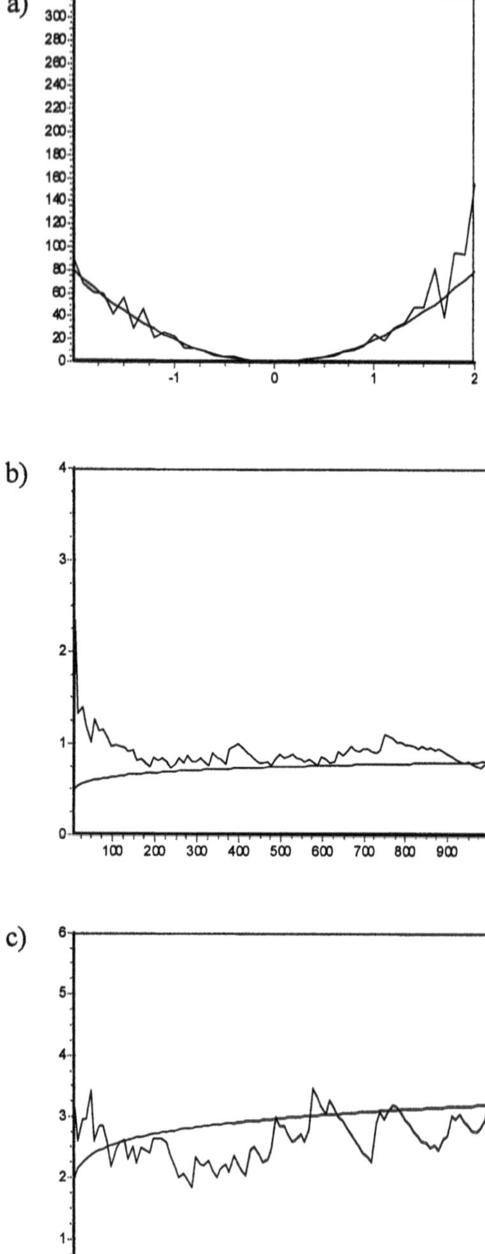

FIGURE 5.6. GRNN for modeling regressions with multiplicative non-stationarity – Simulation 5.1

a)

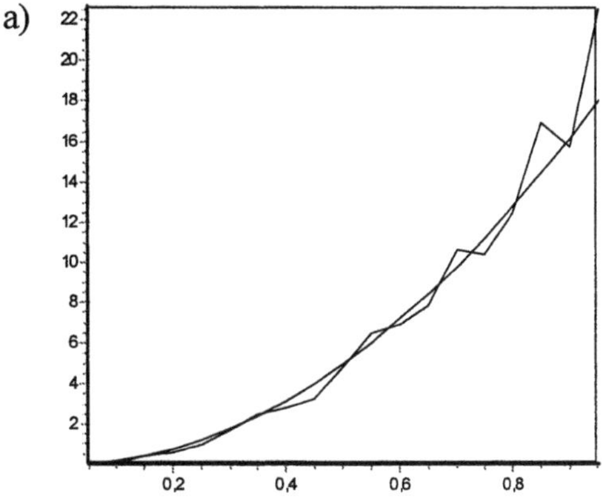

b)

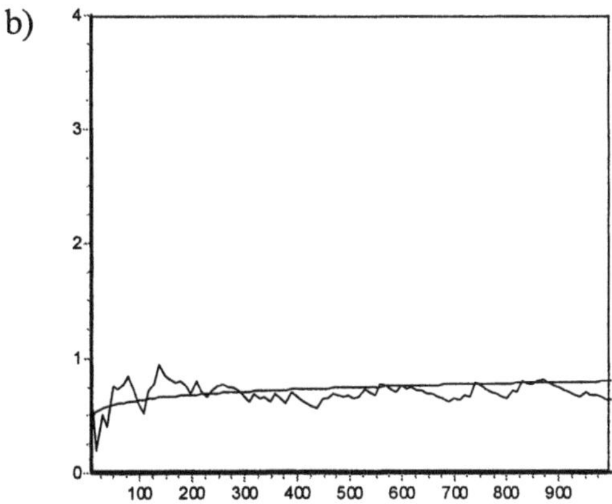

c)

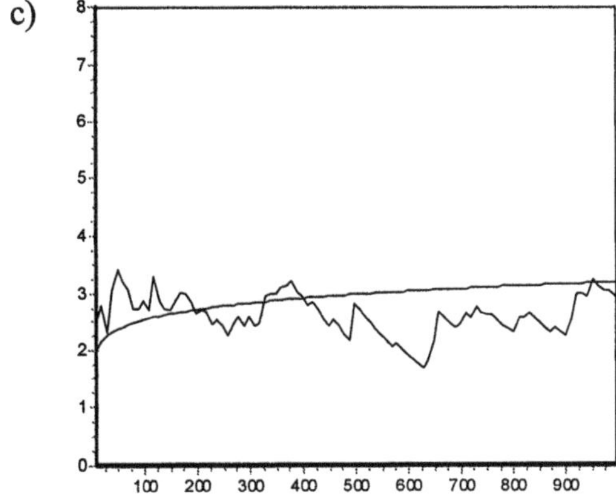

FIGURE 5.7. GRNN for modeling regressions with multiplicative non-stationarity – Simulation 5.2

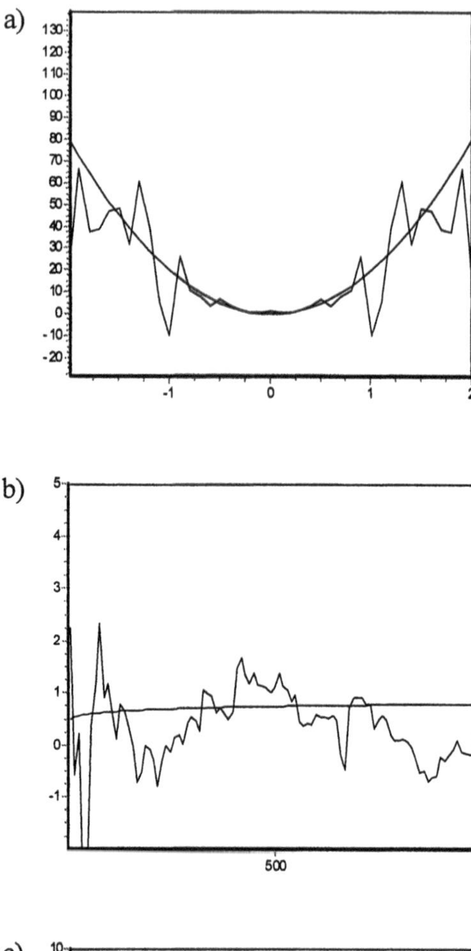

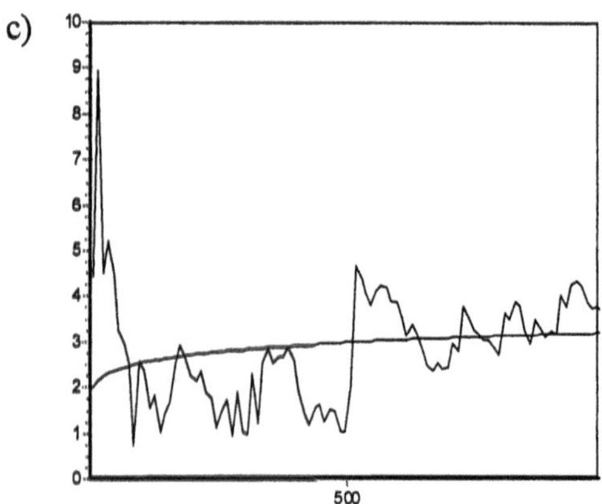

FIGURE 5.8. GRNN for modeling regressions with multiplicative non-stationarity – Simulation 5.3

In Simulations 5.4 and 5.5 we investigate the GRNN applied to track time-varying regressions:

$$\Phi_n^*(x_n) = 10x_n^3 n^{0.2} + z_n$$

Simulation 5.4

We assume, that x_n and z_n are realizations of $N(0,1)$ random variables. The GRNN based on the Parzen kernel has been applied with the following parameters:
$a = 0.7$, $H' = 0.27$, $H'' = 0.5$, $k_1' = k_2'' = 1$.

The results are depicted in Fig. 5.9a, Fig. 5.9b and Fig. 5.9c. In Fig. 5.9a we display a comparison of a true regression and estimated by the GRNN for $n = 1000$. Figure 5.9b and Fig. 5.9c show the tracking of the non-stationary regression with a changing n in points $x = 0.2$ and $x = 1$, respectively.

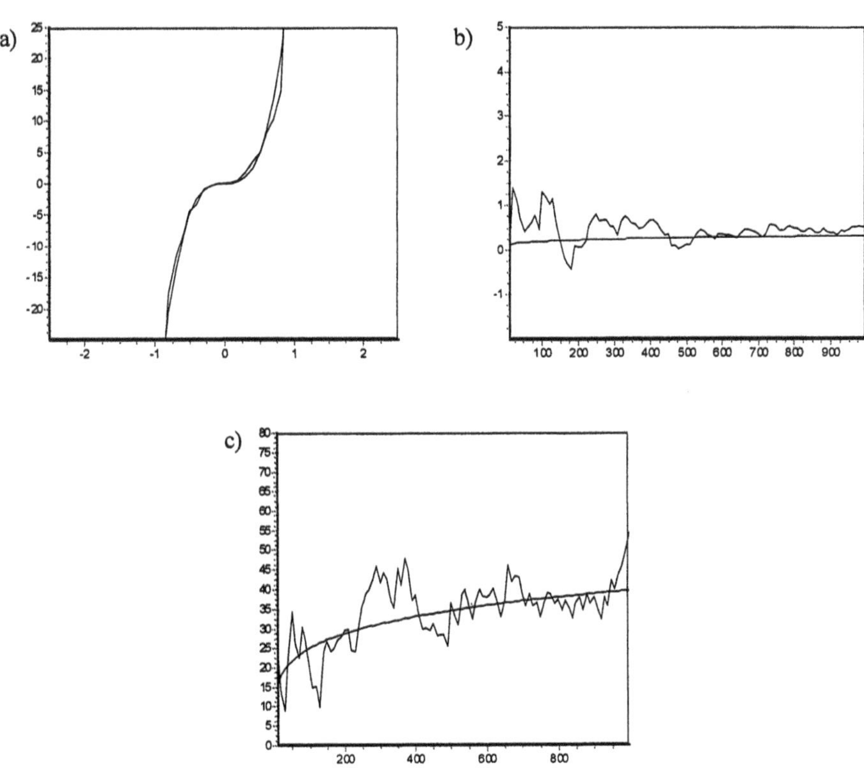

FIGURE 5.9. GRNN for modeling regressions with multiplicative non-stationarity – Simulation 5.4

Simulation 5.5

We assume, that x_n and z_n are realizations of uniformly distributed random variables on $[0, 1]$ and $[-\frac{1}{2}, \frac{1}{2}]$, respectively. The GRNN based on the Parzen triangular kernel has been applied with the following parameters:
$a = 0.7,\ H' = 0.27,\ H'' = 0.5,\ k_1' = k_2'' = 1.$

The results are depicted in Fig. 5.10a, Fig. 5.10b and 5.10c. In Fig. 5.10a we display a comparison of a true regression and estimated by the GRNN for $n = 1000$. Figures 5.10b and 5.10c show the tracking of the non-stationary regression with a changing n in points $x = 0.4$ and $x = 0.8$ respectively.

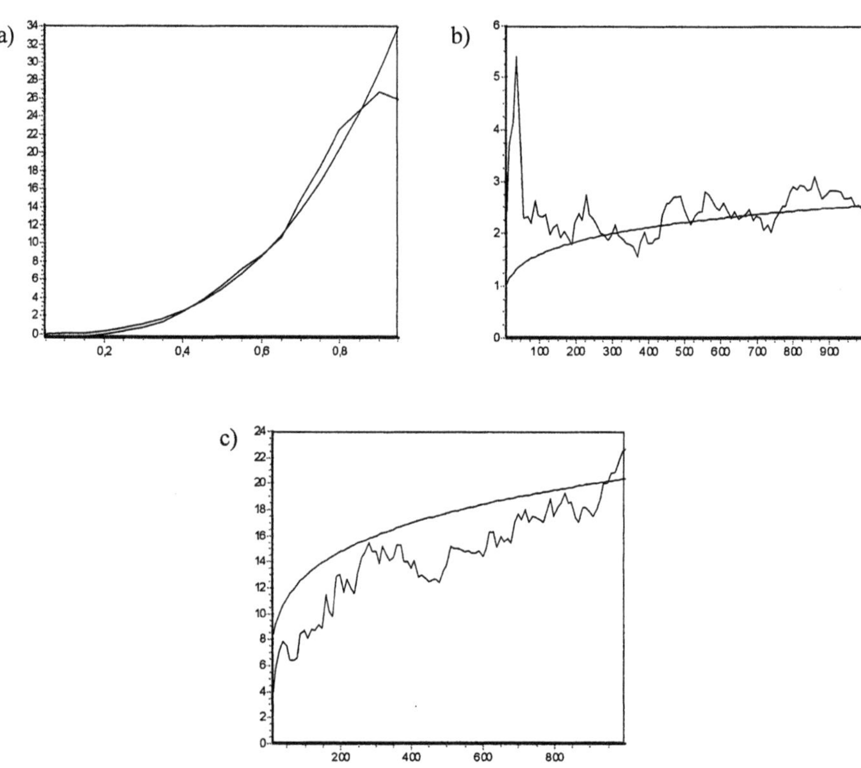

FIGURE 5.10. GRNN for modeling regressions with multiplicative non-stationarity – Simulation 5.5

5.6 Modelling of systems with additive non-stationarity

Let us consider plants described by equation (5.56), where

$$\phi_n^* (x) = \phi (x) + \beta_n, \tag{5.60}$$

β_n is an unknown sequence of numbers, ϕ is an unknown function. Basing on the results of Sections 5.3.1 and 5.3.2, in Tables 5.4a, 5.4b, 5.5a and 5.5b we present conditions that guarantee the convergence of algorithm (5.3) used for tracking regressions with additive non-stationarity. Tables 5.4a and 5.4b show suitable conditions for the algorithm based on the Parzen kernel and Tables 5.5a and 5.5b show similar conditions for the algorithm based on the Fourier and Hermite multidimensional orthogonal series. With reference to conditions presented in both tables, the content of Remark 5.1 applies. Observe that in this case

$$R_n (x) = f (x) \phi (x) + \beta_n f (x)$$

Presently, assumptions (5.22) and (5.26) as well as (5.45) and (5.46) connected with smooth properties of function R_n $(n = 1, 2, ...)$ are replaced by assumptions concerning smooth properties of functions $f\phi$ and f.This fact significantly simplifies the convergence conditions described in Section 5.3 and facilitates the designing of a system that realizes the modelling algorithm (Fig. 5.1).

Remark 5.2

Analysing the conditions of convergence of algorithm (5.3) that are given in Tables 5.4a, 5.4b, 5.5a and 5.5b we may say that they are similar to those that are given in Tables 5.1a, 5.1b, 5.2a and 5.2b and concern the multiplicative non-stationarity. So, as examples of sequences β_n that satisfy conditions given in Tables 5.4a, 5.4b, 5.5a and 5.5b, we may mention sequences specified in examples a) – e) of Section 5.5

Conditions in Tables 5.4a and 5.4b hold under the same conditions as those in (i) – (iv) concerning Tables 5.1a and 5.1b. Moreover, condition (5.34) should be neglected if β_n is a bounded sequence.

TABLE 5.4a. Conditions for weak convergence of the GRNN based on the Parzen kernel – additive non-stationarity

Condition	$\widehat{\phi}_n(x) - \varphi_n^*(x) \xrightarrow{n} 0$ in probability
(5.19)	$a_n h_n^{-p} (\beta_n^2 + 1) \xrightarrow{n} 0$
(5.20)	$a_n h_n^{-2p} (\beta_n^2 + 1) \xrightarrow{n} 0$
(5.21)	$a_n^{-1} \lvert\beta_{n+1} - \beta_n\rvert \xrightarrow{n} 0$
(5.22)	$a_n^{-1} h_n^r (\lvert\beta_n\rvert + 1) \xrightarrow{n} 0$
(5.34)	$\lvert\beta_n\rvert n^{-1} \sum_{i=1}^n h_i^2 \xrightarrow{n} 0$
(5.28), (5.35)	$(\beta_n^2 + 1) n^{-2} \sum_{i=1}^n h_i^{-p} \xrightarrow{n} 0$

TABLE 5.4b. Conditions for strong convergence of the GRNN based on the Parzen kernel – additive non-stationarity

Condition	$\left\lvert\widehat{\phi}_n(x) - \varphi_n^*(x)\right\rvert \xrightarrow{n} 0$ with pr. 1
(5.23)	$\sum_{n=1}^\infty a_n^2 h_n^{-p} (\beta_n^2 + 1) < \infty$
(5.24)	$\sum_{n=1}^\infty a_n^2 h_n^{-2p} (\beta_n^2 + 1) < \infty$
(5.25)	$\sum_{n=1}^\infty a_n^{-1} (\beta_{n+1} - \beta_n)^2 < \infty$
(5.26)	$\sum_{n=1}^\infty a_n^{-1} h_n^{2r} (\beta_n^2 + 1) < \infty$
(5.34)	$\lvert\beta_n\rvert n^{-1} \sum_{i=1}^n h_i^2 \xrightarrow{n} 0$
(5.28), (5.35)	$\sum_{n=1}^\infty (\beta_n^2 + 1) n^{-2} h_n^{-p} < \infty$

TABLE 5.5a. Conditions for weak convergence of the GRNN based on the orthogonal series method – additive non-stationarity

Condition	$\left\|\widehat{\phi}_n(x) - \varphi_n^*(x)\right\| \xrightarrow{n} 0$ in probability
(5.21)	$a_n^{-1}\left\|\beta_{n+1} - \beta_n\right\| \xrightarrow{n} 0$
(5.38)	$a_n q^{(2d+1)p}(n)\left(\beta_n^2 + 1\right) \xrightarrow{n} 0$
(5.39)	$a_n q^{(2d+1)2p}(n)\left(\beta_n^2 + 1\right) \xrightarrow{n} 0$
(5.40)	$a_n^{-1}\left(\left\|\beta_n\right\| + 1\right) q^{-ps}(n) \xrightarrow{n} 0$
(5.47), (5.51)	$\left(\beta_n^2 + 1\right) n^{-2} \sum_{i=1}^{n} q^{(2d+1)2p}(n) \xrightarrow{n} 0$
(5.54)	$\left\|\beta_n\right\| n^{-1} \sum_{i=1}^{n} q^{-ps}(i) \xrightarrow{n} 0$

TABLE 5.5b. Conditions for strong convergence of the GRNN based on the orthogonal series method – additive non-stationarity

Condition	$\left\|\widehat{\phi}_n(x) - \varphi_n^*(x)\right\| \xrightarrow{n} 0$ with pr. 1
(5.25)	$\sum_{n=1}^{\infty} a_n^{-1}\left(\beta_{n+1} - \beta_n\right)^2 < \infty$
(5.41)	$\sum_{n=1}^{\infty} a_n^2 q^{(2d+1)p}(n)\left(\beta_n^2 + 1\right) < \infty$
(5.42)	$\sum_{n=1}^{\infty} a_n^2 q^{(2d+1)2p}(n)\left(\beta_n^2 + 1\right) < \infty$
(5.43)	$\sum_{n=1}^{\infty} a_n^{-1}\left(\beta_n^2 + 1\right) q^{-2ps}(n) < \infty$
(5.48), (5.52)	$\sum_{n=1}^{\infty}\left(\beta_n^2 + 1\right) n^{-2} q^{(2d+1)2p}(n) < \infty$
(5.54)	$\left\|\beta_n\right\| n^{-1} \sum_{i=1}^{n} q^{-ps}(i) \xrightarrow{n} 0$

Conditions in Tables 5.5a and 5.5b hold under the same conditions as those in (i) – (iv) concerning Tables 5.2a and 5.2b. Moreover, condition (5.54) should be neglected if β_n is a bounded sequence.

In Simulations 5.6 and 5.7 we investigate the GRNN applied to track time-varying regressions given by:

$$\Phi_n^*(x_n) = 10\cos(x_n) + n^{0.1} + z_n$$

Simulation 5.6

We assume, that x_n and z_n are realizations of $N(0,1)$ random variables. The GRNN based on the Gaussian kernel has been applied with the following parameters:
$a = 0.8$, $H' = 0.6$, $H'' = 0.5$, $k_1' = k_2'' = 1$.

The results are depicted in Fig. 5.11a, Fig. 5.11b and Fig. 5.11c. In Fig. 5.11a we display a comparison of a true regression and estimated by the GRNN for $n = 1000$. Figure 5.11b and Fig. 5.11c show the tracking of the non-stationary regression with a changing n in points $x = 0.2$ and $x = 1$, respectively.

Simulation 5.7

We assume, that x_n and z_n are realizations of uniformly distributed random variables on $[0,1]$ and $[-\frac{1}{2},\frac{1}{2}]$, respectively. The GRNN based on the Parzen triangular kernel has been applied with the following parameters:
$a = 0.8$, $H' = 0.6$, $H'' = 0.5$, $k_1' = k_2'' = 1$.

The results are depicted in Fig. 5.12a, Fig. 5.12b and 5.12c. In Fig. 5.12a we display a comparison of a true regression and estimated by the GRNN for $n = 1000$. Figure 5.12b and Fig. 5.12c show the tracking of the non-stationary regression with a changing n in points $x = 0.2$ and $x = 0.6$, respectively.

a)

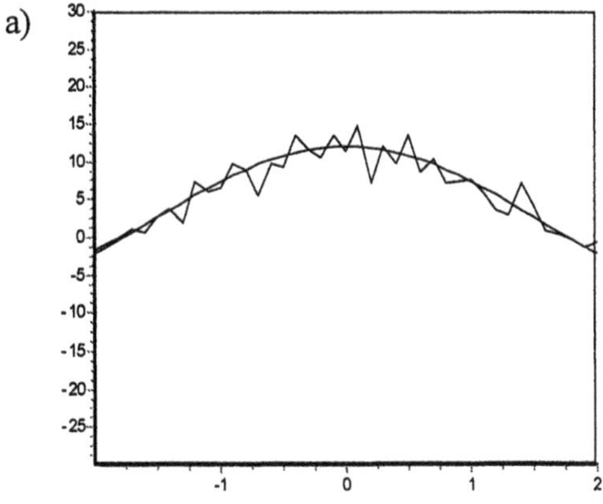

b)

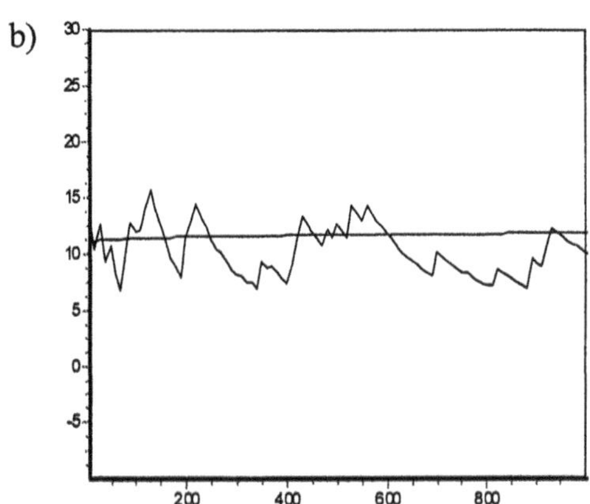

c)

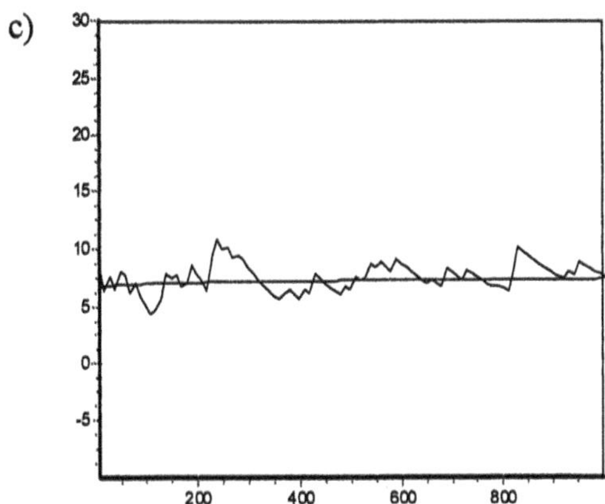

FIGURE 5.11. GRNN for modeling regressions with additive non-stationarity – Simulation 5.6

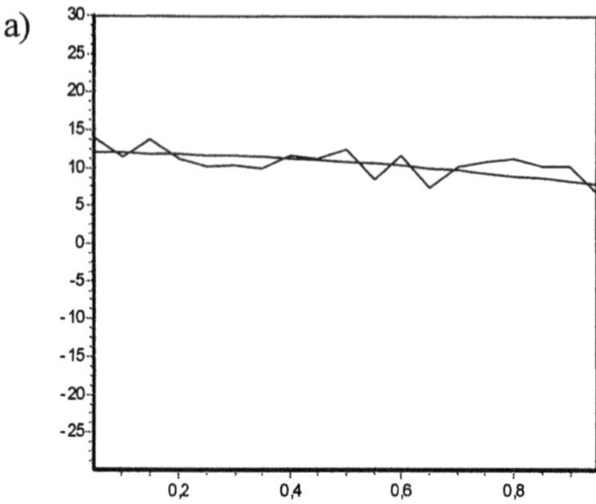

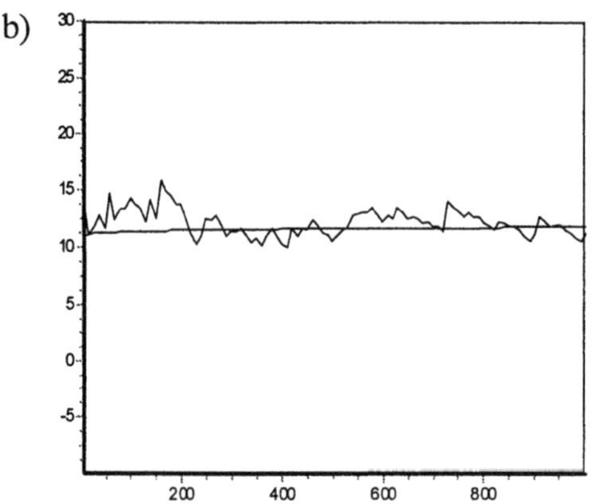

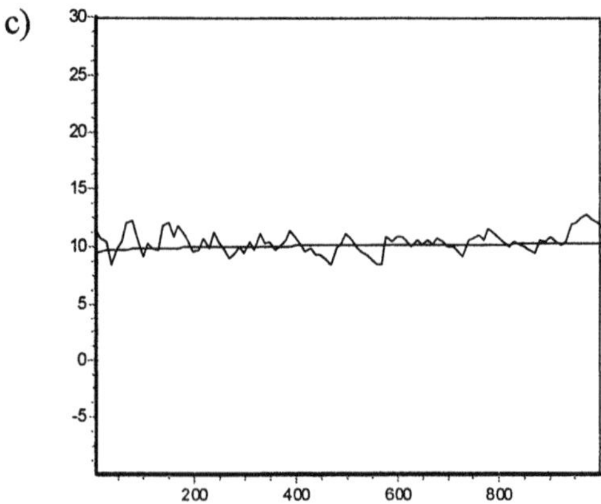

FIGURE 5.12. GRNN for modeling regressions with additive non-stationarity – Simulation 5.7

5.7 Modelling of systems with non-stationarity of the "scale change" and "movable argument" type

Let us consider plants described by equation (5.56), where

$$\phi_n^*(x) = \phi(\omega_n x) \tag{5.61}$$

or

$$\phi_n^*(x) = \phi(x - \lambda_n) \tag{5.62}$$

where:
ω_n – unknown sequence of numbers,
λ_n – unknown sequence of vectors; $\lambda_n = \left[\lambda_n^{(1)}, ..., \lambda_n^{(p)}\right]^T$,
ϕ – unknown function.

With reference to the non-stationarity of plant (5.61) we use the expression "scale change", whereas the non-stationarity of plant (5.62) is referred to as "movable argument". Of course,

$$R_n(x) = f(x)\,\phi(\omega_n x)$$

for plant (5.61) and

$$R_n(x) = f(x)\,\phi(x - \lambda_n)$$

for plant (5.62). In Section 5.5 non-stationary factor α_n and in Section 5.6 non-stationary component β_n were "separable" from function ϕ, significantly simplifying convergence conditions. The present situation is more complicated. Particularly conditions (5.40) and (5.43), taking form (5.45) and (5.46) for the Fourier and Hermite series, now are more complicated. That is why with reference to plants (5.61) and (5.62) we will use algorithm (5.3) based only on the Parzen kernel. With the help of results of Section 5.3.1, in Tables 5.6a and 5.6b are shown conditions ensuring the convergence of algorithm (5.3) tracking changing characteristics (5.61) and (5.62). Conditions (5.19) – (5.22) and (5.23) – (5.26) concern the selection of sequence h_n' and conditions (5.28) and (5.30) concern the selection of sequence h_n'' (according to symbols in Fig. 5.2).

TABLE 5.6a. Conditions for weak convergence of the GRNN based on the Parzen kernel – non-stationarity of the type "scale change" and "movable argument"

Condition	$\lvert\phi(x)-\varphi_n^*(x)\rvert \xrightarrow{n} 0$ in probability
(5.19)	$a_n h_n^{-p} \xrightarrow{n} 0$
(5.20)	$a_n h_n^{-2p} \longrightarrow 0$
(5.21)	$\begin{cases} a_n^{-1}\lvert\omega_{n+1}-\omega_n\rvert \xrightarrow{n} 0 \\ a_n^{-1}\lvert\lambda_{n+1}-\lambda_n\rvert \xrightarrow{n} 0 \end{cases}$
(5.22)	$\begin{cases} a_n^{-1}h_n^r\left(\omega_n^2+1\right) \xrightarrow{n} 0 \\ a_n^{-1}h_n^r \xrightarrow{n} 0 \end{cases}$
(5.28)	$n^{-2}\sum_{i=1}^{n} h_i^{-p} \xrightarrow{n} 0$

TABLE 5.6b. Conditions for strong convergence of the GRNN based on the Parzen kernel – non-stationarity of the type "scale change" and "movable argument"

Condition	$\lvert\phi_n(x)-\varphi_n^*(x)\rvert \xrightarrow{n} 0$ with pr. 1
(5.23)	$\sum_{n=1}^{\infty} a_n^2 h_n^{-p} < \infty$
(5.24)	$\sum_{n=1}^{\infty} a_n^2 h_n^{-2p} < \infty$
(5.25)	$\begin{cases} \sum_{n=1}^{\infty} a_n^{-1}\left(\omega_{n+1}-\omega_n\right)^2 < \infty \\ \sum_{n=1}^{\infty} a_n^{-1}\left(\lambda_{n+1}-\lambda_n\right)^2 < \infty \end{cases}$
(5.26)	$\begin{cases} \sum_{n=1}^{\infty} a_n^{-1}h_n^{2r}\left(\omega_n^4+1\right) < \infty \\ \sum_{n=1}^{\infty} a_n^{-1}h_n^{2r} < \infty \end{cases}$
(5.30)	$\sum_{n=1}^{\infty} n^{-2}h_n^{-p} < \infty$

Conditions in Tables 5.6a and 5.6b hold under the following assumptions:
(i) f and ϕ are bounded
(ii) ϕ satisfies Lipschitz condition
(iii) f and ϕ have bounded partial derivatives up to the r–th order
(iv) f has continuous derivatives up to the third order
Let us now assume that sequences ω_n and λ_n are of the following type:
(i) $\omega_n = k_1 n^t$, $t > 0$
(ii) $\lambda_n = k_2 n^t$, $t > 0$

Employing Theorem 5.2 and arguments similar to those in Section 5.5, we obtain the following expressions defining the speed of convergence of algorithm (5.3):
(i) If $\omega_n = k_1 n^t$, $t > 0$ then

$$P\left(\left|\widehat{\phi}_n(x) - \phi_n^*(x)\right| > \varepsilon\right)$$
$$\leq \left(\frac{\varepsilon + 2}{\varepsilon\ f(x)}\right)^2 \left(c_1 n^{2r(t-H')} + c_2 n^{-r_1}\right. \tag{5.63}$$
$$\left. + c_3 n^{-4H''} + c_4 n^{-(1-H''p)}\right)$$

where

$$r_1 = \min\left[a - H'p,\ 2(1-a-t),\ 2\left(r - H' - rt - a\right)\right]$$

Presently, algorithm (5.3) has tracking properties if

$$t + \frac{a}{r} < H' < \frac{a}{p},\ a < 1 - t,\ 0 < H'' < \frac{1}{p}$$

It means that parameter t should be contained within the range

$$t \in \left(0,\ 1 - \frac{1}{1 + \frac{1}{p} - \frac{1}{r}}\right)$$

where $r > p$. In the one-dimensional case

$$t \in \left(0,\ 1 - \frac{r}{2r-1}\right)$$

Along with an increase of parameter r specifying smooth properties of functions f and ϕ, the range in which parameter t is contained

widens, not exceeding the interval $(0, 1/2)$. The increase of dimension p results in the decrease of the above mentioned range.
(ii) If $\lambda_n = k_2 n^t$, $t > 0$ then

$$P\left(\left|\widehat{\phi}_n(x) - \phi_n^*(x)\right| > \varepsilon\right)$$
$$\leq \left(\frac{\varepsilon + 2}{\varepsilon\, f(x)}\right)^2 \left(c_1 n^{-2rH'} + c_2 n^{-r_1}\right. \tag{5.64}$$
$$\left.+ c_3 n^{-4H''} + c_4 n^{-(1-H''p)}\right)$$

where

$$r_1 = \min\left[a - H'p,\ 2(1 - a - t),\ 2(rH' - a)\right]$$

In other words, algorithm (5.3) has tracking properties if for $r > p$,

$$\frac{a}{r} < H' < \frac{a}{p}, \quad a < 1 - t, \quad 0 < H'' < \frac{1}{p}$$

Assuming that $t \in (0, 1)$ it is possible to select such parameters H' and H'' in algorithm (5.3) that would satisfy the above inequalities.

In Simulations 5.8 and 5.9 we investigate the GRNN applied to track time-varying regressions given by:

$$\Phi_n^*(x_n) = 10\cos(x_n n^{0.1}) + z_n$$

Simulation 5.8
We assume, that x_n and z_n are realizations of $N(0, 1)$ random variables. The GRNN based on the Gaussian kernel has been applied with the following parameters:
$a = 0.6$, $H' = H'' = 0.5$, $k_1' = k_2'' = 1$.

The results are depicted in Fig. 5.13a, Fig. 5.13b and Fig. 5.13c. In Fig. 5.13a we display a comparison of a true regression and estimated by the GRNN for $n = 1000$. Figures 5.13b and Fig. 5.13c show the tracking of the non-stationary regression with a changing n in points $x = 0.2$ and $x = 0.8$, respectively.

a)

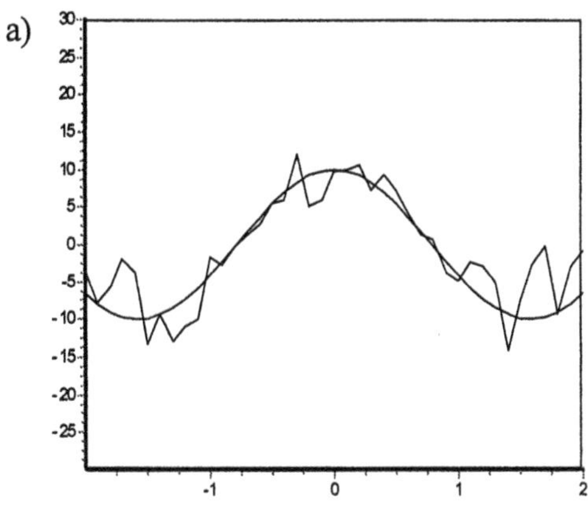

b)

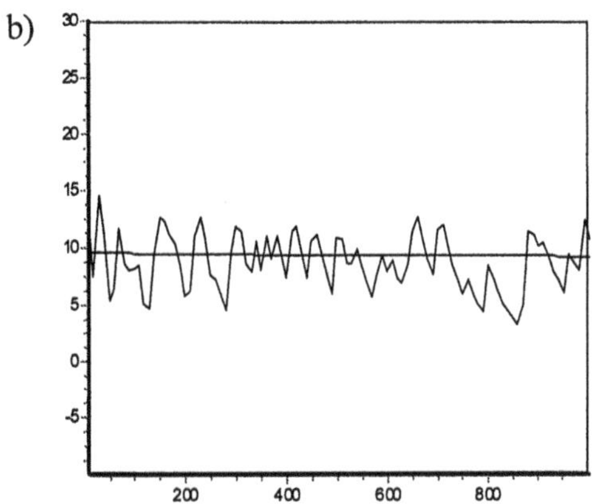

c)

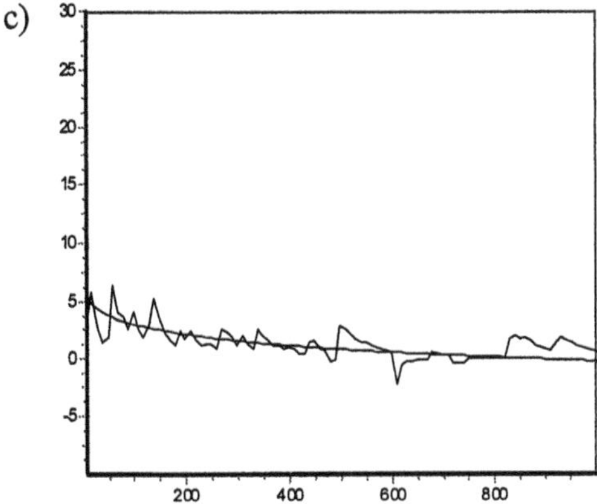

FIGURE 5.13. GRNN for modeling regressions with non-stationarity of the type "scale change" – Simulation 5.8

Simulation 5.9

We assume, that x_n and z_n are realizations of uniformly distributed random variables on $[0, 1]$ and $[-\frac{1}{2}, \frac{1}{2}]$, respectively. The GRNN based on the Parzen triangular kernel has been applied with the following parameters:

$a = 0.6$, $H' = H'' = 0.5$, $k_1' = k_2'' = 1$.

The results are depicted in Fig. 5.14a, Fig. 5.14b and Fig. 5.14c. In Fig. 5.14a we display a comparison of a true regression and estimated by the GRNN for $n = 1000$. Figures 5.14b and Fig. 5.14c show the tracking of the non-stationary regression with a changing n in points $x = 0.2$ and $x = 0.8$, respectively.

In Simulations 5.10 and 5.11 we investigate the GRNN applied to track time-varying regressions:

$$\Phi_n^*(x_n) = 10 \sin(x_n - n^{0.1}) + z_n$$

Simulation 5.10

We assume, that x_n and z_n are realizations of $N(0, 1)$ random variables. The GRNN based on the Gaussian kernel has been applied with the following parameters:

$a = 0.8$, $H' = H'' = 0.5$, $k_1' = k_2'' = 1$.

The results are depicted in Fig. 5.15a, Fig. 5.15b and Fig. 5.15c. In Fig. 5.15a we display a comparison of a true regression and estimated by the GRNN for $n = 1000$. Figures 5.15b and Fig. 5.15c show the tracking of the non-stationary regression with a changing n in points $x = 0.2$ and $x = 0.4$, respectively.

Simulation 5.11

We assume, that x_n and z_n are realizations of uniformly distributed random variables on $[0, 1]$ and $[-\frac{1}{2}, \frac{1}{2}]$, respectively. The GRNN based on the Parzen triangular kernel has been applied with the following parameters:

$a = 0.8$, $H' = H'' = 0.5$, $k_1' = k_2'' = 1$.

The results are depicted in Fig. 5.16a, Fig. 5.16b and Fig. 5.16c. In Fig. 5.16a we display a comparison of a true regression and estimated by the GRNN for $n = 1000$. Figures 5.16b and Fig. 5.16c show the tracking of the non-stationary regression with a changing n in points $x = 0.2$ and $x = 0.4$, respectively.

a)

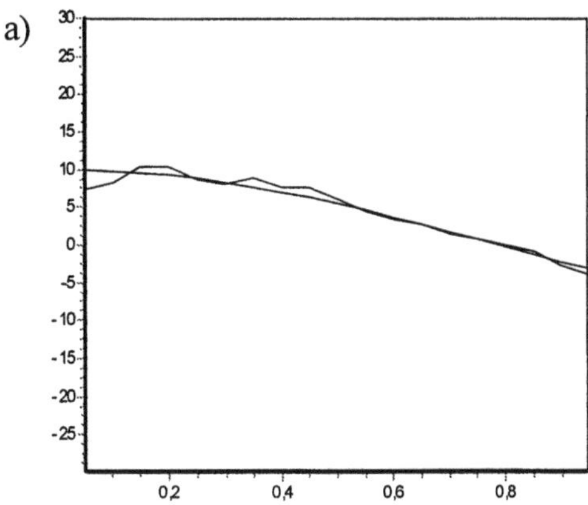

b)

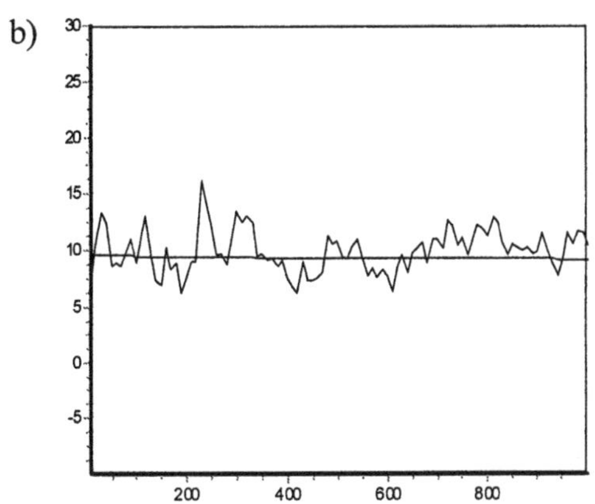

c)

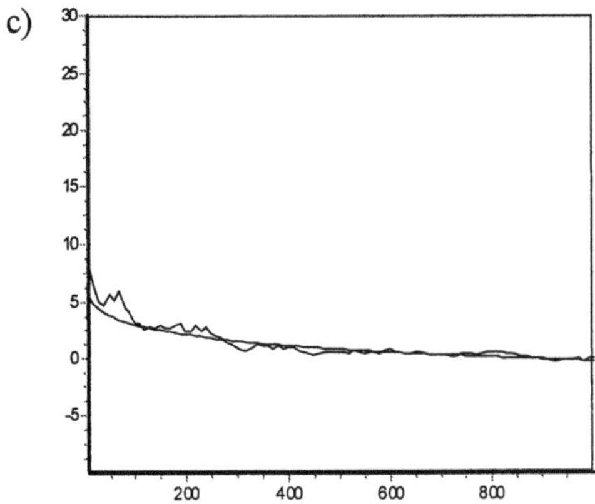

FIGURE 5.14. GRNN for modeling regressions with non-stationarity of the type "scale change" – Simulation 5.9

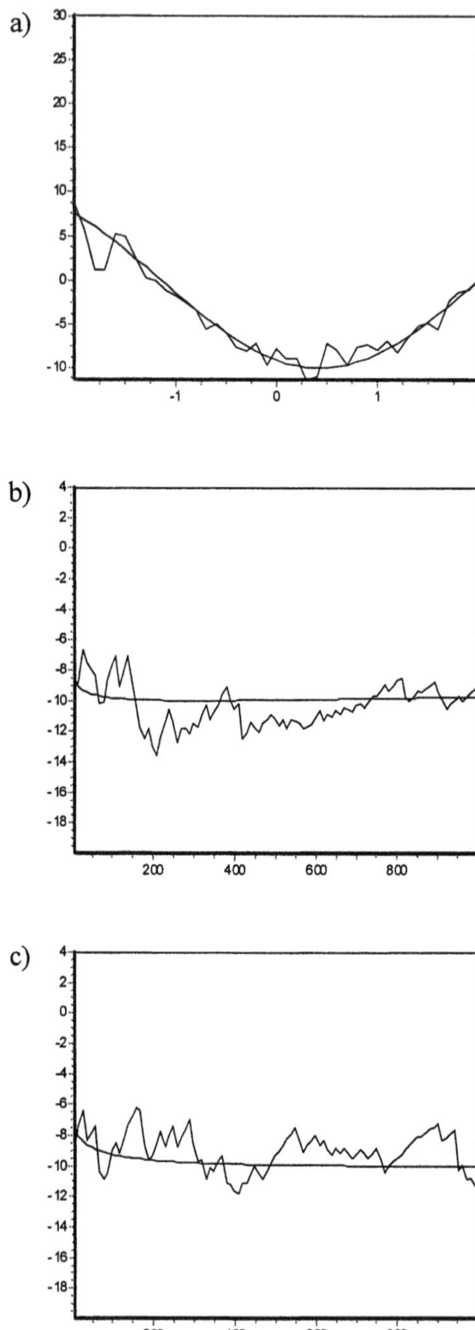

FIGURE 5.15. GRNN for modeling regressions with non-stationarity of the type "movable argument" – Simulation 5.10

a)

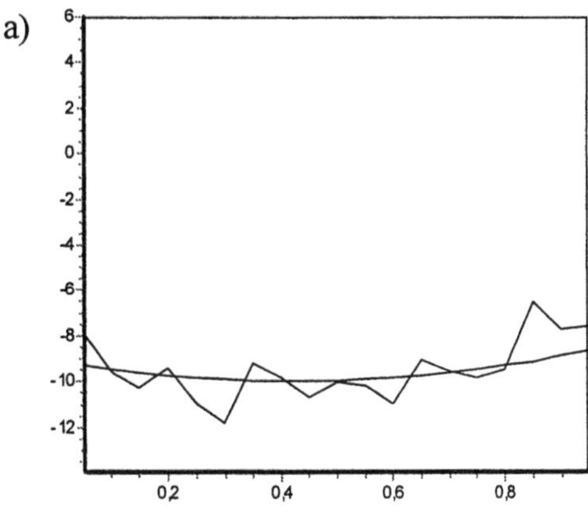

b)

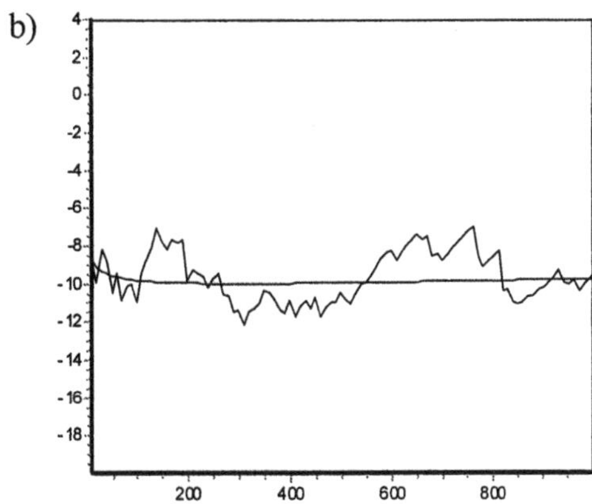

c)

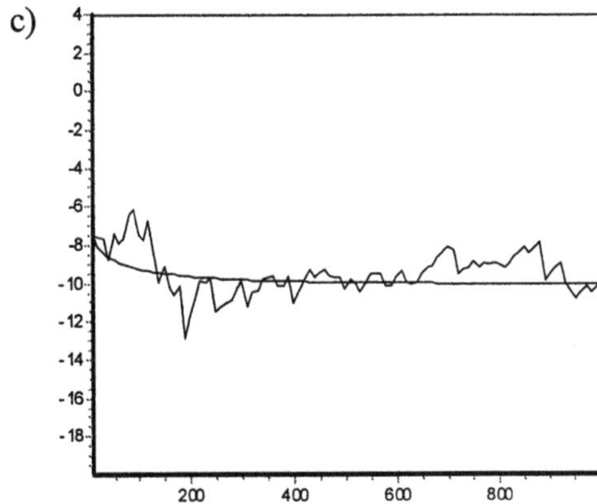

FIGURE 5.16. GRNN for modeling regressions with non-stationarityof the type "movable argument" – Simulation 5.11

In Simulations 5.12 and 5.13 we investigate the GRNN applied to track time-varying regressions:

$$\Phi_n^*(x_n) = 10\cos(x_n + n^{0.2}) + z_n$$

Simulation 5.12
We assume, that x_n and z_n are realizations of $N(0,1)$ random variables. The GRNN based on the Gaussian kernel has been applied with the following parameters:
$a = 0.6$, $H' = 0.37$, $H'' = 0.45$, $k_1' = k_2'' = 1$.

The results are depicted in Fig. 5.17a, Fig. 5.17b and Fig. 5.17c. In Fig. 5.17a we display a comparison of a true regression and estimated by the GRNN for $n = 1000$. Figure 5.17b and Fig. 5.17c show the tracking of the non-stationary regression with a changing n in points $x = 0.6$ and $x = 0.8$, respectively.

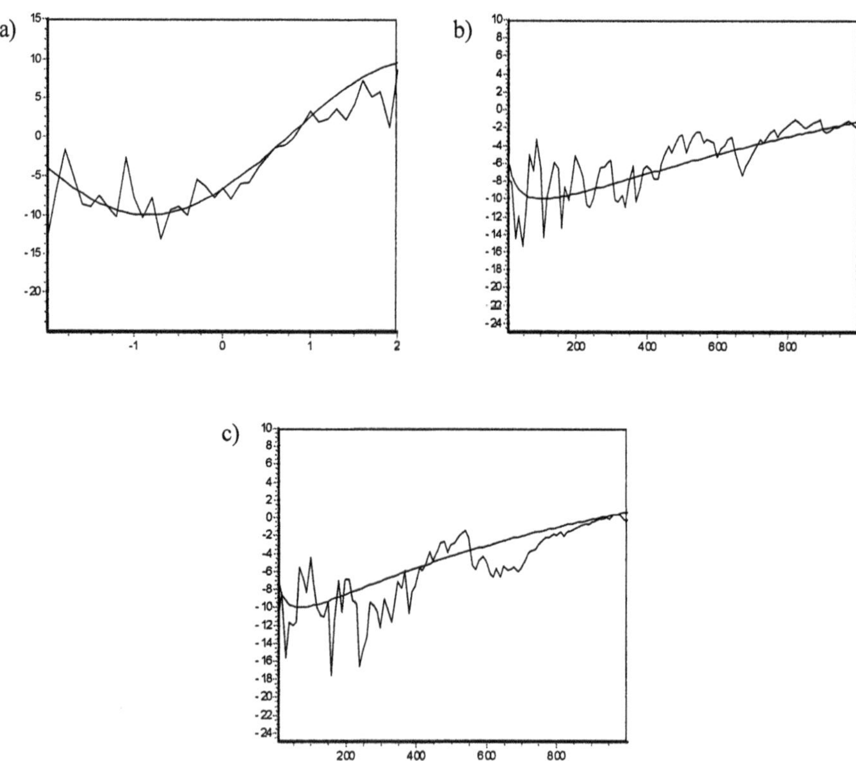

FIGURE 5.17. GRNN for modeling regressions with non-stationarity of the type "movable argument" – Simulation 5.12

Simulation 5.13

We assume, that x_n and z_n are realizations of uniformly distributed random variables on $[0,1]$ and $[-\frac{1}{2},\frac{1}{2}]$, respectively. The GRNN based on the Parzen triangular kernel has been applied with the following parameters:
$a = 0.6$, $H' = 0.37$, $H'' = 0.45$, $k_1' = k_2'' = 1$.

The results are depicted in Fig. 5.18a, Fig. 5.18b and Fig. 5.18c. In Fig. 5.18a we display a comparison of a true regression and estimated by the GRNN for $n = 1000$. Figure 5.18b and Fig. 5.18c show the tracking of the non-stationary regression with a changing n in points $x = 0.2$ and $x = 0.8$ respectively.

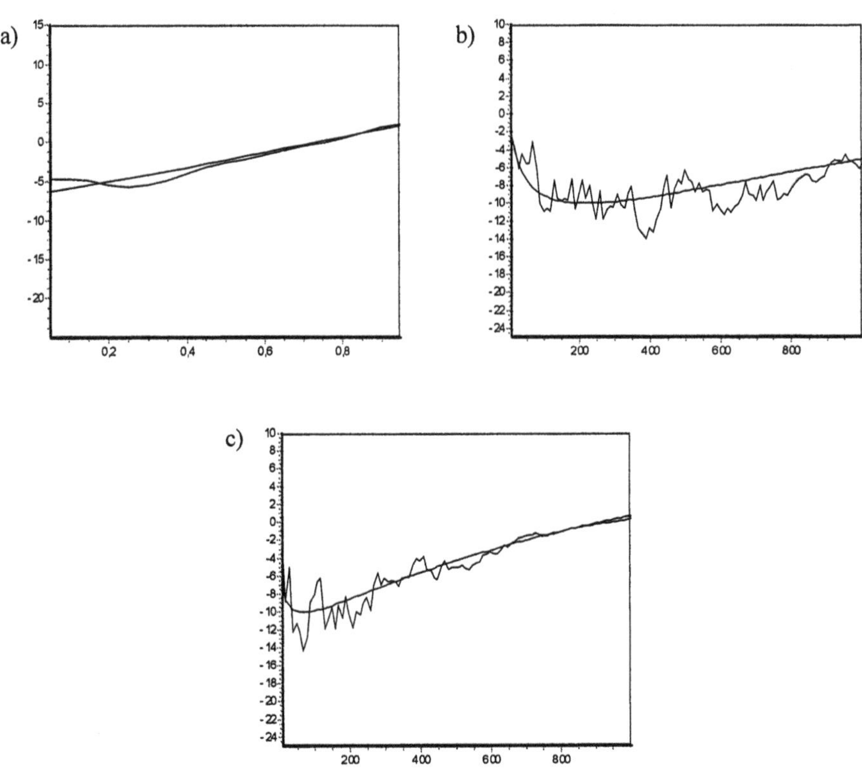

FIGURE 5.18. GRNN for modeling regressions with non-stationarity of the type "movable argument" – Simulation 5.13

5.8 Modelling of systems with a diminishing non-stationarity

Now we will consider a certain special case of non-stationary plants, namely such a one when non-stationarity diminishes as $n \to \infty$. In other words we will assume that

$$\phi_n^*(x) \xrightarrow{n} F(x) \tag{5.65}$$

where F – unknown function. Convergence (5.65) can describe the effect of the ageing of elements or reaching the proper operation regime after the trial period.

In the considered situation, it would be possible to employ all the results of Sections 5.3 and 5.4. An important reason for distinguishing the above case is the possibility of a significant simplification of the convergence conditions that were described above. On the basis of the results from Section 4.5 we will present weaker conditions ensuring the convergence of algorithm (5.3). Like before, a detailed discussion will be carried out for procedures using the Parzen kernel and the orthogonal series method.

The conditions formulated in Corollaries 5.5 and 5.6 concern the selection of sequences h_n' and $q'(n)$ (structural Schemes 5.2 and 5.3). As regards the selection of sequences h_n'' and $q''(n)$ connected with the satisfaction of condition B, the corollaries refer to Sections 5.3.1b and 5.3.2b.

A) Algorithms constructed on the basis of the Parzen kernel.

The following result is a conclusion from general Theorem 4.7.

Corollary 5.5

(pointwise convergence of algorithm (5.3) in a quasi-stationary case). In procedure (4.3), let us choose $a_n = n^{-1}$. Moreover, let us assume that function K satisfies conditions (5.27) and (5.31), $h_n \xrightarrow{n} 0$ and one of the three following conditions holds:

$$\sup |f(x)(\phi_n^*(x) - F(x))| \xrightarrow{n} 0 \tag{5.66}$$

$$h_n^{-p} \int f(x) |\phi_n^*(x) - F(x)| \, dx \xrightarrow{n} 0 \tag{5.67}$$

$$h_n^{-p} \int f^2(x) (\phi_n^*(x) - F(x))^2 \, dx \xrightarrow{n} 0 \tag{5.68}$$

a) If the weak version of condition B (Section 5.3.1) is satisfied and

$$n^{-2}\sum_{i=1}^{n} h_i^{-p} m_i' \xrightarrow{n} 0 \tag{5.69}$$

then

$$\left|\widehat{\phi}_n(x) - \phi_n^*(x)\right| \xrightarrow{n} 0, \quad \text{in probability} \tag{5.70}$$

b) If the strong version of condition B (Section 5.3.1) is satisfied and

$$\sum_{n=1}^{\infty} n^{-2} h_n^{-p} m_n' < \infty \tag{5.71}$$

then

$$\left|\widehat{\phi}_n(x) - \phi_n^*(x)\right| \xrightarrow{n} 0 \tag{5.72}$$

with pr.1. Convergence (5.70) and (5.72) hold in continuity points of functions Ff and f.

Remark 5.3
Replacing condition (5.31) by (5.32) and additionally assuming $Ff \in L_1$, we obtain convergence (5.70) and (5.72) in almost all points x (Stein [260], Wheeden and Zygmund [299]).
It should be noted that conditions (5.69) and (5.71) can be replaced with

$$n^{-2}\sum_{i=1}^{n} h_i^{-2p} m_i'' \xrightarrow{n} 0 \tag{5.73}$$

and

$$\sum_{n=1}^{\infty} n^{-2} h_n^{-2p} m_n'' < \infty \tag{5.74}$$

Example 5.2
Let us consider a plant described by dependence (5.56) where

$$\phi_n^*(x) = c_n \phi(x), c_n \xrightarrow{n} c$$

Let us point out that in this case $F(x) = c\phi(x)$. Thus, conditions (5.66) – (5.68) take the form
a)

$$|c_n - c| \xrightarrow{n} 0, \ \sup |f(x)\phi(x)| < \infty$$

b)

$$h_n^{-p} |c_n - c| \xrightarrow{n} 0, \int f(x) |\phi(x)| \, dx < \infty$$

c)

$$h_n^{-p} (c_n - c)^2 \xrightarrow{n} 0, \quad \int f^2(x) \phi^2(x) \, dx < \infty$$

It is easily seen that conditions (5.69) and (5.71) reduce to

$$n^{-2} \sum_{i=1}^{n} h_i^{-p} \xrightarrow{n} 0, \quad \sum_{n=1}^{\infty} n^{-2} h_n^{-p} < \infty \tag{5.75}$$

if

$$\sup f(x) < \infty \quad \text{and} \quad \sup \left| f(x) \phi^2(x) \right| < \infty$$

Analysing the above conditions, we see that the satisfaction of condition a) does not require additional assumptions imposed on the sequence h_n. In other words, during the designing of a system realizing algorithm (5.3), the non-stationarity effect is not taken into consideration at all. This remark is not true in relation to conditions b) and c). However, if we assume that

$$h_n = n^{-H}, \; H > 0$$

and

$$c_n = c + n^{-t}, \; t > 0$$

then conditions b) and c) take the form

$$n^{Hp-t} \xrightarrow{n} 0, \quad \int f(x) |\phi(x)| \, dx < \infty$$

and

$$n^{Hp-2t} \xrightarrow{n} 0, \quad \int f^2(x) \phi(x)^2 \, dx < \infty$$

Moreover, conditions (5.75) can be written as

$$n^{Hp-1} \xrightarrow{n} 0, \quad \sum_{n=1}^{\infty} n^{Hp-2} < \infty \tag{5.76}$$

On the other hand, conditions (5.75) ensure the convergence of algorithm (5.3) in a stationary situation (e.g. [99]). So if $t \geq 1$ for condition b) or $t \geq \frac{1}{2}$ for condition c) then while designing a system realizing algorithm (5.3), we select sequence h_n in the same way as in the case of a stationary plant. Recapitulating:

1) meeting of condition (5.66) does not require additional assumptions as regards sequence h_n regardless of the speed of convergence of characteristics ϕ_n^* to function F,

2) meeting of conditions (5.67) and (5.68) does not require additional assumptions as regards sequence h_n only when characteristic ϕ_n^* convergences fast enough.

B) Algorithms constructed on the basis of the orthogonal series.

For $n = 1, 2, ...,$we will define a function

$$S_n(x) = \sum_{|\underline{j}| \leq q} \Psi_{\underline{j}}(x) D_{\underline{j}} \tag{5.77}$$

where

$$D_{\underline{j}} = \int F(x) f(x) \Psi_{\underline{j}}(x) dx.$$

The following result is a corollary from Theorem 4.7.

Corollary 5.6

(pointwise convergence of algorithm (5.3) in a quasi-stationary case). In procedure (4.3), let us choose $a_n = n^{-1}$. Let us assume that $q(n) \xrightarrow{n} \infty$ and one of the following conditions is satisfied:

$$\sup |\phi_n^*(x) - F(x)| \left(\sum_{j=0}^{q(n)} G_j^2 \right)^p \xrightarrow{n} 0 \tag{5.78}$$

$$\int f(x) |\phi_n^*(x) - F(x)| dx \left(\sum_{j=0}^{q(n)} G_j^2 \right)^p \xrightarrow{n} 0, \tag{5.79}$$

$$\int f^2(x) (\phi_n^*(x) - F(x))^2 dx \left(\sum_{j=0}^{q(n)} G_j \right)^{2p} \xrightarrow{n} 0, \tag{5.80}$$

a) If the weak version of condition b (Section 5.3.2b) is satisfied and

$$n^{-2} \sum_{i=1}^{n} \left(\sum_{j=0}^{q(i)} G_j^2 \right)^p m_i' \xrightarrow{n} 0, \tag{5.81}$$

then

$$\left|\widehat{\phi}_n(x) - \phi_n^*(x)\right| \xrightarrow{n} 0. \tag{5.82}$$

in probability.

b) If the strong version of condition b (Section 5.3.2b) is satisfied and

$$\sum_{n=1}^{\infty} n^{-2} \left(\sum_{j=0}^{q(n)} G_j^2 \right)^p m_n' < \infty \tag{5.83}$$

then

$$\left|\widehat{\phi}_n(x) - \phi_n^*(x)\right| \xrightarrow{n} 0. \tag{5.84}$$

with pr.1. The convergence (5.82) and (5.84) hold in points x at which

$$S_n(x) \xrightarrow{n} F(x) f(x) \tag{5.85}$$

and condition (5.49) is met.

The problem of convergence of (5.49) and (5.85) for various orthogonal series is discussed in detail in Section 5.3.2b. Observe that conditions (5.81) and (5.83) can be replaced by

$$n^{-2} \sum_{i=1}^{n} \left(\sum_{j=0}^{q(i)} G_j^2 \right)^{2p} m_i'' \xrightarrow{n} 0, \tag{5.86}$$

and

$$\sum_{n=1}^{\infty} n^{-2} \left(\sum_{j=0}^{q(n)} G_j^2 \right)^{2p} m_i'' < \infty \tag{5.87}$$

Analysing conditions (5.78) – (5.80) we have to state that they all require meeting additional assumptions as regards sequence $q(n)$ unless characteristic ϕ_n^* converges fast enough. Let us get back to Example 5.2. For the Fourier system, conditions (5.78) – (5.80) take the same form

$$|c_n - c|\, q^p(n) \xrightarrow{n} 0 \tag{5.88}$$

with various assumptions as regards functions f and ϕ.
For $c_n = c + n^{-t}$, the above condition can be written as

$$n^{-t} q^p(n) \xrightarrow{n} 0$$

Since

$$n^{-1} q^p(n) \xrightarrow{n} 0$$

implies convergence in a stationary situation (it can be shown if we weaken results presented in [209]), then the non-stationarity effect is neglected while designing a system realizing algorithm (5.3) for $t \geq 1$.

5.9 Concluding remarks

The general learning procedure (4.3) that was described in Chapter 4 enabled the construction of non-parametric algorithms for the modelling of non-stationary systems. The convergence of these algorithms was shown through the formulation of corollaries from general theorems on convergence of procedure (4.3). In Section 5.3, conditions for convergence of algorithm (5.3) tracking changes of optimal characteristic ϕ_n^*, $n = 1, 2, \ldots$, were given.

Since the type of non-stationarity was not specified, the results obtained have on the one hand a universal nature (they can be used for the modelling of plants with various types of non-stationarity), but on the other hand they are less transparent. However, if the non-stationarity is specified, e.g. multiplicative or additive, the conditions given in tables of this chapter are clear. In order to illustrate the possibilities of the application of algorithm (5.3), let us assume sequences α_n, β_n, ω_n and λ_n (connected with plants investigated in this chapter) as sequences of a type n^t, i.e. let us consider the following plants:

$$y_n = c_1 n^t \phi(x_n) + z_n$$

$$y_n = \phi(x_n) + c_2 n^t + z_n$$

$$y_n = \phi\left(x_n c_3 n^t\right) + z_n$$

$$y_n = \phi\left(x_n - c_4 n^t\right) + z_n$$

Using tables given in Sections 5.5, 5.6 and 5.7 it is possible to design the GRNN, i.e. to find sequences a_n, h_n or q_n that will guarantee tracking properties despite the fact that we do not know parameter t and function ϕ.

In this chapter, two types of algorithms were used:

a) GRNN based on the Parzen kernel,

b) GRNN based on the orthogonal series method.

Generally, we should say that the convergence of algorithm b) is connected with both the convergence of the orthogonal series and with the speed of this convergence (assessment of the "tail" of the series). The problem of convergence of the orthogonal series is less complicated in the scalar case (e.g. Sansone [237]), but more complicated in the multidimensional case, that was investigated in more detail only for the Fourier series (e.g. Sjölin [246]). An additional problem is the examination of the orthogonal series speed of convergence because even in the one-dimensional case, appropriate results can be obtained with quite complicated assumptions as regards expanded functions which should be "smooth" enough. The selection of a particular series depends on both boundedness or unboundedness of the input signal and on the dimensionality of the problem; if $A = R^p$, $p \geq 1$ then the proper series is the Hermite series. If $A \subset R^p$, $p > 1$, $\mu(A) < \infty$, then it is reasonable to employ the Fourier series because its properties are well known. In the scalar case, when the input signal is bounded, it is possible to use, among others, the Fourier, the Haar and the Legendre series. The last series is particularly interesting because the assumptions concerning smooth properties connected with the application of this series are much weaker than in the case of application of the Fourier series (Sansone [237]). The above considerations suggest that we should rather use algorithm a) that is based on the Parzen's kernel. However, the orthogonal series method has a very desirable advantage: if e.g. sequence α_n (or β_n) is of type n^t, then the maximum value of t at which the algorithm still has tracking properties does not depend on dimension p (contrary to the algorithm based on the Parzen kernel). Moreover, simulations that were conducted do not discredit the orthogonal series method, especially when it is used for the identification of plants with multiplicative and additive non-stationarity. A certain problem in the case of applying the algorithm based on the Parzen kernel may be the selection of function H meeting conditions (5.14) – (5.17). This problem will arise for some types of non-stationarity when a high degree of smooth properties of functions R_n, $n = 1, 2, ...$ will be required. In this chapter examples of function H were given for which parameter r connected with smooth properties of $R_n, n = 1, 2, ...$ takes the value of 2 and 4.

In the stationary case, the pointwise convergence of non-parametric identification algorithms depends on local properties of an unknown characteristic of the best model ϕ^*. The constructed algorithms are asymptotically optimal in the points of convergence of the orthogonal series (see Rutkowski [209]). Comparing it with the non-stationary case, we must point out that the pointwise convergence of algorithm (5.3) depends on global properties of functions f and $f\ \phi^*$ (conditions (5.22) and (5.26) as well as (5.45) and (5.46)), which relates to all types of non-stationarity considered in Sections 5.5 – 5.7.

The examples of particular types of non-stationarity given in Sections 5.5 – 5.7 do not exhaust all the possibilities of application of algorithm (5.3). In particular, the results can be used for the identification of plants with non-stationarity that is a combination of cases discussed in this chapter, i.e.

$$\phi_n^*(x) = \alpha_n \phi(\omega_n x - \lambda_n) + \beta_n$$

It is well known that identification problems are usually closely connected with control tasks in real-time systems. Taking into consideration the requirements of control algorithms, identification algorithms should have the capability to predict characteristics of the best model some time in advance. In other words, on the basis of observations

$$(X_1, Y_1), ..., (X_n, Y_n)$$

we should predict

$$\phi_{n+k}^*(x) = E[Y_{n+k} | X_{n+k} = x], \ k \geq 1.$$

Using the results from Section 4.6 it is possible to say that Corollaries 5.1 – 5.4 (given in Section 5.3) ensure the convergence of algorithm (5.3) in the sense

$$\left| \widehat{\phi}_n(x) - \phi_{n+k}^*(x) \right| \xrightarrow{n} 0$$

in probability (with pr.1). The speed of convergence of algorithm (5.3) in the sense

$$P\left(\left| \widehat{\phi}_n(x) - \phi_{n+k}^*(x) \right| > \varepsilon \right)$$

is a result of inequality (5.55). Of course, the higher the value k, the lower the speed of convergence (the right side of inequality (5.55) takes on a higher value).

For plants with a multiplicative and additive type of non-stationarity, it is possible to use the identification algorithm based on the modified procedure (4.3), expressed by formula (4.71). The selection of a sequence of functions W_n in procedure (4.71) depends directly on the form of sequences α_n and β_n in plants (5.57) and (5.60). In particular, if α_n = const. n^t, $t > 0$, then we can take

$$W_n\left(R_n\left(x\right)\right) = \left(1 + \frac{1}{n+1}\right) R_n\left(x\right)$$

Condition (4.72) will be reduced to

$$a_n^{-1}\left|\alpha_{n+1} - \alpha_n\left(1 + \frac{1}{n+1}\right)\right| \xrightarrow{n} 0.$$

The above condition along with a set of the remaining assumptions from corollaries 5.1 and 5.3 is met when $0 < t < \frac{1}{2}$. So the application of the modified procedure slightly widens the range in which parameter t is contained (in comparison with results from Tables 5.3a and 5.3b). However, as we mentioned in Section 4.7, this "widening" is not very large.

The fundamental results of this chapter were partly presented in [211], [218] – [221], [224], [225] and [235]. In the future research it would be interesting to describe non-stationary changes linguistically and apply flexible neuro-fuzzy systems (see Rutkowski and Cpałka [232], [233]) for their modelling.

6

Probabilistic Neural Networks for Pattern Classification in a Time-Varying Environment

6.1 Introduction

In Chapter 3 we presented probabilistic neural networks working in a stationary environment. However, as we mentioned in Chapter 5, a lot of phenomena have a non-stationary character, e.g. the conventer-oxygen process of steelmaking, the change of catalyst properties in an oil refinery or in the process of carbon dioxide conversion. In particular, a large group of problems from the domain of technology, biology and physics is described by non-stationary probability densities, e.g. of the "movable argument" type – $f_n(x) = f(x - c_n)$ [196]. The problem of pattern classification in a time-varying environment was analyzed in only a few works: de Figueiredo [84] and Tzypkin [277] used the parametric approach approximating probability densities that change with time by means of linear combinations of a certain fixed set (base) of functions. Appropriate coefficients were estimated by making use of dynamic stochastic approximation algorithms. Of course, such an approach does not ensure an asymptotic optimality of the classification rules. Rutkowski [217] considered the asymptotically optimal classification rules in a quasi-stationary environment when class conditional densities are convergent to a finite limit.

In this chapter we propose a new class of probabilistic neural networks working in a non-stationary environment. The novelty of this chapter is summarized as follows:

1. We formulate the problem of pattern classification in a non-stationary environment as a prediction problem and design a probabilistic neural network to classify patterns having time-varying probability distributions. We note that the problem of pattern classification in the non-stationary case is closely connected with the problem of prediction because on the basis of a learning sequence of length n, a pattern in the moment $n+k$, $k \geq 1$ should be classified.

2. We present, for the first time in literature, definitions of optimality of PNN in a time-varying environment. Moreover, we prove that our PNN asymptotically approach the Bayes optimal (time-varying) decision surface. Time-varying discriminant functions are estimated by means of a general learning procedure presented in Chapter 3 and the convergence of algorithms is a consequence of theorems presented in that chapter.

3. We investigate the speed of convergence of the constructed PNN.

4. We design in detail PNN based on the Parzen kernels and multivariate Hermite series.

It should be emphasized that the design of the PNN in a time-varying environment is much more difficult than in the stationary case. In order to design PNN approaching the Bayes optimal decision surfaces (time-varying), we should pick up not only a smoothing parameter (denoted in this book by h_n for the Parzen kernel and by $q(n)$ for the orthogonal series kernel) but also a learning sequence a_n which should satisfy conditions typical for stochastic approximation procedures (see e.g. [7]). In this context, it is worth to quote Bendat and Piersol [32], who believed that the problem of estimation of a non-stationary probability density requires the possession of many realizations of the stochastic process. We would like to emphasize that the PNN constructed in this chapter allow to track time-varying discriminant functions (in particular, tracking time-varying probability densities) with use of only one realization

of the stochastic process – subsequent observations of a learning sequence. For illustration of the capability of our PNN we mention that having a sequence $\{X_n\}$ of independent random variables with probability densities $f_n(x) = f(x - n^t)$, we are able to estimate time-varying densities despite the fact that both f and parameter t $(0 < t < 1)$ are unknown. Consequently, we are able to estimate time-varying discriminant functions and corresponding classification rules. The PNN studied in this chapter are adaptive in the sense that they adopt to changes of the time-varying environment.
Undoubtedly, one of the most important problems in pattern classification is the selection of features. These problems were widely discussed in literature (e.g. [89], [90]). Therefore in this chapter we assume that the features that interest us have already been selected and we will focus on the non-stationarity of the problem.

6.2 Problem description and presentation of classification rules

Let (X_n, V_n), $n = 1, 2, ...$ be a sequence of independent pairs of random variables. Random variable X_n has an interpretation of the pattern connected with a given class and takes values in space A, $A \subset R^p$. Random variable V_n takes values in set $\{1, ..., M\}$ called the set of classes, specifying the class number.
A priori probabilities of occurrence of class m in moment n $(m = 1, ..., M, n = 1, 2, ...)$ will be represented by p_{mn}, i.e. $p_{mn} = P(V_n = m)$. It is assumed that there are conditional probability densities f_{mn} of random variable X_n on condition that $V_n = m$. These densities are called densities in classes.
The classification rule is a measurable mapping $\varphi_n : A \to \{1, ..., M\}$. The measure of quality of the rule φ_n is the probability of missclassification

$$P(\varphi_n(X_n) \neq V_n) \stackrel{df}{=} L_n(\varphi_n) \tag{6.1}$$

The rule that minimizes the above index is called the Bayes' rule. The Bayes' rule in moment n is denoted by φ_n^* and the value of $L_n(\varphi_n^*)$ is denoted by L_n^*, i.e.

$$L_n(\varphi_n^*) = L_n^* \tag{6.2}$$

We will define the following function

$$d_{mn}(x) = p_{mn} f_{mn}(x) \tag{6.3}$$

This function will be called the discriminant function of class m in moment n. Generalizing considerations for the stationary case (e.g. [75]), it is easily seen that the rule φ_n^* has the form

$$\boxed{\varphi_n^*(X_n) = m, \text{ if } d_{mn}(X_n) > d_{in}(X_n)} \tag{6.4}$$

for $i \neq m$, $i = 1, ..., M$, $n = 1, 2, ...$. We assume that both a priori probabilities p_{mn} and densities in classes f_{mn}, $m = 1, ..., M$, $n = 1, 2, ...$, are completely unknown. For this reason we use empirical classification rules based on estimators of discriminant functions.

The problem of non-parametric pattern classification in a non-stationary case boils down to constructing empirical classification rules that on the basis of the learning sequence

$$(X_1, V_1), ..., (X_n, V_n) \tag{6.5}$$

would classify pattern X_{n+k}, $k \geq 1$. It is, of course, an issue of prediction of patterns having non-stationary probability densities. In the case of complete probabilistic information, i.e. the knowledge of discriminant functions

$$d_{m,n+k}(x) = p_{m,n+k}\, f_{m,n+k}(x), \tag{6.6}$$

pattern X_{n+k} could be classified by means of rule (6.4).

Let $\widehat{d}_{mn}$ be an estimator – constructed on the basis of the learning sequence (6.5) – of function $d_{m,n+k}$, $k \geq 1$. In this chapter, we will consider the empirical rules of the form

$$\boxed{\widehat{\varphi}_n(X_{n+k}) = m, \text{ if } \widehat{d}_{mn}(X_{n+k}) > \widehat{d}_{in}(X_{n+k})} \tag{6.7}$$

for $i \neq m$, $i = 1, ..., M$. The sequence of empirical rules $\widehat{\varphi}_n$ is called the classification learning algorithm. The rule $\widehat{\varphi}_n$ is a function of the learning sequence $(X_1, V_1), ..., (X_n, V_n)$ and classified pattern X_{n+k}, $k \geq 1$.

We will now construct an estimator of function (6.6). We will first show that procedure (4.3) that was presented in Chapter 4 can be used for estimation of time-varying discriminant functions (6.3). Let

$$T_{mn} = \begin{cases} 1 & \text{if} \quad V_n = m \\ 0 & \text{if} \quad V_n \neq m \end{cases} \tag{6.8}$$

Discriminant function (6.3) can be presented as

$$d_{mn}(x) = f_n(x) E[T_{mn} | X_n = x] \tag{6.9}$$

where

$$f_n(x) = \sum_{m=1}^{M} p_{mn} f_{mn}(x). \tag{6.10}$$

Comparing (6.9) and (4.3), setting $Y_n = T_{mn}$ for the fixed m, we use procedure (4.3) for the estimation of discriminant functions (6.3):

$$\begin{gathered} \widehat{d}_{m,n+1}(x) = \widehat{d}_{mn}(x) + a_{n+1} \\ \cdot \left(T_{m,n+1} K_{n+1}(x, X_{n+1}) - \widehat{d}_{mn}(x) \right) \\ d_{m,0}(x) = 0 \;\; \text{for} \;\; m = 1, ..., M, \;\; n = 0, 1, 2, ... \end{gathered} \tag{6.11}$$

On the basis of considerations of Section 4.6, $\widehat{d}_{mn}$ can be used not only for the estimation of d_{mn}, but also for the estimation of $d_{m,n+k}$, $k \geq 1$.

The structural scheme of a system that realizes the classification algorithm with the use of procedure (6.11) is presented in Fig. 6.1.

Sequences $\{K_n\}$ and $\{a_n\}$ on which this procedure is based, should generally satisfy different conditions depending on the number of class m. That is why in Fig. 6.1, symbols $K_n^{(m)}$ and $a_n^{(m)}$, $m = 1, ..., M$ are used. We should point out that in order to classify pattern X_{n+k}, $k \geq 1$, it is necessary to store the whole learning set of length n. Next, when pattern X_{n+k} to be classified appears, procedure (6.11) is activated starting from $n = 0$ and putting $x = X_{n+k}$.

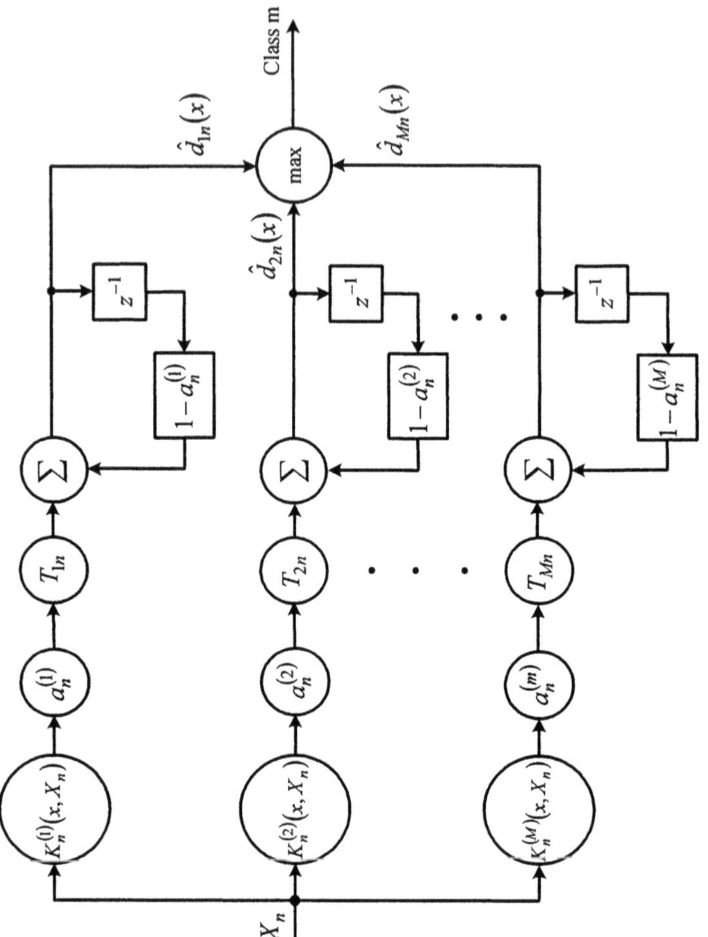

FIGURE 6.1. The PNN for pattern classification in a time-varying environment

6.3 Asymptotic optymality of classification rules

As was mentioned in the introduction to this chapter, the concept of asymptotic optimality of classification rules in the non-stationary case has not been studied in literature yet. In this section we will present appropriate definitions and show that when the length of learning set (6.5) increases, classification learning algorithms become increasingly similar to optimal algorithm (6.4) which could be determined when a priori probabilities p_{mn} and densities in classes f_{mn} are known, $m = 1, ..., M$, $n = 1, 2, ...$. In Section 6.4 we will discuss the issue of the speed of convergence of empirical classification rules.

The global performance measure of classification rule (6.4), classifying X_{n+k}, is the probability of missclassification in moment $n+k$:

$$L_{n+k}\left(\varphi_{n+k}^{*}\right) = P\left(\varphi_{n+k}^{*}\left(X_{n+k}\right) \neq V_{n+k}\right) \tag{6.12}$$

As a performance measure of empirical rule (6.7) we take

$$L_{n+k}\left(\widehat{\varphi}_{n}\right) = P\left(\widehat{\varphi}_{n}\left(X_{n+k}\right) \neq V_{n+k} \left| X_{1}, V_{1}, ..., X_{n}, V_{n}\right.\right), \tag{6.13}$$

i.e. the probability of misclassification of pattern X_{n+k} determined on the basis of rule (6.7) and learning sequence (6.5).

Definition 6.1
Classification algorithm $\widehat{\varphi}_n$ defined by (6.7) is weakly asymptotically optimal when

$$EL_{n+k}\left(\widehat{\varphi}_{n}\right) - L_{n+k}\left(\varphi_{n+k}^{*}\right) \xrightarrow{n} 0. \tag{6.14}$$

Definition 6.2
Classification algorithm $\widehat{\varphi}_n$ defined by (6.7) is strongly asymptotically optimal when

$$L_{n+k}\left(\widehat{\varphi}_{n}\right) - L_{n+k}\left(\varphi_{n+k}^{*}\right) \xrightarrow{n} 0. \tag{6.15}$$

with probability 1.

The following theorem ensures asymptotic optimality of the rule (6.7) if estimator $\widehat{d}_{mn}$ (expressed by formula (6.11) "follows" the changes of discriminant function $d_{m,n+k}$, when $n \to \infty$.

Theorem 6.1
Let χ_n, $n = 1, 2, ...$, be a sequence of sets in R^p such that for $\varepsilon > 0$ the following conditions are satisfied

$$\int_{\chi_n} f_{n+k}(x)\, dx \geq 1 - \varepsilon/2, \;\; n = 1, 2, ..., \tag{6.16}$$

and

$$\mu\left(\chi_n\right) \leq \text{const.}, \; n = 1, 2, ..., \tag{6.17}$$

where

$$f_n(x) = \sum_{m=1}^{M} p_{mn} f_{mn}(x)$$

a) If

$$E \int \left(\widehat{d}_{mn}(x) - d_{m,n+k}(x) \right)^2 dx \xrightarrow{n} 0, \tag{6.18}$$

then the pattern classification rule (6.7) is weakly asymptotically optimal.
b) If

$$\int \left(\widehat{d}_{mn}(x) - d_{m,n+k}(x) \right)^2 dx \xrightarrow{n} 0, \tag{6.19}$$

with pr.1, then the pattern classification rule (6.7) is strongly asymptotically optimal.

Remark 6.1
It is always possible to select sequence χ_n in such a way so that condition (6.16) could be met. However, it does not mean that the condition (6.17) is automatically satisfied. For instance, if densities in classes are of the exponential type

$$f_{mn}(x) = \lambda_{mn} e^{-\lambda_{mn} x}, \ x \geq 0 \tag{6.20}$$

and

$$\lambda_{mn} \xrightarrow{n} 0, \ m = 1, ..., M, \tag{6.21}$$

then there does not exist sequence χ_n that satisfies conditions (6.16) and (6.17) at the same time. However, if the densities in the classes are of the "movable argument" type

$$f_{mn}(x) = f_m(x - c_{mn}), \tag{6.22}$$

then it is possible to take (in the scalar case)

$$\chi_n = [c_{mn} - \nu, \ c_{mn} + \nu] \tag{6.23}$$

for a sufficiently large ν.

6.4 Speed of convergence of classification rules

The speed of convergence of procedure (6.11) in the sense

$$E\int\left(\widehat{d}_{mn}(x)-d_{m,n+k}(x)\right)^2 dx \xrightarrow{n} 0$$

can be evaluated by means of the general Theorem 4.6 taking into account Corollary 4.2 concerning prediction. For this purpose, it is necessary to specify constants A_2, B_2 and C_2 that are present in the assumptions of Theorem 4.6. We will do it in Section 6.7, considering a particular type of non-stationarity. Now, we will connect the speed of convergence of (6.19) with the speed of convergence of (6.14), i.e.

$$EL_{n+k}\left(\widehat{\varphi}_n\right)-L_{n+k}\left(\varphi^*_{n+k}\right) \xrightarrow{n} 0$$

As we know (Definition 6.1), convergence (6.14) ensures a weak asymptotic optimality of rule (6.7).

Let us denote

$$t_{m,n+k}=\int\ldots\int\left|x^{(1)}\right|^s\ldots\left|x^{(p)}\right|^s f_{m,n+k}(x)\,dx,\ \ s>0 \tag{6.24}$$

and

$$t_{n+k}=\sum_{m=1}^{M} t_{m,n+k}, \tag{6.25}$$

where $t_n<\infty$, $n=1,2,\ldots$.

Theorem 6.2
Let us assume that

$$E\int\left(\widehat{d}_{mn}(x)-d_{m,n+k}(x)\right)^2 dx = 0\left(u_n\right) \tag{6.26}$$

a) if sequence t_n is bounded, i.e.

$$t_n\le \text{const.},\ n=1,2,\ldots, \tag{6.27}$$

then

$$EL_{n+k}\left(\widehat{\varphi}_n\right)-L_{n+k}\left(\varphi^*_{n+k}\right)=0\left(u_n^{s/2s+1}\right) \tag{6.28}$$

b) if sequence t_n is not bounded, i.e.

$$t_n\ge \text{const.}>0,\ \ n>n_0 \tag{6.29}$$

then

$$EL_{n+k}\left(\widehat{\varphi}_n\right) - L_{n+k}\left(\varphi^*_{n+k}\right) = 0\left(t_{n+k}u_n^{s/2s+1}\right) \tag{6.30}$$

From the proof of Theorem 6.2 it follows that if non-stationarity densities in classes are of the "movable argument" type (6.22), i.e.

$$f_{mn}(x) = f_m(x - c_{mn})$$

then conclusion a) is true when

$$\int \cdots \int \left|x^{(1)}\right|^s \cdots \left|x^{(p)}\right|^s f_m(x)\, dx < \infty \tag{6.31}$$

for a certain $s > 0$.

In the next sections we will consider procedures of type (6.7) constructed on the basis of the Parzen kernel and the orthogonal series method. Using general Theorems 4.3 and 4.4 and Theorem 6.1, we will present conditions ensuring the convergence of algorithm (6.7).

6.5 Classification procedures based on the Parzen kernels

In Fig. 6.2 we present the PNN based on the Gaussian kernel for pattern classification in a time-varying environment ($M = 2$). We assume the normalization of vectors x and X_i.

As in Chapter 5, appropriate conditions for the convergence of the classification algorithm shown in Fig. 6.2 will depend on smooth properties of the density function f_{mn} ($m = 1, ..., M$, $n = 1, 2, ...$). We define

$$\delta^{\underline{i}}_{mn} = \int \left[\frac{\partial^r}{\partial x^{(i_1)} ... \partial x^{(i_r)}} f_{mn}(x)\right]^2 dx, \tag{6.32}$$

where $\underline{i} = (i_1, ..., i_r)$, $i_k = 1, ..., p$, $k = 1, ..., r$.

The following result is a corollary from Theorems 4.3 and 6.1:

Corollary 6.1

If function K satisfies conditions (5.13) – (5.17), assumptions (6.16) and (6.17) hold, $h_n \to 0$, and

$$a_n h_n^{-p} \xrightarrow{n} 0, \tag{6.33}$$

$$a_n^{-2} p_{mn}^2 h_n^{2r} \delta_{mn}^i \xrightarrow{n} 0, \tag{6.34}$$

$$a_n^{-2} \left| p_{m,n+1} - p_{mn} \right| \int f_{mn}^2 (x)\, dx \xrightarrow{n} 0, \tag{6.35}$$

$$a_n^{-2} p_{mn}^2 \int \left(f_{m,n+1} (x) - f_{mn} (x) \right)^2 dx \xrightarrow{n} 0, \tag{6.36}$$

then the pattern classification rule (6.7) is weakly asymptotically optimal.

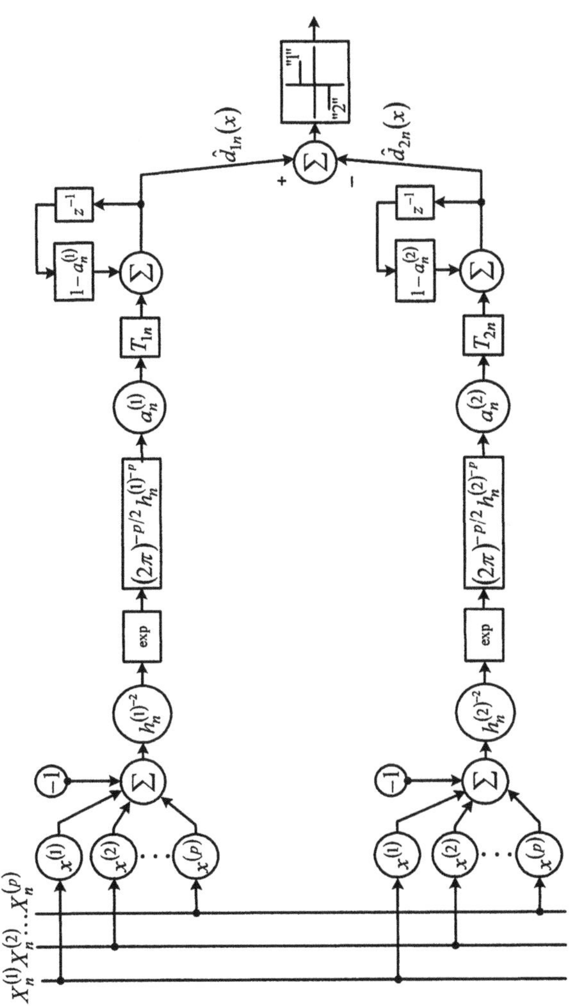

FIGURE 6.2. The PNN based on the Parzen kernel for pattern classification in a time-varying environment

Corollary 6.2. is a consequence of the general Theorem 4.4 and Theorem 6.1.

Corollary 6.2
If function K satisfies conditions (5.13) – (5.17), assumptions (6.16) and (6.17) hold, $h_n \to 0$, and

$$\sum_{n=1}^{\infty} a_n^2 h_n^{-p} < \infty, \tag{6.37}$$

$$\sum_{n=1}^{\infty} a_n^{-1} h_n^{2r} \delta_{mn}^{\underline{i}} < \infty, \tag{6.38}$$

$$\sum_{n=1}^{\infty} a_n^{-1} \left(p_{m,n+1} - p_{mn}\right)^2 \int f_{mn}^2 (x)\, dx < \infty, \tag{6.39}$$

$$\sum_{n=1}^{\infty} a_n^{-1} p_{mn}^2 \int \left(f_{m,n+1} (x) - f_{mn} (x)\right)^2 dx < \infty, \tag{6.40}$$

then the pattern classification rule (6.7) is strongly asymptotically optimal.

6.6 Classification procedures based on the orthogonal series

In Fig 6.3 we present the PNN based on orthogonal series for pattern classification in a time-varying environment.

Let us denote

$$S_{mn} = \int \left(\sum_{|\underline{j}| \le q} l_{\underline{j}n}^m \Psi_{\underline{j}} (x) - f_{mn} (x) \right)^2 dx \tag{6.41}$$

where

$$l_{\underline{j}n}^m = \int f_{mn} \Psi_{\underline{j}} (x)\, dx \tag{6.42}$$

The following result is a corollary from Theorems 4.3 and 6.1.

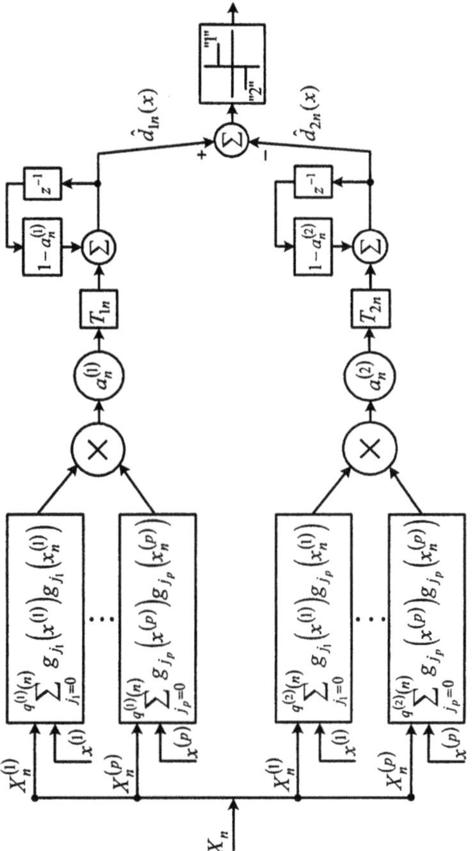

FIGURE 6.3. The PNN based on orthogonal series for pattern classification in a time-varying environment

Corollary 6.3
If conditions (6.35) and (6.36) are satisfied, $q(n) \to \infty$, and

$$a_n \left(\sum_{j=0}^{q(n)} G_j^2 \right)^p \xrightarrow{n} 0, \tag{6.43}$$

$$a_n^{-2} p_{mn}^2 S_{mn} \xrightarrow{n} 0, \tag{6.44}$$

then the pattern classification rule (6.7) is weakly asymptotically optimal.

Corollary 6.4 is a consequence of the general Theorem 4.4 and Theorem 6.1.

Corollary 6.4
If conditions (6.39) and (6.40) are satisfied, $q(n) \to \infty$,and

$$\sum_{n=1}^{\infty} a_n^2 \left(\sum_{j=0}^{q(n)} G_j^2 \right)^p < \infty \tag{6.45}$$

$$\sum_{n=1}^{\infty} a_n^{-1} p_{mn}^2 S_{mn} < \infty \tag{6.46}$$

then the pattern classification rule (6.7) is strongly asymptotically optimal.
Conditions (6.44) and (6.46) take a more concrete form depending on the smooth properties of function f_{mn} and the orthogonal series used. As an example, we will use the multidimensional Hermite series.

Let us assume that

$$D_{mn}^l \left(x; f_{mn} \right) \in L_2 \left(R^p \right), \tag{6.47}$$

where

$$D_{mn}^l \left(x; f_{mn} \right) = \prod_{j=1}^{P} \left(x^{(j)} - \frac{\partial}{\partial x^j} \right)^l f_{mn} \left(x \right), \; l > 1.$$

Then

$$S_{mn} \leq \left|\left| D_{mn}^l \left(x; f_{mn} \right) \right|\right|_{L_2}^2 q^{-pl} \left(n \right) \tag{6.48}$$

The above inequality is a generalization to the multidimensional and non-stationary case of the result obtained by Walter [289]. By means of this inequality, conditions (6.44) and (6.46) can be expressed as

$$a_n^{-2} p_{mn}^2 q^{-pl} \left(n \right) \left|\left| D_{mn}^l \right|\right|_{L_2}^2 \xrightarrow{n} 0 \tag{6.49}$$

and

$$\sum_{n=1}^{\infty} a_n^{-1} p_{mn}^2 q^{-pl} \left(n \right) \left|\left| D_{mn}^l \right|\right|_{L_2}^2 < \infty. \tag{6.50}$$

As we will see in Section 6.7, these conditions take a simple form for a particular type of non-stationarity of function f_{mn}.

6.7 Non-stationarity of the "movable argument" type

In order to simplify our considerations, let us examine the one-dimensional case and assume that a priori probabilities p_{mn} do not change with time. As regards density in classes, we assume that they are of the form

$$f_{mn}(x) = f_m(x - c_{mn})$$

for $n = 1, 2, ..., m = 1, ..., M$, where $x \in R^1$. The above case occurs most often in practice [196]. In Tables 6.1a, 6.1b, 6.2a and 6.2b, we present conditions that ensure the weak and strong asymptotic optimality of procedure (6.7) constructed on the basis of the Parzen kernel and the Hermite orthonormal series.

TABLE 6.1a. Conditions for weak convergence of the PNN based on the Parzen kernel

Condition	$L_{n+k}(\widehat{\varphi}) - L_{n+k}(\varphi^*_{n+k}) \overset{n}{\longrightarrow} 0$ in probability	Assumptions
(6.33)	$a_n h_n^{-1} \overset{n}{\longrightarrow} 0$	
(6.34)	$a_n^{-1} h_n^2 \overset{n}{\longrightarrow} 0$	$f''_m \in L_2,\ r = 2$
(6.36)	$a_n^{-1}\left\|c_{m,n+1} - c_{mn}\right\| \overset{n}{\longrightarrow} 0$	$f_m, f'_m, f''_m \in L_2$

TABLE 6.1b. Conditions for strong convergence of the PNN based on the Parzen kernel

Condition	$L_{n+k}(\widehat{\varphi}) - L_{n+k}(\varphi^*_{n+k}) \overset{n}{\longrightarrow} 0$ with pr. 1	Assumptions
(6.37)	$\sum_{n=1}^{\infty} a_n^2 h_n^{-1} < \infty$	
(6.38)	$\sum_{n=1}^{\infty} a_n^{-1} h_n^4 < \infty$	$f''_m \in L_2,\ r = 2$
(6.40)	$\sum_{n=1}^{\infty} a_n^{-1}(c_{m,n+1} - c_{mn})^2 < \infty$	$f_m, f'_m, f''_m \in L_2$

TABLE 6.2a. Conditions for weak convergence of the PNN based on the orthogonal series

Condition	$L_{n+k}(\widehat{\varphi}) - L_{n+k}(\varphi^*_{n+k}) \xrightarrow{n} 0$ in probability	Assumptions
(6.43)	$a_n q^{5/6}(n) \xrightarrow{n} 0$	$G_j = \text{const.}\,(j+1)^{-\frac{1}{12}}$ $l = 2$
(6.44)	$a_n^{-2}\left(c_{mn}^4 + 1\right) q^{-2}(n) \xrightarrow{n} 0$	$f_m, f_m', f_m'' \in L_2$ $\int x^4 f_m^2(x)\,dx < \infty$
(6.36)	$a_n^{-1}\left\lvert c_{m,n+1} - c_{mn}\right\rvert \xrightarrow{n} 0$	$\int x^2 f_m^2(x)\,dx < \infty$

TABLE 6.2b. Conditions for strong convergence of the PNN based on the orthogonal series

Condition	$L_{n+k}(\widehat{\varphi})$ $-L_{n+k}(\varphi^*_{n+k}) \xrightarrow{n} 0$ with. pr. 1	Assumptions
(6.45)	$\sum_{n=1}^{\infty} a_n^2 q^{5/6}(n) < \infty$	$G_j =$ $\text{const.}(j+1)^{-\frac{1}{12}}$
(6.46)	$\sum_{n=1}^{\infty} a_n^{-1}\left(c_{mn}^4 + 1\right) q^{-2}(n) < \infty$	$l = 2$ $f_m, f_m', f_m'' \in L_2$
(6.40)	$\sum_{n=1}^{\infty} a_n^{-1}\left(c_{m,n+1} \quad c_{mn}\right)^2 < \infty$	$\int x^4 f_m^2(x)\,dx < \infty$ $\int x^2 f_m^2(x)\,dx < \infty$

As regards smooth properties of function f_m, $r = 2$ (conditions (6.34) and (6.38)) is assumed in the case of the use of the Parzen kernel and $m = 2$ (conditions (6.49) and (6.50)) in the case of the use of the Hermite series. However, as follows from the tables, the use of the orthogonal series method requires more assumptions as regards function f_m and its derivatives. For instance, let us assume that sequence c_{mn} representing density function non-stationarity is of the type

$$c_{mn} = n^{t_m}, \quad t_m > 0, \quad m = 1, ..., M \quad n = 1, 2, ... \tag{6.51}$$

Analysing all conditions specified in Tables 6.1a, 6.1b, 6.2a and 6.2b, it is possible to establish within what limits parameters t_m, $m =$

$1, ..., M$ should be contained so that Corollaries 6.1 – 6.4 could be true. The results of such an analysis are presented in Table 6.3.

TABLE 6.3. Conditions imposed on parameter t_m – non-stationarity of the "movable argument" type

Method	Weak convergence	Strong convergence
Parzen	$0 < t_m < 1$	$0 < t_m < \frac{1}{7}$
Hermite orthogonal series	$0 < t_m < \frac{1}{11}$	conditions of Corollary 6.4 are not satisfied

It is easily seen that the use of the Hermite orthogonal series requires much more strict assumptions as regards the range within which parameters t_m can be contained.

Using Theorem 4.6 and Corollary 4.2 we will now evaluate the speed of convergence of algorithms (6.7) and (6.11). In procedure (6.11), let us select sequences h_n, $q(n)$ and a_n of the following type:

$$h_n = k_1 n^{-H}, \ k_1 > 0, \ H > 0, \tag{6.52}$$

$$q(n) = \left[k_2 n^Q\right], \ k_2 > 0, \ Q > 0, \tag{6.53}$$

$$a_n = \frac{k}{n^a}, \ k > 0, \ a > 0, \tag{6.54}$$

a) Speed of convergence of algorithms based on the Parzen kernel

With reference to the symbols from Theorem 4.6, we obtain:

$$A_2 = H, \ B_2 = 2(1 - t_m), \ C_2 = 4H \tag{6.55}$$

Consequently,

$$\begin{aligned} & E \int \left(\widehat{d}_{mn}(x) - d_{m,n+k}(x)\right)^2 dx \\ \leq \ & l_3 n^{-4H} + l_4 n^{-r_2} + l''(k) n^{-2(1-t_m)}, \end{aligned} \tag{6.56}$$

where

$$r_2 = \min\left[a - H,\ 2\left(1 - t_m - a\right),\ 2\left(2H - a\right)\right]. \tag{6.57}$$

If

$$\int |x|^s f_m(x)\, dx < \infty, \quad s > 0, \tag{6.58}$$

then, from Theorem 6.2, we obtain

$$EL_{n+k}\left(\widehat{\varphi}_n\right) - L_{n+k}\left(\varphi^*_{n+k}\right) = 0\left(n^{-As/s+1}\right), \tag{6.59}$$

where

$$A = \min\left[4H,\ r_2,\ 2\left(1 - t_m\right)\right]. \tag{6.60}$$

b) Speed of convergence of algorithms based on the Hermite orthonormal series

In this case we have

$$A_2 = \frac{5}{6}Q,\ B_2 = 2\left(1 - t_m\right),\ C_2 = 2\left(Q - 2t_m\right) \tag{6.61}$$

From Theorem 4.6 it follows that

$$\begin{aligned} & E\int\left(\widehat{d}_{mn}(x) - d_{m,n+k}(x)\right)^2 dx \\ \leq\ & l_3 n^{2(2t_m - Q)} + l_4 n^{-r_2} + l''(k) n^{-2(1-t_m)}, \end{aligned} \tag{6.62}$$

where

$$r_2 = \min\left[a - \frac{5}{6}Q,\ 2\left(1 - t_m - a\right),\ 2\left(Q - 2t_m - a\right)\right]. \tag{6.63}$$

If condition (6.58) holds, then

$$EL_{n+k}\left(\widehat{\varphi}_n\right) - L_{n+k}\left(\varphi^*_{n+k}\right) = 0\left(n^{-Bs/s+1}\right), \tag{6.64}$$

where

$$B = \min\left[2\left(Q - t_m\right),\ r_2,\ 2\left(1 - t_m\right)\right] \tag{6.65}$$

Analysing formulas (6.59) and (6.64) we see that the influence of parameters t_m on the speed of convergence of both algorithms is much more significant in the case of the use of the algorithm based on the Hermite series (it results in a decrease of this speed).

In all the above considerations, the same degree of smooth properties of function f_m was assumed: $r = 2$ for the algorithm based on the Parzen kernel and $l = 2$ for the algorithm based on the Hermite orthogonal series method. It is easy to prove that for $r > 2$, the range within which parameters t_m, $m = 1, ..., M$, are contained and which ensures weak asymptotic optimality of the algorithm does not widen. For $l \geq 2$ (with additional assumptions as regards function f_m and its derivatives up to the l-th order), the following inequality holds

$$E \int \left(\widehat{d}_{mn}(x) - d_{m,n+k}(x) \right)^2 dx \leq l_3 n^{l(2t_m - Q)} + l_4 n^{-r_2} + l''(k) n^{-2(1-t_m)}, \tag{6.66}$$

where

$$r_2 = \min \left[a - \frac{5}{6}Q,\ 2(1 - t_m - a),\ Ql - 2lt_m - 2a \right]. \tag{6.67}$$

From the last inequality it follows that

$$0 < t_m < \frac{3l - 5}{8l - 5} \approx \frac{3}{8} \tag{6.68}$$

for a sufficiently large l. In other words, for the algorithm based on the Hermite series, a significant increase of smooth properties of function f_m allows the widening of the range within which parameters t_m are contained, but this "widened" range is nevertheless significantly narrower than in the case of the use of the algorithm based on the Parzen kernel ($0 < t_m < 1$).

6.8 Classification in the case of a quasi-stationary environment

We will show that the previously presented conditions ensuring asymptotic optimality of rule (6.7) can be weakened if

$$p_{m,n} \xrightarrow{n} p_m \tag{6.69}$$

and

$$f_{mn}(x) \xrightarrow{n} f_m(x) \tag{6.70}$$

for $m = 1, ..., M$, where functions f_m are probability densities defined on A. Presently, if

$$\left|\widehat{d}_{mn}(x) - d_{m,n+k}(x)\right| \xrightarrow{n} 0, \ m = 1, ..., M \tag{6.71}$$

in probability (with pr.1) for almost all points x then the pattern classification learning algorithm (6.7) is weakly (strongly) asymptotically optimal. This result can be obtained through a slight modification of Greblicki's result [98] for the stationary case. Using results from Section 4.5 (which are true if $a_n = n^{-1}$ in procedure (6.7)), we will prove convergence (6.71) for algorithms based on the Parzen kernel and the orthogonal series method.

a) Algorithms based on the Parzen kernel

The following result is a corollary from the general Theorem 4.7.

Corollary 6.5

Let us select $a_n = n^{-1}$ in procedure (6.11). Let us assume that $h_n \xrightarrow{n} 0$, function K satisfies conditions (5.31), the convergence

$$|p_{mn} - p| \, h_n^{-p} \xrightarrow{n} 0 \tag{6.72}$$

holds and one of the following conditions is satisfied:

$$\sup |f_{mn}(x) - f_m(x)| \xrightarrow{n} 0, \tag{6.73}$$

$$h_n^{-p} \int |f_{mn}(x) - f_m(x)| \, dx \xrightarrow{n} 0, \tag{6.74}$$

$$h_n^{-p} \int (f_{mn}(x) - f_m(x))^2 \, dx \xrightarrow{n} 0, \tag{6.75}$$

i) If

$$n^{-2} \sum_{i=1}^{n} h_i^{-p} \xrightarrow{n} 0, \tag{6.76}$$

then the pattern classification algorithm defined by (6.7) is weakly asymptotically optimal.

ii) If

$$\sum_{n=1}^{\infty} n^{-2} h_n^{-p} < \infty, \tag{6.77}$$

then the pattern classification algorithm defined by (6.7) is strongly asymptotically optimal.

Let us point out that the satisfaction of condition (6.73) does not require additional assumptions as regards sequence h_n. If $p_{mn} = p$ then, taking into consideration (6.73), while designing a system realizing algorithm (6.7), the non-stationarity effect is not taken into consideration at all, regardless of the speed of the diminishment of the non-stationarity. Asymptotic optimality of the algorithm is then ensured by conditions (6.76) and (6.77), identical as in the stationary case [67]. Observe that

$$n^{-2}\sum_{i=1}^{n} h_i^{-p} \leq n^{-1}\sum_{i=1}^{n} i^{-1}h_i^{-p} \xrightarrow{n} 0 \tag{6.78}$$

if

$$h_n^{-p}n^{-1} \xrightarrow{n} 0, \tag{6.79}$$

Thus, conditions (6.74) and (6.75) are implied by condition (6.76) when

$$\int \left|f_{mn}(x) - f_m(x)\right| dx = 0\left(n^{-1}\right), \tag{6.80}$$

or

$$\int \left(f_{mn}(x) - f_m(x)\right)^2 dx = 0\left(n^{-1}\right). \tag{6.81}$$

In other words, while designing a system realizing algorithm (6.7), the non-stationarity effect is not taken into consideration also when the non-stationarity diminishes fast enough. In the above considerations, we assumed that $p_{mn} = p$.

b) Algorithms based on the orthogonal series method

The following corollary is a consequence of the general Theorem 4.7.

Corollary 6.6

In procedure (6.11) let us select $a_n = n^{-1}$. Let us assume that $q(n) \xrightarrow{n} \infty$, and

$$\left|p_{mn} - p\right| \left(\sum_{j=0}^{q(n)} G_j^2\right)^2 \xrightarrow{n} 0, \tag{6.82}$$

$$\sum_{|\underline{j}|\leq q} B_{\underline{j}}^{m}\Psi_{\underline{j}}(x) \xrightarrow{n} f_m(x) \tag{6.83}$$

for almost all x, where $B_j^m = \int \Psi_{\underline{j}}(x) f_m(x)\, dx$ and one of the following conditions is satisfied:

$$\left(\sum_{j=0}^{q(n)} G_j^2\right)^p \int |f_{mn}(x) - f_m(x)|\, dx \xrightarrow{n} 0, \tag{6.84}$$

$$\left(\sum_{j=0}^{q(n)} G_j^2\right)^{2p} \int (f_{mn}(x) - f_m(x))^2\, dx \xrightarrow{n} 0, \tag{6.85}$$

i) If

$$n^{-2} \sum_{i=1}^{n} \left(\sum_{j=0}^{q(i)} G_j^2\right)^{2p} \xrightarrow{n} 0 \tag{6.86}$$

then the pattern classification algorithm defined by (6.7) is weakly asymptotically optimal.
ii) If

$$\sum_{n=1}^{\infty} n^{-2} \left(\sum_{j=0}^{q(n)} G_j^2\right)^{2p} < \infty \tag{6.87}$$

then the pattern classification learning algorithm defined by (6.7) is strongly asymptotically optimal.

Conditions that ensure convergence (6.83) for various orthogonal series are given in Section 5.3.2. Let us assume that $p_{mn} = p$. Now, in contrast to the algorithm based on the Parzen kernel, the non-stationarity effect is not taken into consideration only when the non-stationarity diminishes fast enough. It is then possible to show that conditions (6.84) and (6.85) are implied by (6.86). It should be noticed that condition (6.86) ensures asymptotic optimality in the stationary case (Rutkowski [208]).

6.9 Simulation results

6.9.1 PNN for estimation of a time-varying probability density

Let $\{X_n\}$ be a sequence of independent random variables with probability densities $f_n(x) = f(x - n^t)$. If $f(x) = N(0,1)$, then

$$f_n(x) = \frac{1}{\sqrt{2\pi}} e^{-\frac{(x-n^t)^2}{2}} \tag{6.88}$$

It is easily seen that if

$$p_{mn} = \begin{cases} 1 & \text{when} \quad m = 1 \\ 0 & \text{when} \quad m \neq 1 \end{cases}$$

then algorithm (6.11) can be used for non-parametric learning of time-varying probability densities f_n. Procedure (6.11) applied for non-parametric estimation of time-varying probability densities takes the form

$$\widehat{f}_{n+1}(x) = \widehat{f}_n(x) + a_{n+1}\left(K_{n+1}(x, X_{n+1}) - \widehat{f}_n(x)\right) \tag{6.89}$$

which becames

$$\widehat{f}_{n+1} = \widehat{f}_n + a_{n+1}\left[\frac{1}{\sqrt{2\pi}} h_{n+1}^{-1} e^{-\frac{1}{2}\left(\frac{x - X_{n+1}}{h_{n+1}}\right)^2} - \widehat{f}_n(x)\right] \tag{6.90}$$

for the Gaussian kernel. Let us choose

$$\begin{aligned} h_n &= kn^{-H}, \quad k > 0, \quad H > 0 \\ a_n &= n^{-a} \end{aligned} \tag{6.91}$$

Depending on the parameter t in model $f_n(x) = f(x - n^t)$ we pick up parameters a and H such that

$$a < 1 - t, \quad \tfrac{a}{2} < H < a \tag{6.92}$$

ensuring the convergence of the PNN.

Simulation 6.1

In this simulation we apply the recursive PNN given by (6.90) for the estimation of the time-varying normal distribution given by (6.88). In Fig. 6.4 we show the results for a) $t = 0.1$, $H = 0.5$, $a = 0.8$, $k = 2$, b) $t = 0.2$, $H = 0.45$, $a = 0.7$, $k = 2$, c) $t = 0.3$, $H = 0.4$, $a = 0.6$, $k = 2$, d) $t = 0.4$, $H = 0.35$, $a = 0.5$, $k = 2$. In each case $n = 100$.

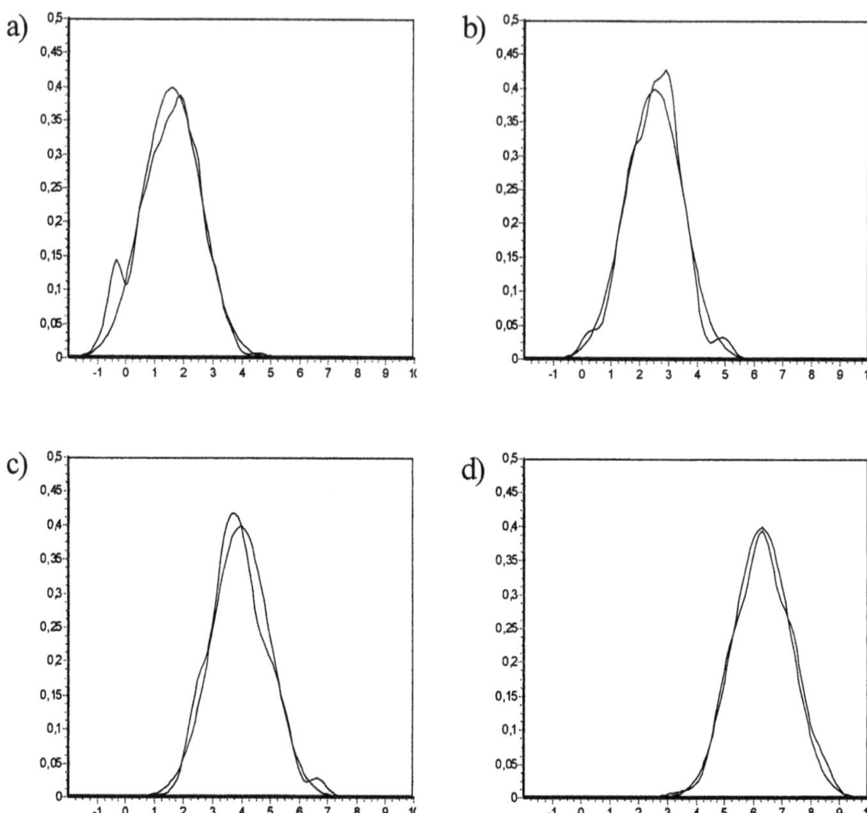

FIGURE 6.4. Estimation of time-varying probability density – Simulation 6.1

Simulation 6.2

In this simulation we investigate the influence of parameter a on the results of the simulations. We apply the recursive PNN given by (6.90) for the estimation of the time-varying normal distribution. In Fig. 6.5 we show the results for $n = 100$, $t = 0.1$, $H = 0.5$, $k = 2$ and a varying a: a) $a = 0.5$, b) $a = 0.6$, c) $a = 0.8$.

Simulation 6.3

In this simulation we would like to show the influence of the sample size n on the results of the simulations. We apply the recursive PNN given by (6.90) for the estimation of the time-varying normal distribution. In Fig. 6.6 we show the results for $t = 0.4$, $H = 0.3$, $a = 0.7$, $k = 2$ and a varying n: a) $n = 10$, b) $n = 100$, c) $n = 200$.

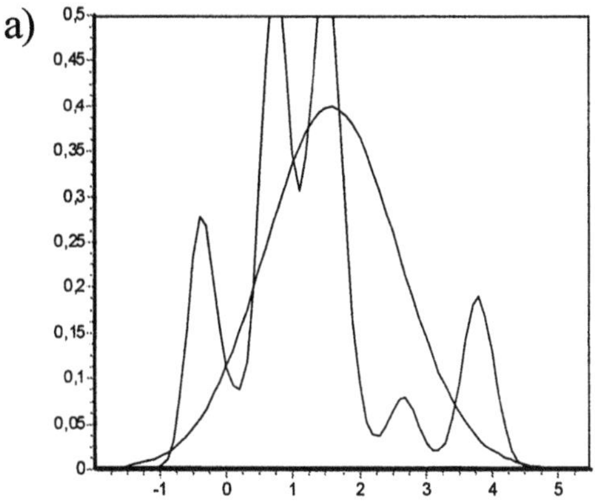

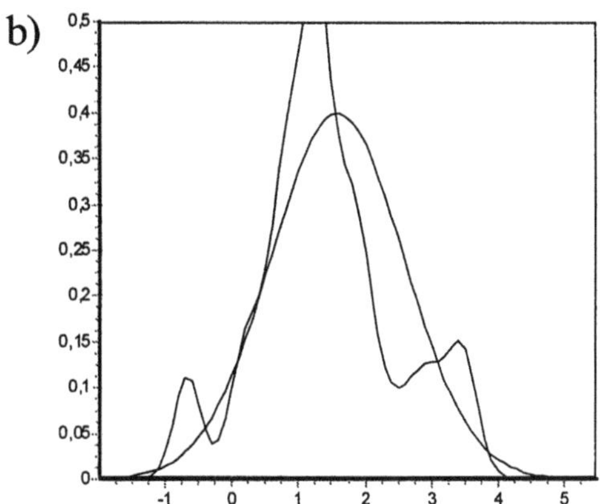

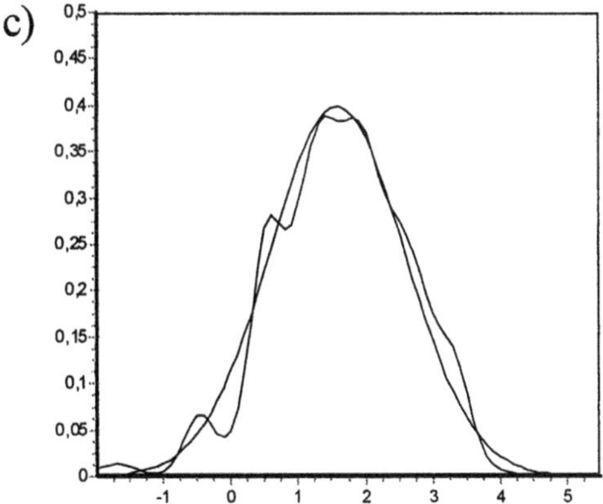

FIGURE 6.5. Estimation of time-varying probability density – Simulation 6.2

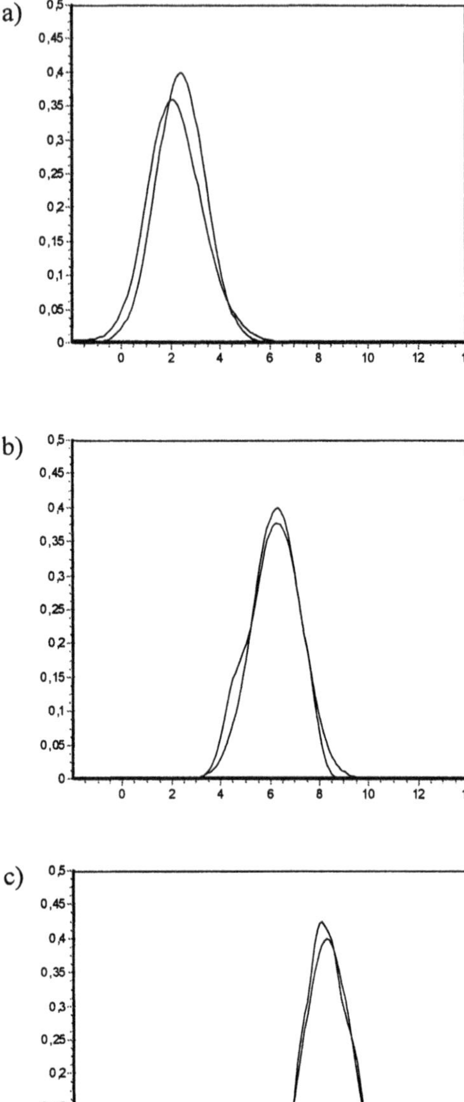

FIGURE 6.6. Estimation of time-varying probability density – Simulation 6.3

Simulation 6.4

In this simulation we apply the recursive PNN given by (6.90) for the estimation of the time-varying normal distribution. In Fig. 6.7 we present the results for $n = 100$, $t = 0.2$, $H = 0.45$, $a = 0.7$, $k = 2$ and the varying Parzen kernels given in Table 2.1: a) uniform, b) triangular, c) Gaussian d) Mexican hat.

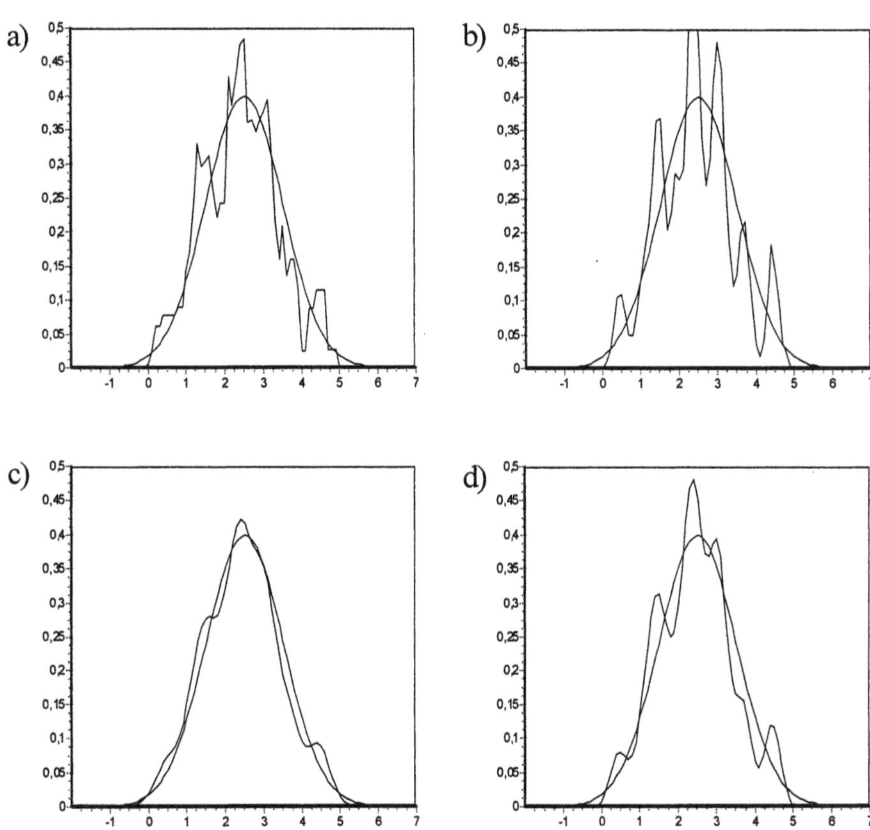

FIGURE 6.7. Estimation of time-varying probability density – Simulation 6.4

Simulation 6.5

In this simulation we apply the recursive PNN given by (6.90) for the estimation of the time-varying normal distribution. In Fig. 6.8 – Fig. 6.12 we show the results for $k = 2$ and a) $n = 1000$, b) $n = 5000$. Parameters t, H and a satisfy conditions (6.92).

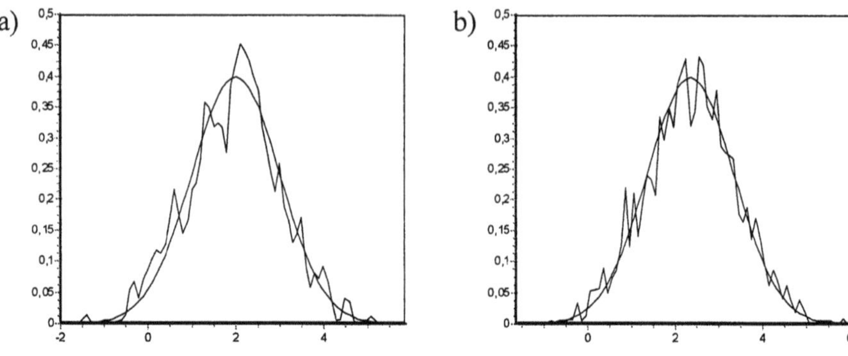

FIGURE 6.8. Estimation of time-varying probability density – Simulation 6.5 ($t = 0.1$, $H = 0.5$, $a = 0.8$)

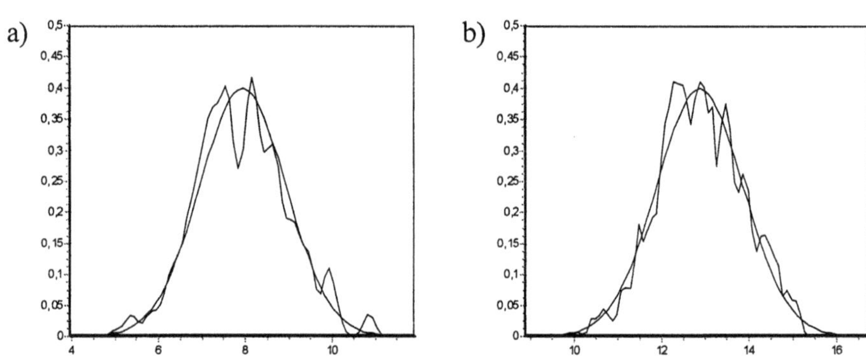

FIGURE 6.9. Estimation of time-varying probability density – Simulation 6.5 ($t = 0.3$, $H = 0.4$, $a = 0.6$)

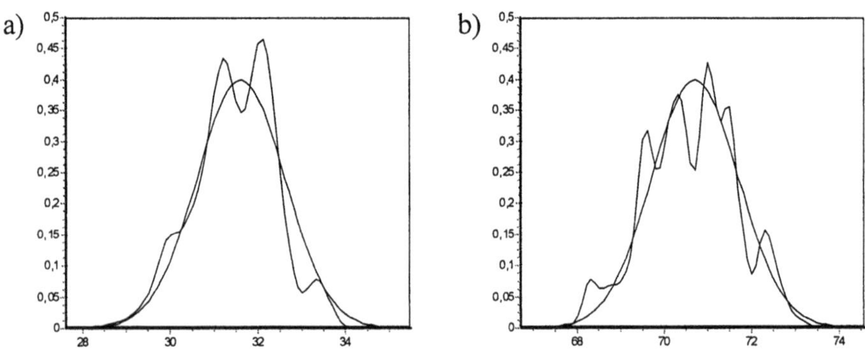

FIGURE 6.10. Estimation of time-varying probability density – Simulation 6.5 ($t = 0.5$, $H = 0.3$, $a = 0.4$)

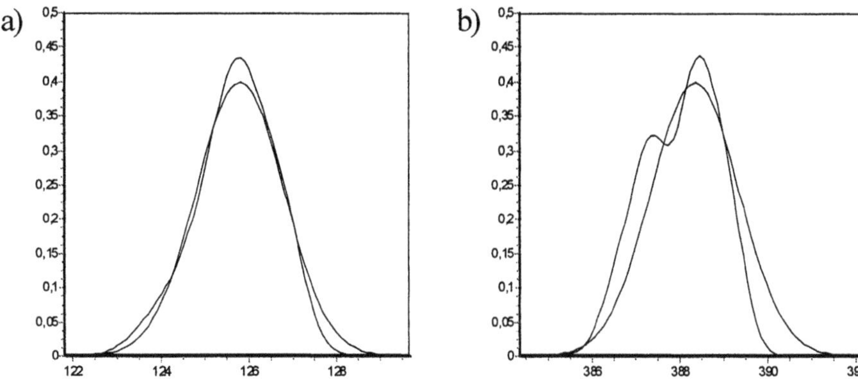

FIGURE 6.11. Estimation of time-varying probability density – Simulation 6.5 ($t = 0.7$, $H = 0.2$, $a = 0.2$)

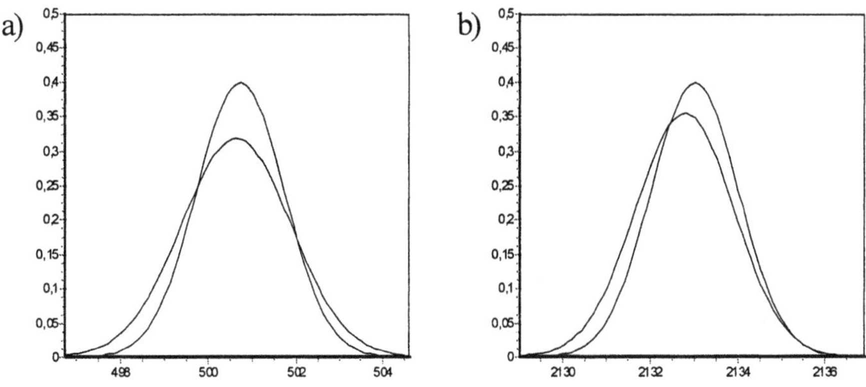

FIGURE 6.12. Estimation of time-varying probability density – Simulation 6.5 ($t = 0.9$, $H = 0.07$, $a = 0.05$)

Simulation 6.6

In this simulation we apply the recursive PNN given by (6.89) based on the trigonometric orthogonal series for the estimation of the time-varying normal distribution. In Fig. 6.13 we show the results for $n = 100$ and a) $t = 0.1$, $Q = 0.45$, $a = 0.8$, b) $t = 0.2$, $Q = 0.4$, $a = 0.75$.

a)

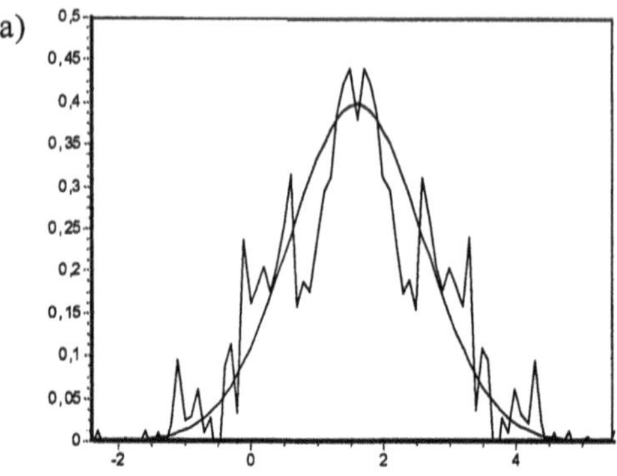

b)

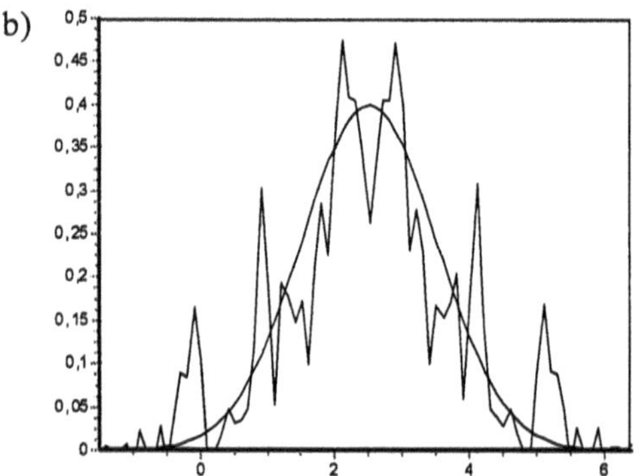

FIGURE 6.13. Esitmation of time-varying probability density – Simulation 6.6

6.9.2 PNN for classification in a time-varying environment

Simulation 6.7

Consider a two-category classification problem with $p_{1n} = p_{2n} = \frac{1}{2}$ and

$$f_{1n}(x) = f_1(x - n^t), \quad f_{2n}(x) = f_2(x - n^t)$$

where

$$f_1(x) = N(0,1) \quad f_2(x) = N(2,1)$$

In this case the minimum probability of error is given by (see [75], page 73)

$$P_e = \frac{1}{\sqrt{2\pi}} \int_1^\infty e^{-u^2/2} du = 0.159 \tag{6.93}$$

We will use procedures (6.7) and (6.11) for classification and show that the empirical probability of the misclassification approaches the minimum probability (6.93). In Fig. 6.14a and Fig. 6.14b we present the results of simulations for $n = 10000$ (for each class) and $k = 5$. Two cases are considered: $t = 0.1$, $a = 0.7$, $H = 0.5$ (Fig.6.14a) and $t = 0.3$, $a = 0.4$, $H = 0.3$ (Fig. 6.14b). We observe that the decision boundary is almost perfectly estimated. Tables 6.1 and 6.2 show empirical probability of misclassification tracking the minimum probability of error P_e as n grows large. For each n we test the PNN on 5000 samples from class 1 and 5000 from class 2.

TABLE 6.1. Empirical probability of misclassification for $t = 0.1$, $a = 0.7$, $H = 0.5$

n	Error
1000	0.1559
2000	0.1602
3000	0.1573
4000	0.1691
5000	0.1566
6000	0.1598
7000	0.1562
8000	0.1607
9000	0.1561
10000	0.1572

TABLE 6.2. Empirical probability of misclassification for $t = 0.3$, $a = 0.4$, $H = 0.3$

n	Error
1000	0.1639
2000	0.1648
3000	0.1687
4000	0.1649
5000	0.1564
6000	0.1611
7000	0.1558
8000	0.1586
9000	0.1569
10000	0.1562

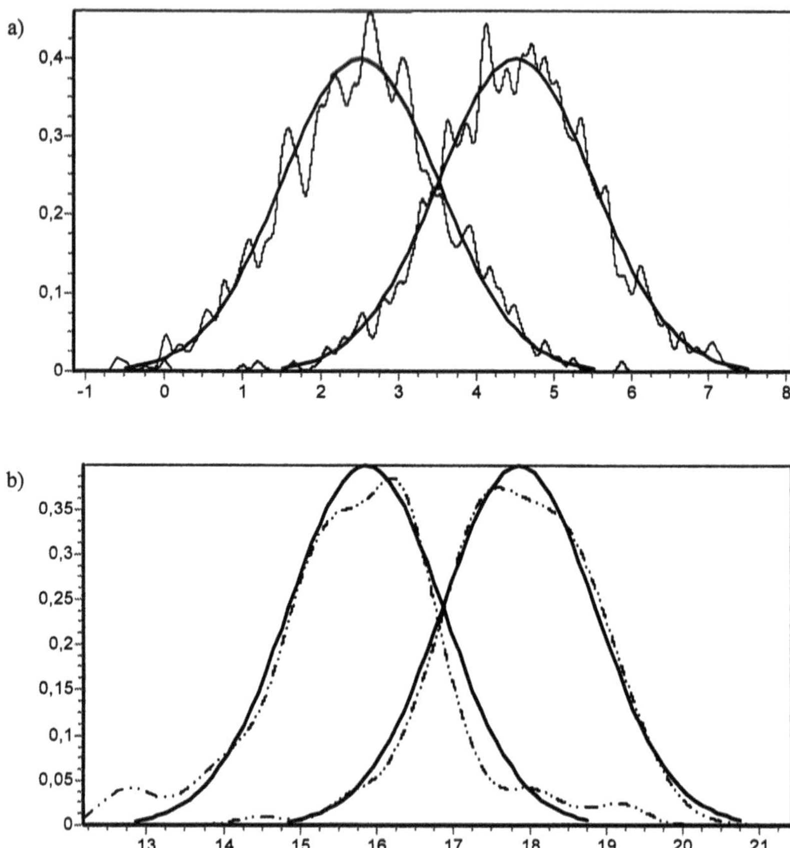

FIGURE 6.14. PNN for classification in a time-varying environment

6.10 Concluding remarks

The presented pattern classification procedures are asymptotically optimal in the sense of Definitions 6.1 and 6.2. These properties are true with certain assumptions concerning densities in classes f_{mn}, $m = 1, ..., M$, $n = 1, 2, ...$.

In the stationary case, analogous properties are true with no assumptions concerning densities in classes (e.g. Devroye and Wagner [64], Greblicki and Rutkowski [101]), but we should remember that the problems considered in this chapter are much more difficult.

In Corollaries 6.1 – 6.4 concerning the asymptotic optimality of rule (6.7), the type of non-stationarity was not specified; which enables us to use the obtained results for classifications of patterns characterized by various types of non-stationarity, but at the cost of the clarity of the respective conditions. These conditions, as it was shown in Tables 6.1a, 6.1b, 6.2a, 6.2b, 6.3, are clear and understandable in the case of the "movable argument" – type of non-stationarity. By the use of these tables it is possible to design a system realizing the pattern classification algorithm (6.7), i.e. properly select sequences a_n, h_n or $q(n)$, when non-stationary densities in classes are of the "movable argument" type, i.e.

$$f_{mn}(x) = f_m(x - c_{mn}), \quad m = 1, ..., M, \quad n = 1, 2, ...$$

For example, if $c_{mn} = n^{t_m}$ then neither the knowledge of function f_m nor the knowledge of parameters t_m is necessary in order to design algorithm (6.7). In spite of this our algorithm will possess asymptotically optimal properties in the sense of Definitions 6.1 and 6.2.

The comparison of algorithms based on Parzen's kernel with algorithms constructed on the basis of the orthogonal series method, carried out in Section 6.7, was undoubtedly more favorable to the former. Their application requires weaker assumptions concerning smooth properties of the density function and they allow the tracking of more significant changes of these functions (Table 6.3). As we remember, in Chapter 5 the algorithm based on the Parzen's kernel was also preferred in the case of plants with the non-stationarity of the "movable argument" type.

In Section 6.9 we have noticed that if

$$p_{mn} = \begin{cases} 1 & \text{when} \quad m = 1 \\ 0 & \text{when} \quad m \neq 1 \end{cases}$$

then algorithm (6.11) can be used for tracking time-varying probability densities. The convergence of algorithm (6.11) applied to nonparametric learning of time-varying probability densities follows from Corollaries 6.1 – 6.4. This problem was also investigated in works [217] and [280] but the authors assumed then that the sequence of non-stationary probability density functions is convergent in a specified sense to a finite limit.

The material presented in this chapter was partially published by the author in [215] – [217] and [234].

Part II

Soft Computing Techniques for Image Compression

7

Vector Quantization for Image Compression

7.1 Introduction

Vector quantization (VQ) is one of the most often discussed methods in scientific literature on image data compression [95], [103], [126], [171], [194], [329]. The idea of applying VQ to image compression relies on replacing separable image blocks with some class of model blocks which are the most representative blocks for this image. Since human eye perception is based on seeing bigger image parts, changes made by the vector quantization method may not be detected at all. In this chapter we first explain how to prepare an image for vector quantization. Next the image compression problem is explained and the VQ algorithm based on neural networks is described.

7.2 Preprocessing

In order to apply the vector quantization algorithm, for a given digitized image we have to fix separable blocks of the same size. These blocks will be represented by vectors created as a result of setting a mapping between the successive pixels from a block and each vector component. A method of selecting pixels for creating vectors can be a succession-line method, that is, pixels which are lying in consecu-

tive block rows are picked up from the left side, row after row. To compress a given image with the help of the system described above, preprocessing should be done first.
First, every pixel of a digital image is treated as an element of matrix $\mathbf{D} = [d(i, j)]$ according to the schema presented in Fig. 7.1.

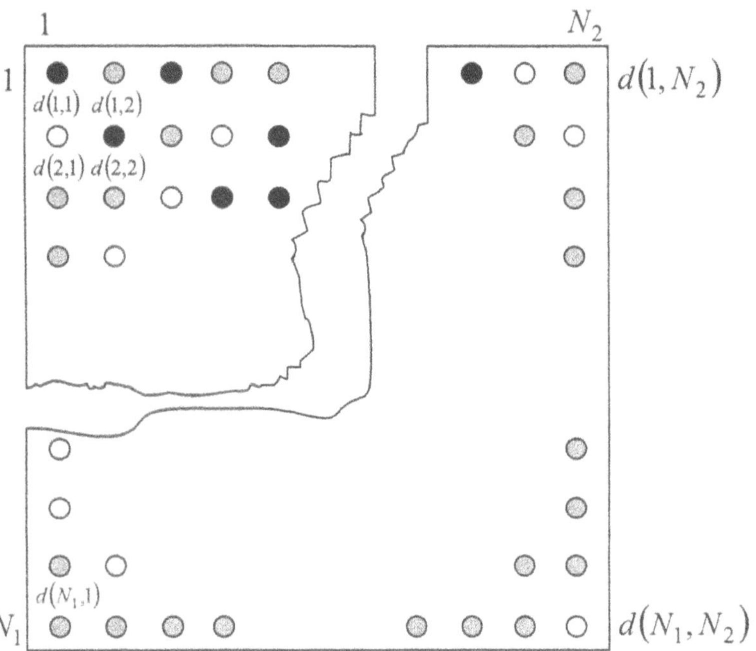

FIGURE 7.1. Pixel reprezentation of an image

The value of every element $d(i, j)$, where $i = 1, 2, ..., N_1$, $j = 1, 2, ..., N_2$ is the luminance value of the pixel corresponding to this element. The luminance value was coded by means of the binary code on the p-bytes basis, i.e. through the help of 2^p quantization levels: $\mathbf{D} \in \Omega^{N_1 \cdot N_2}$, $\Omega = \{0, 1, ..., 2^p - 1\}$, $d(i, j) \in \Omega$.
The next thing that should be done with the image is to separate blocks $n_1 \times n_2$ of pixels from a given picture of dimension $N_1 \times N_2$ as it is shown in Fig. 7.2.

Every pixel in a currently considered block is treated as an element of matrix $\mathbf{Y}(m, n)$, $m = 1, 2, ..., N_1/n_1$, $n = 1, 2, ..., N_2/n_2$. Matrices $\mathbf{Y}(m, n)$ are created as a result of the following transformation:

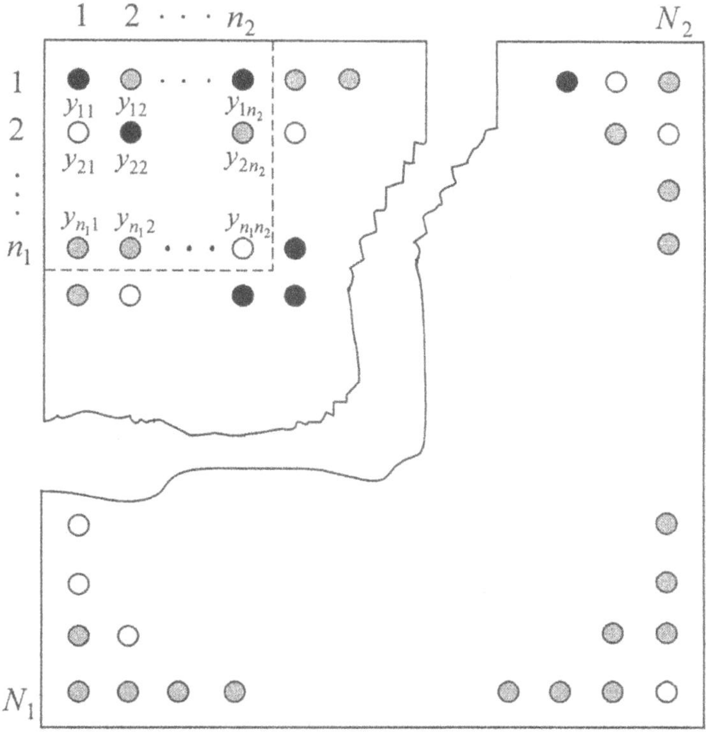

FIGURE 7.2. Separation of a block from an image

$$\mathbf{D} = \begin{bmatrix} d(1,1) & \cdots & d(1,N_2) \\ \vdots & \ddots & \vdots \\ d(N_1,1) & \cdots & d(N_1,N_2) \end{bmatrix}$$

$$\xrightarrow{\xi} \begin{bmatrix} y_{11}(m,n) & \cdots & y_{1,n_2}(m,n) \\ \vdots & \ddots & \vdots \\ y_{n_1,1}(m,n) & \cdots & y_{n_1,n_2}(m,n) \end{bmatrix} \tag{7.1}$$

$$= \mathbf{Y}(m,n) \in \Omega^{n_1 \times n_2},$$

where

$$\left(\frac{N_1}{n_1}\right) \text{mod } 1 = 0 \text{ and } \left(\frac{N_2}{n_2}\right) \text{mod } 1 = 0$$

The course of variability of the parameters m and n is explained in Fig. 7.3.

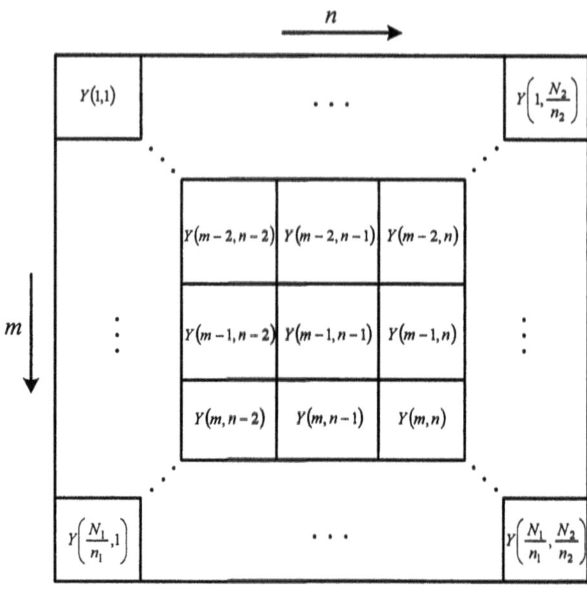

FIGURE 7.3. Block representation of an image

Note that in books on image compression most often we can see 2×2 and 4×4 pixel dimension blocks, more seldom 3×3, 5×5 or 10×10 pixels. Matrices $\mathbf{Y}(m,n)$ formed in such a way, representing image blocks from the original picture, are subject to the following transformation:

$$\mathbf{Y}(m,n) = \begin{bmatrix} y_{11}(m,n) \cdots y_{1,n_2}(m,n) \\ \vdots \quad \ddots \quad \vdots \\ y_{n_1,1}(m,n) \cdots y_{n_1,n_2}(m,n) \end{bmatrix}$$

$$\xrightarrow{\psi} [e_1(m,n), e_2(m,n), ..., e_q(m,n)]^T \tag{7.2}$$

$$= \mathbf{E}(m,n) \in \Omega^q$$

where $q = n_1 \cdot n_2$. Such a form of vectors representing a sequence of image blocks is useful when we construct a predictor. The equation (7.2) means that pixels in a sequence of image blocks are transformed into vector components in a sequential-linear order, i.e. pixels lying in a sequence of rows of blocks are chosen from the left side, row after row.

7.3 Problem description

Let $\mathbf{G} = [\mathbf{g}_0, \mathbf{g}_1, ..., \mathbf{g}_J]$ be the set of prototypes (codewords or code-vectors) such that $\mathbf{g}_j = [g_{1,j}, g_{2,j}, ..., g_{q,j}]^T$. The set $\mathbf{G} \subset \mathbf{R}^q$ is commonly called the code-book.
The vector quantization algorithm is based on the VQ transformation:

$$\mathrm{VQ} : \Re^q \xrightarrow{\mathrm{VQ}} \Omega, \tag{7.3}$$

where
$\Re^q$ – q-dimension vector space spanned on the vectors
$\mathbf{E}(m, n) = [e_1(m, n), e_2(m, n), ..., e_q(m, n)]^T$ representing image blocks,
$m = 1, 2, ..., N_1/n_1$ – vertical image block coordinate,
$n = 1, 2, ..., N_2/n_2$ – horizontal image block coordinate,
$N_1/n_1, N_2/n_2$ – number of image blocks, in vertical and horizontal directions, respectively
$\Omega = \{j, j = 0, 1, ..., J\}$ – set of indices in the code-book,
$J + 1$ – size of the code-book.

Remark 7.1
From the above description it follows that the image is represented by $\frac{N_1 \cdot N_2}{q}, q = n_1 \cdot n_2$, vectors in the form

$$\mathbf{E}(1,1), (1,2), ..., \mathbf{E}\left(\frac{N_1}{n_1}, \frac{N_2}{n_2}\right) \tag{7.4}$$

which represents a learning sequence for determination of the code-book. Sometimes it is more convenient to use the following substitutions:

$$\begin{aligned} \mathbf{E}(1,1) &= \mathbf{E}(1) \\ \mathbf{E}(1,2) &= \mathbf{E}(2) \\ &\vdots \\ \mathbf{E}\left(\frac{N_1}{n_1}, \frac{N_2}{n_2}\right) &= \mathbf{E}\left(\frac{N_1 \cdot N_2}{q}\right) \end{aligned}$$

Therefore, the learning sequence is alternatively represented by $\mathbf{E}(t)$, $t = 1, 2, ..., \frac{N_1 \cdot N_2}{q}$, i.e.

$$\mathbf{E}(1), \ \mathbf{E}(2), ..., \mathbf{E}\left(\frac{N_1 \cdot N_2}{q}\right) \tag{7.5}$$

We will use both notations interchangeably, depending on the situation. Moreover, it will be convenient to denote $\frac{N_1 \cdot N_2}{q} = M$.
The VQ algorithm can be written as a composition of two functions ϕ and φ, transforming vector $\mathbf{E}(m,n)$ in the following way:

$$\mathrm{VQ} : \mathbf{E}\left(m,n\right) \xrightarrow{\phi} \mathbf{g}_{j^0}\left(m,n\right) \xrightarrow{\phi} j^0\left(m,n\right), \tag{7.6}$$

where
$\mathbf{g}_{j^0}(m,n) = \left[g_{1j^0}, g_{2j^0}, ..., g_{qj^0}\right]^T \in \mathbf{G}$ – vector chosen from the code-book $\mathbf{G}$ as a representation of a given vector $\mathbf{E}(m,n)$ based on the closest similarity method,
$j^0\left(m,n\right) \in \Omega$ – label of a prototype vector chosen to represent a given image block, which is the number for this prototype vector in the code-book $\mathbf{G}$.

The selection of a representative pattern vector for vector $\mathbf{E}\left(m,n\right)$ is based on the closest similarity method, which means picking up such a prototype vector from the code-book that:

$$d\left[\mathbf{E}\left(m,n\right), g_{j^0}\right] = \min_{0 \le j \le J} \{d\left[\mathbf{E}\left(m,n\right), g_j\right]\}, \tag{7.7}$$

where measure d can be defined in several ways, e.g. as the Euclidean distance

$$d\left[\mathbf{E}\left(m,n\right), \mathbf{g}_j\right] = \sqrt{\sum_{i=1}^{q} [e_i - g_{ij}]^2} \tag{7.8}$$

or as the taxi metric (also known as the Manhattan-type distance)

$$d\left[\mathbf{E}\left(m,n\right), \mathbf{g}_j\right] = \sum_{i=1}^{q} |e_i - g_{ij}| \tag{7.9}$$

or as a scalar product of vectors $\mathbf{E}\left(m,n\right)$ and $\mathbf{g}_j$

$$d\left[\mathbf{E}\left(m,n\right), \mathbf{g}_j\right] = 1 - \mathbf{E}\left(m,n\right) \circ \mathbf{g}_j = 1 - \sum_{i=1}^{q} e_i g_{ij} \tag{7.10}$$

In literature, the term "contest winner" is often used to denote the prototype vector $\mathbf{g}_{j^0}$. A set of labels $j^0\left(m,n\right)$ which corresponds to the prototype vectors $\mathbf{g}_{j^0}$, selected as "contest winners" for each vector $\mathbf{E}\left(m,n\right)$ is an information which is coded and transmitted or stored on the disc. This ends the image compression process.

The decompression process is performed according to an inverse transformation of the VQ technique:

$$\mathrm{VQ}^{-1}: \Omega \xrightarrow{\mathrm{VQ}^{-1}} \Re^{q} \tag{7.11}$$

For label $j^0(m,n)$, the decompression process in the VQ method proceeds in the following way:

$$\mathrm{VQ}^{-1}: j^0(m,n) \rightarrow \mathbf{g}_{j^0}(m,n) = \mathbf{E}'(m,n), \tag{7.12}$$

where $\mathbf{E}'(m,n) = [e_1'(m,n), e_2'(m,n), ..., e_n'(m,n)]^T$ – is a set of vectors that represents image blocks reconstructed after compression. Thus, the error that occurred in the whole image can be expressed as:

$$MSE = \frac{\sum_{n=1}^{N_1/n_1} \sum_{m=1}^{N_2/n_2} \sum_{i=1}^{q} [e_i(m,n) - e_i'(m,n)]^2}{N_1 N_2} \tag{7.13}$$

Of course, the size of the code-book $J+1$ is much smaller than the number of possible pixel combinations constituting one image block. Each of the prototype vectors $\mathbf{g}_j$, represents a certain image block class. These blocks are in some way similar to each other, but only within one class. Vector $\mathbf{g}_j$ can be compared with the mid-point of the image blocks from a given class. In this way, we can obtain the best representative for each vector $\mathbf{E}(m,n)$.

The crucial problem in the vector quantization method is to choose such prototype vectors $\mathbf{g}_j$ from the code-book, so that the total mean square error measured as e.g.

$$MSE = \frac{\sum_{m=1}^{N_1/n_1} \sum_{n=1}^{N_2/n_2} \sum_{i=1}^{q} \left[e_i(m,n) - g_{ij^0}\right]^2}{N_1 N_2}, \tag{7.14}$$

attains as low a value as possible. To determine the optimal code-book in the sense of a minimal error as in (7.14), usually Lloyd's algorithm was used. However, the algorithms for determining an optimal code-book that are based on neural networks are becoming increasingly popular.

7.4 VQ algorithm based on neural network

The neural network shown in Fig. 7.4. is designed to determine the code-book $\mathbf{G}$ in the process of learning. The weights of the net correspond to prototype vectors $\mathbf{g}_j$, $j = 0, 1, ..., J$.

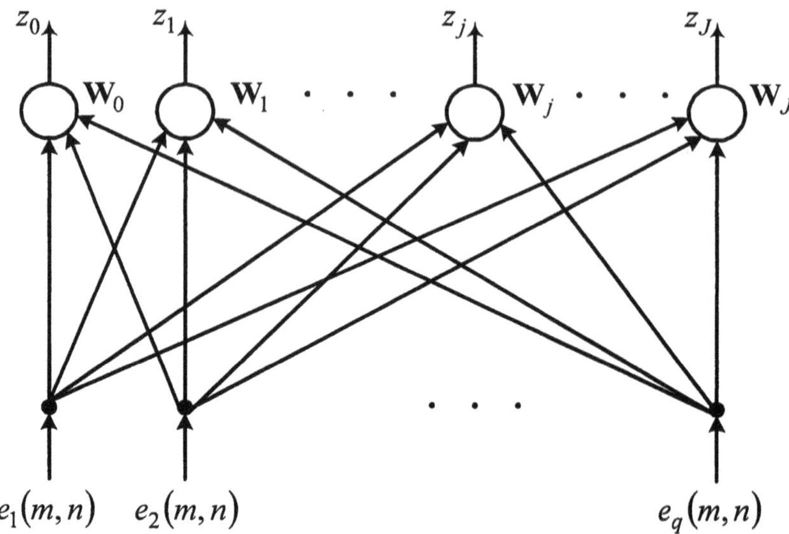

FIGURE 7.4. Neural network for vector quantization

Each separate symbol in this figure means:
$\mathbf{E}(m,n) = [e_1(m,n), e_2(m,n), ..., e_q(m,n)]^T$ – neural network input signal, $\mathbf{W} = [\mathbf{W}_0, \mathbf{W}_1, ..., \mathbf{W}_J]$ – weights matrix of the connections between input nodes and network neurons, $\mathbf{W}_j = [w_{1j}, w_{2j}, ..., w_{qj}]$ – weights vector determining connection weights to the j-th neuron from all q input nodes, $\mathbf{Z} = [z_0, z_1, ..., z_J]^T$ – vector determining the state of neurons in the network:

$$z_j = \begin{cases} 1 & \text{for} \quad j = j^0 \\ 0 & \text{for} \quad j \neq j^0 \end{cases}$$

where $j = 0, 1, ..., J$, j^0– label of the contest "winner".

An application of the neural network in the vector quantization algorithm can be divided into two stages: the network learning stage and the network practical use stage. In the network learning stage, we provide the network with vectors $\mathbf{E}(m,n)$ coming from the learning set. The learning set is a sequence of all vectors representing image blocks of a given image. Each time when vector $\mathbf{E}(m,n)$ enters the network, a contest "winner" is chosen, according to the principle of the smallest distance between a given input vector $\mathbf{E}(m,n)$ and the prototype vector $\mathbf{W}_{j^0}$. All other weight vectors $\mathbf{W}_j$ are "beaten". As a measure of the distance between vectors $\mathbf{E}(m,n)$ and $\mathbf{W}_j$ we can use formulas (7.8), (7.9) or (7.10). For the "winner" neuron value 1 is assigned to its output, and for the rest of them value 0. Next, the neuron weights are modified according to the learning rule. The neural network learning algorithm will be described in detail along with the PVQ compression algorithm in Chapter 12. The operation of providing the network input with vectors $\mathbf{E}(m,n)$ and the modification of network weights is repeated until the learning set has been exhausted. At the end of the network learning process, the finally formed weights vectors $\mathbf{W}_j$, $j = 0, 1, ..., J$, are identified with the prototype vectors from the code-book in the following way:

$$\mathbf{g}_j = \mathbf{W}_j \quad \text{for} \quad j = 0, 1, ..., J. \tag{7.15}$$

The stage of practical use of a neural network for the vector quantization method begins when the learning stage ends. Then, the possibility of modifying the network weights becomes blocked and the feeding of the set of vectors $\mathbf{E}(m,n)$ begins. These vectors are formed from the consecutive image blocks. Every time, the selection of the "winner" (similarly as in the learning stage) takes place, the number j^0 in the code-book will be used as an equivalent for the input vector $\mathbf{E}(m,n)$. In this way, we receive a sequence of labels which are coded, e.g. with the help of a binary code or Huffman's code. In this form, a sequence of labels is sent by a transmission channel or stored on the disc.

Decompression is performed as described above, but in a reverse order and is based on setting a position in the code-book with the use of sequentially read labels j^0. The sequence of the prototype vectors $\mathbf{g}_{j^0}(m,n)$ obtained in this way is treated as a sequence of vectors connected with the reproduced image blocks.

7.5 Concluding remarks

In this chapter we have described the VQ technique for image compression. Another area of applications of VQ is speech recognition and speech compression [52], [140]. An extensive discussion of vector quantization techniques and applications is given in [95].

8
The DPCM Technique

8.1 Introduction

The Differential Pulse Code Modulation (DPCM) method was used with a large success in the image compression [125], [153], [159]. This technique is based on removing redundant information, existing between consecutive image sequences, and coding only new information, which is passed with new image elements.

In this chapter various aspects of the prediction technique used for image compression will be presented:
- general schema of prediction technique in the scalar case,
- prediction technique in the multi-dimensional case,
- application of neural networks to non-linear prediction.

8.2 Scalar case

Before entering into discussion on the DPCM algorithm we assume that we have samples of signals (e.g. samples of speech or images) represented by the sequence

$$\{v\,(n)\} \in \Re, n = 1, 2, ..., n_{\max}, n_{\max} \text{ – number of samples.}$$

The scheme of the image compression with the use of the DPCM technique is shown in Fig. 8.1.

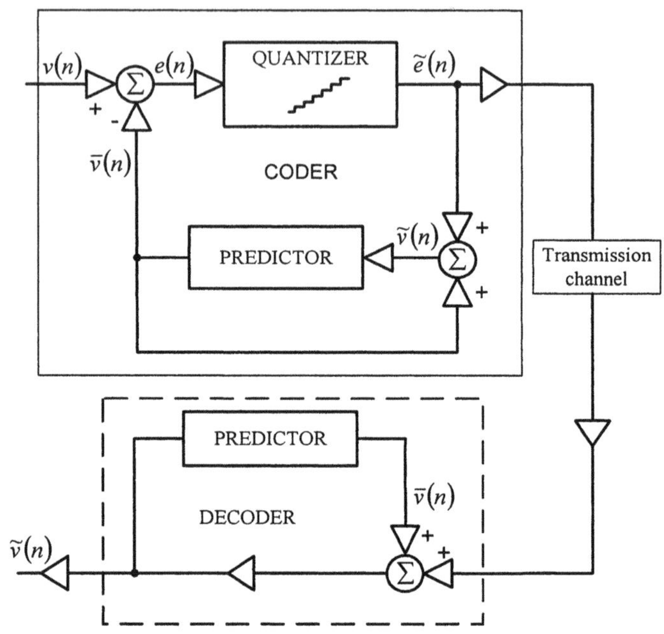

FIGURE 8.1. Scalar DPCM algorithm block diagram

The system realizing the DPCM algorithm consists of three parts:
a) encoder
b) transmission channel
c) decoder
The encoder's task is such a conversion of the input signal so that the load of the transmission channel (disk memory) is reduced. The decoder's task is to reproduce an image coming from the transmission channel (disk). To discuss the DPCM algorithm it is convenient to concentrate on the presentation of equations describing two summary nodes shown in Fig. 8.1:
(i) system input node,
(ii) predictor input node.
The equation of the system input node is expressed by the following formula:

$$e(n) = v(n) - \overline{v}(n), \tag{8.1}$$

where $e(n) \in \Re$ – prediction error, the value of which is quantized $\overline{v}(n) = F[\widetilde{v}(n-1), \widetilde{v}(n-2), ..., \widetilde{v}(n-K)]$ – predicted value of the signal currently given on system input, determined based on the K-previously reproduced pixels, where K-predictor order, F – predictor function which can be linear or non-linear.

For the linear predictor function, equation (8.1) takes the form:

$$e(n) = v(n) - \sum_{k=1}^{K} a_k \, \widetilde{v}(n-k), \tag{8.2}$$

where $a_1, a_2, ..., a_K$ are linear predictor coefficients.
For the non-linear predictor function, we can use the AR model [159]:

$$\overline{v}(n) = \sum_i a_i \, \widetilde{v}(n-i) + \sum_i \sum_j a_{ij} \, \widetilde{v}(n-i) \, \widetilde{v}(n-j) \tag{8.3}$$

$$+ \sum_i \sum_j \sum_k a_{ijk} \, \widetilde{v}(n-i) \, \widetilde{v}(n-j) \, \widetilde{v}(n-k) + ...$$

The equation of the predictor input node is given by:

$$\widetilde{v}(n) = \overline{v}(n) + \widetilde{e}(n), \tag{8.4}$$

where $\widetilde{e}(n)$ is the quantized prediction error value. The quantized prediction error is treated as the output signal of the coder and is transmitted by a transmission channel (or is stored on disk) to the decoder. The decoder contains an identical loop as in the coder. The loop is composed of the input node of the predictor and the predictor itself. The equation of the input node of the predictor in the decoder is expressed by formula (8.4), and the signal $\widetilde{v}(n), n = 1, 2, ..., n_{\max}$, is the output signal of the decoder representing the reproduced image.
The error caused by the DPCM technique for image compression is given by

$$d(n) = |v(n) - \widetilde{v}(n)| = |e(n) - \widetilde{e}(n)|, \tag{8.5}$$

From equation (8.5) it follows that the overall reproduction error in predictive vector quantization is equal to the error in quantizing the difference signal presented to the vector quantizer. However, the average value of the prediction error $e(n)$ is, for all samples, less than the average value of the original signal $v(n)$. Maintaining the error level determined by the length of the quantization intervals, we can use less quantization levels for the $e(n)$ signal than in the case of the $v(n)$ signal. A lower number of quantization levels means a lower number of bytes necessary for encoding the sequence of the signal samples.

8.3 Vector case

In the case where the input signal of the predictor is a sequence of vectors (multidimensional case), the prediction idea does not change. Only the way of signal interpretation and coefficients in the DPCM system change. The signals in the DPCM vector system are the following vectors:

$\mathbf{V}(n) = [v_1(n), v_2(n), ..., v_q(n)]^T \in \Re^q$ – input vector

$\widetilde{\mathbf{V}}(n) = [\widetilde{v}_1(n), \widetilde{v}_2(n), ..., \widetilde{v}_q(n)]^T \in \Re^q$ – reproduced vector

$\overline{\mathbf{V}}(n) = [\overline{v}_1(n), \overline{v}_2(n), ..., \overline{v}_q(n)]^T \in \Re^q$ – predicted input vector

$\mathbf{E}(n) = [e_1(n), e_2(n), ..., e_q(n)]^T \in \Re^q$ – residual (difference) vector

where q is the dimension of all vectors. Note that the prediction coefficients are, in such a case, matrices $\mathbf{A}_k \in \Re^{q \cdot q}$, $k = 1, 2, ..., K$, where K is the predictor order. In an analogous manner to equations (8.1) and (8.4), the equations of nodes will be given by, respectively:

$$\mathbf{E}(n) = \mathbf{V}(n) - \widetilde{\mathbf{V}}(n), \tag{8.6}$$

$$\widetilde{\mathbf{V}}(n) = \overline{\mathbf{V}}(n) + \widetilde{\mathbf{E}}(n) \tag{8.7}$$

In the linear case equation (8.2) takes the form

$$\mathbf{E}(n) = \mathbf{V}(n) - \sum_{k=1}^{K} \mathbf{A}_k \widetilde{\mathbf{V}}(n-k) \tag{8.8}$$

8.4 Application of neural network

The DPCM system with a neural network [159] used for the realization of a non-linear predictor is shown in Fig. 8.2.

In the above system, the neural network approximates the non-linear predictor, the function of which is given by expression (8.3), where the predictor order $K = 3$. With such assumptions, the following symbols in the figure are used:

$w_i = a_i,$

$w_{ij} = a_{ij},$

$w_{ijki} = a_{ijki},$

$y_i = \widetilde{v}\,(n-i)$ for $i = 1...3,$

$y_{ij} = \widetilde{v}\,(n-i)\,\widetilde{v}\,(n-j)\,,(i,j) \in \{(1,2)\,,(1,3)\,,(2,3)\},$

$y_{ijk} = \widetilde{v}\,(n-i)\,\widetilde{v}\,(n-j)$

$\widetilde{v}\,(n-k)\,, i = 1, j = 2, k = 3.$

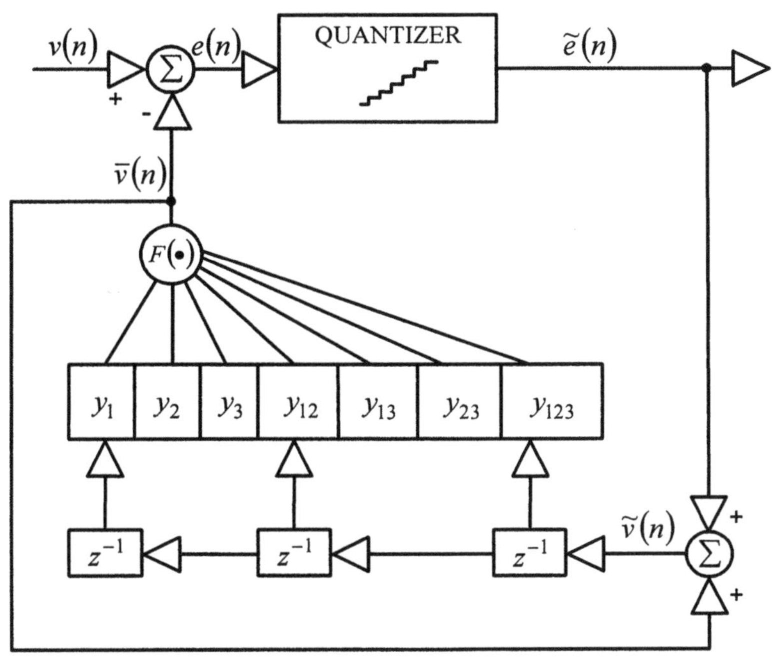

FIGURE 8.2. Scalar DPCM algorithm block diagram

The predicted value of the input signal will be given by:

$$\widetilde{v}\,(n) = F\,(w_1 y_1 + w_2 y_2 + w_3 y_2 + w_{12} y_{12} \tag{8.9}$$

$$+ w_{13} y_{13} + \; w_{23} y_{23} + w_{123} y_{123} + b)$$

where

$F\,(z+b) = \frac{1}{1+e^{-(z+b)}}$ – sigmoidal function, b – threshold value

$z = \sum_{i=1}^{3} w_i y_i + w_{12} y_{12} + w_{13} y_{13} + w_{23} y_{23} + w_{123} y_{123},$

The DPCM system operation with the use of a neural network can be divided into two phases:
- learning phase,
- practical operation phase.

In the learning phase, network weights marked here by the variable w with indices i, j and k, are learned in such a way so that the error defined as:

$$e = \frac{1}{n_{\max}} \sum_{n=1}^{n_{\max}} \left(v\left(n\right) - \widetilde{v}\left(n\right)\right)^2, \tag{8.10}$$

is as small as possible, e.g. with the help of the back-propagation error algorithm. After the setting of the neural network weights value, the phase of the practical operation of the DPCM system follows, and the neural network works as a predictor.

8.5 Concluding remarks

In this chapter we have described the basic idea of the DPCM algorithm. In the past, when designing the predictor in the DPCM schema, the most popular was the linear predictor due to its well known statistical properties. In the last decade predictors based on neural networks have received considerable attention and outperformed linear predictors.

9
The PVQ Scheme

9.1 Introduction

The VQ DPCM algorithm, in short PVQ (Predictive Vector Quantization), was proposed [47], [48], [164], [200], [231] as a new idea of image compression and was supposed to combine the advantages of both the VQ and DPCM methods and to remove their disadvantages in the image compression process. In this chapter the PVQ algorithm is described in detail.

9.2 Description of the PVQ scheme

A major drawback of the VQ method is the requirement of the usage of sizable code-books, which is to guarantee an acceptable level of image distortion after decompression. A remedy for this problem can be the use of the compression algorithm based on the connection of two methods: VQ described in Chapter 7 and DPCM described in Chapter 8. The new method created from the combination of the previously mentioned two is called VQ DPCM or in short PVQ. Schematically, the system is depicted in Fig. 9.1.

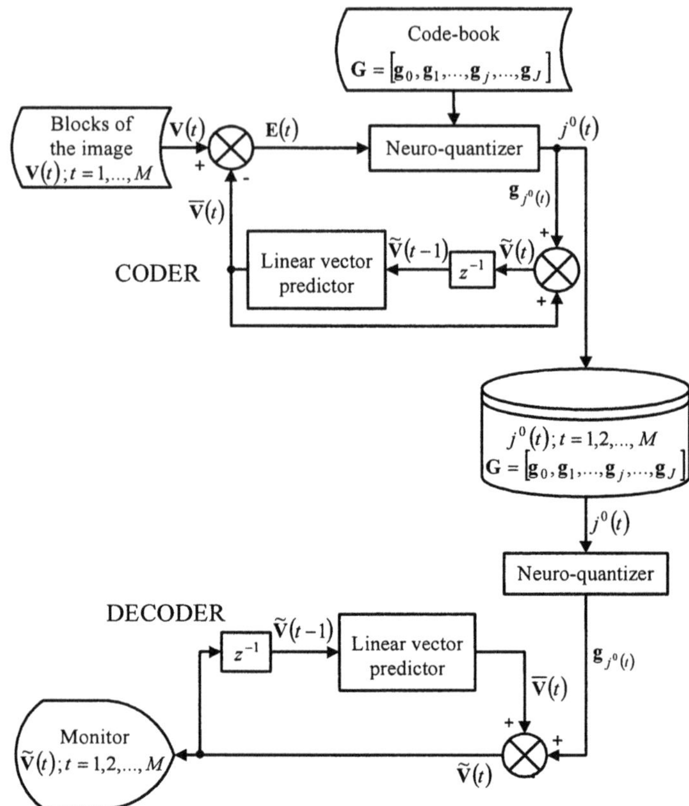

FIGURE 9.1. The PVQ block diagram with linear predictor

The system in Fig. 9.1. consists of three basic parts: coder (transmitter), memory module or transmission channel and decoder (receiver). The coder's task is the reduction of the volume of information without losing the important parts of this information (compression with losses). The coder is located in the system starting from the input summer and ends on the signal line leading to the memory module or to the transmission channel. The second system element is the memory module or transmission channel. The third system element is the receiver, whose task is to reproduce a coded image based on the information received from the transmission channel or read from the memory. The decoder is located between the signal line coming from the memory or the transmission channel and ends at the system output from which the information about the reproduced image is read.

The input $\mathbf{V}(m,n)$ or equivalently $\mathbf{V}(t)$ to the PVQ scheme is prepared in the same way as sequence $\mathbf{E}(m,n)$ given by (7.4) or equivalently $\mathbf{E}(t)$ given by (7.5), see Remark 7.1. Vectors $\mathbf{V}(m,n)$ obtained in this way are passed to the PVQ compression system input, as shown in Fig. 9.1. In this system, from each component of the consecutively given vectors $\mathbf{V}(m,n)$, components of the vectors $\overline{\mathbf{V}}(m,n) = [\overline{v}_1(m,n), \overline{v}_2(m,n), ..., \overline{v}_q(m,n)]^T \in \Re^q$ are subtracted. On the output of this summary node, we receive the difference signal (or residual vector) $\mathbf{E}(m,n) = [e_1(m,n), e_2(m,n), ..., e_q(m,n)]^T$ given by the equation:

$$\mathbf{E}(m,n) = \mathbf{V}(m,n) - \overline{\mathbf{V}}(m,n) \tag{9.1}$$

The task of this node originates from the DPCM algorithm. It has to reduce the amplitude of the components of input vectors $\mathbf{V}(m,n)$ and consequently reduce the code-book size (VQ algorithm), maintaining the appropriate distribution of vectors from the code-book in space of difference vectors $\Re^q$.

The difference vector $\mathbf{E}(m,n)$ is applied to the input of the vector quantizer based on the neural network (see Chapter 11 for details). From among prototype vectors in the code-book $\mathbf{G} = [\mathbf{g}_0, \mathbf{g}_1, ..., \mathbf{g}_J]$, where $J+1$ is the code-book size, the prototype vector $\mathbf{g}_{j^0}(m,n)$ is chosen on the principle of the greatest similarity to the one currently applied to the quantizer input. The variable that will finally represent the input vector $\mathbf{V}(m,n)$ is label $j^0(m,n)$ of the corresponding prototype vector. This process can be formally expressed by the VQ transformation:

$$\text{VQ: } \Re^q \xrightarrow{\text{VQ}} j^0 \tag{9.2}$$

Prototype vectors $\mathbf{g}_j \in \Re^q$ should represent the distribution of vectors $\mathbf{E}(m,n)$ in a given image as well as possible. Label $j^0(m,n)$ of a selected prototype vector is either sent by the transmission channel to the receiver or stored in the memory as an equivalent of the given vector $\mathbf{E}(m,n)$. The transformation of the signal in the receiver will be discussed below.

To create vector $\overline{\mathbf{V}}(m,n)$, we have to close the feedback loop directing vector $\mathbf{g}_{j^0}$ to the input of the summary node that is described by the following equation:

$$\widetilde{\mathbf{V}}(m,n) = \overline{\mathbf{V}}(m,n) + \mathbf{g}_{j^0}(m,n) \tag{9.3}$$

On the output of this node, vector $\widetilde{\mathbf{V}}(m,n)$ is retrieved, being the reproduced vector corresponding to vector $\mathbf{V}(m,n)$, applied at that time on the summary node in the encoder. Vector $\widetilde{\mathbf{V}}(m,n)$ is the input signal for the predictor, whose construction will be described in Chapter 10. The equation of the vector predictor in this case is given by:

$$\overline{\mathbf{V}}(m,n) = \sum_{k=0}^{K}\sum_{l=0}^{L} \mathbf{A}_{kl}\widetilde{\mathbf{V}}(m-k,n-l), \tag{9.4}$$

where

$\mathbf{A}_{kl} \in \Re^{q\times q}$ – prediction coefficients in the form of matrix,

$(k,l) \neq (0,0)$,

K – predictor order in horizontal direction,

L – predictor order in vertical direction,

$\widetilde{\mathbf{V}}(m-k,n-l)$ – vectors reproduced in previous steps of image transformation.

The predictor should have $K \cdot L$ memory cells that store vectors reproduced in the preceding steps of the image transformation. Basing on equation (9.4), the predictor generates vector $\overline{\mathbf{V}}(m,n)$ with regard to the existence of dependencies on respective components of input vectors $\widetilde{\mathbf{V}}(m-k,n-l)$ in the vertical direction (parameter m) and in the horizontal direction (parameter n). The selection of the optimal predictor, i.e. prediction coefficients $\mathbf{A}_{kl}$, ensures a reduction of the amplitude of components of the difference vectors. Methods of choosing the optimal predictor will be described in Chapter 10.

As was mentioned earlier, the output signal of the coder (transmitter), which is based on the PVQ algorithm, is label $j^0(m,n)$ which is the label of the prototype vector $\mathbf{g}_{j^0}(m,n)$ coming from the code-book. These labels are stored in the memory or transmitted to the receiver. The size of that code-book is $J+1$ and should be chosen from sequence 2^p, $p = 1, 2, \ldots$, which will guarantee an optimal use of p-bytes intended for the coding of labels. The image compression process enables to save memory space, e.g. on the disk, or to shorten transmission time.

The discussion of the decompression process and part of the system from Fig. 9.1 called the decoder (receiver) will be considerably shortened because all of its elements are described along with the presentation of the coder. The decoder consists of an identical vector quantizer as in the case of the coder and the loop that has a predictor and summary node, described by equation (9.3). A signal in the form of labels $j^0\,(m,n)$, transmitted or read from the memory, is directed to the decoder input. An assigning for every label j^0 corresponding prototype vector $\mathbf{g}_{j^0}\,(m,n)$ takes place just like in the coder. Formally it is done according to the transformation:

$$\mathrm{VQ}^{-1} : j^0 \overset{\mathrm{VQ}^{-1}}{\longrightarrow} \Re^q \tag{9.5}$$

The prototype vector $\mathbf{g}_{j^0}\,(m,n)$ is the input signal for the summary node described by equation (9.3). The loop with this node and predictor is identical to the loop with the same elements in the coder. As was mentioned by the way of describing the coder, vector $\widetilde{\mathbf{V}}\,(m,n)$ represents a retrieved image block and is also the output signal of the whole PVQ system. In order to retrieve an image block we have to perform the following transformations:

$$\begin{aligned} \widetilde{\mathbf{V}}\,(m,n) &= \left[\tilde{v}_1\,(m,n)\,, ..., \tilde{v}_q\,(m,n)\right]^T \\ &\overset{\psi^{-1}}{\longrightarrow} \begin{bmatrix} \tilde{y}_{11}\,(m,n) \cdots \tilde{y}_{1,n_2}\,(m,n) \\ \vdots \quad \ddots \quad \vdots \\ \tilde{y}_{n_1,1}\,(m,n) \cdots \tilde{y}_{n_1,n_2}\,(m,n) \end{bmatrix} \\ &= \widetilde{\mathbf{Y}}\,(m,n) \in \Omega^{n_1 \times n_2}, \end{aligned} \tag{9.6}$$

$$\begin{aligned} \widetilde{\mathbf{Y}}\,(m,n) &= \begin{bmatrix} \widetilde{y}_{11}\,(m,n) & \cdots & \widetilde{y}_{1,n_2}\,(m,n) \\ \vdots & \ddots & \vdots \\ \widetilde{y}_{n_1,1}\,(m,n) & \cdots & \widetilde{y}_{n_1,n_2}\,(m,n) \end{bmatrix} \\ &\overset{\xi^{-1}}{\longrightarrow} \begin{bmatrix} \widetilde{d}\,(1,1) & \cdots & \widetilde{d}\,(1,N_2) \\ \vdots & \ddots & \vdots \\ \widetilde{d}\,(N_1,1) & \cdots & \widetilde{d}\,(N_1,N_2) \end{bmatrix} \\ &= \widetilde{D} \in \Omega^{n_1 \times n_2}, \end{aligned} \tag{9.7}$$

Of course, pixels in the retrieved image have to correspond to pixels in the original image $\widetilde{d}\,(i,j) \leftrightarrow d\,(i,j)$, which ensures the correct location of pixels in the retrieved image. Since, vector quantization is a loss-involving technique, we can write $\widetilde{d}\,(i,j) \approx d\,(i,j)$.

Methods measuring errors created after the image compression are listed as follows:

a) MSE (Mean Square Error):

$$\mathrm{MSE} = \frac{1}{N_1 N_2} \sum_{i=1}^{N_1} \sum_{j=1}^{N_2} \left[d(i,j) - \tilde{d}(i,j) \right]^2,$$

b) PMSE (Peak Mean Square Error):

$$\mathrm{PMSE} = \frac{\sum_{i=1}^{N_1} \sum_{j=1}^{N_2} \left[d(i,j) - \tilde{d}(i,j) \right]^2}{\left[\max\{d(i,j)\}\right]^2},$$

c) NMSE (Normalized Mean Square Error):

$$\mathrm{NMSE} = \frac{\sum_{i=1}^{N_1} \sum_{j=1}^{N_2} \left[d(i,j) - \tilde{d}(i,j) \right]^2}{\sum_{i=1}^{N_1} \sum_{j=1}^{N_2} \left[d(i,j) \right]^2}$$

d) SNR (Signal to Noise Ratio):

$$\mathrm{SNR} = 10 \log_{10} \left(\frac{\sum_{i=1}^{N_1} \sum_{j=1}^{N_2} \left[d(i,j) \right]^2}{\sum_{i=1}^{N_1} \sum_{j=1}^{N_2} \left[d(i,j) - \tilde{d}(i,j) \right]^2} \right)$$

e) PSNR (Peak Signal to Noise Ratio)

$$\mathrm{PSNR} = 10 \log_{10} \left(\frac{\left[\max\{d(i,j)\}\right]^2}{\sum_{i=1}^{N_1} \sum_{j=1}^{N_2} \left[d(i,j) - \tilde{d}(i,j) \right]^2} \right)$$

9.3 Concluding remarks

The difficulties in constructing an image compression system based on the PVQ algorithm consist in, among other things, a lack of any information needed for initializing the parameter of one system element (predictor or vector quantizer) together with a lack of the initial value of the second system element. Among the schemes of determining predictor parameters and the vector quantizer used in the PVQ algorithm, three approaches will be distinguished in Chapter 12:

(i) with "open loop",
(ii) with "closed loop",
(iii) with "modified closed loop".

In the future work it would be interesting to combine the PVQ schemes with the new method of automatic understanding of images [271].

10

Design of the Predictor

10.1 Introduction

An essential part of the PVQ scheme is the predictor. In this chapter we discuss problems concerned with determining the optimal vector predictor's coefficients $\mathbf{A}_{kl}$, $k = 0, 1, ..., K$, $l = 0, 1, ..., L$, where K – predictor order in the horizontal direction, L – predictor order in the vertical direction. The autocorrelation and the covariance method will be described to estimate matrix coefficients. Moreover, we present a neural architecture for the vector prediction.

10.2 Optimal vector linear predictor

We will consider a q-dimensional stationary discrete stochastic process $\left\{\widetilde{\mathbf{X}}\left(m,n\right)\right\}$ with the finite moment of the second order, i.e.

$$\mathbf{E}\left(\left\|\widetilde{\mathbf{X}}\left(m,n\right)\right\|^{2}\right) < \infty \tag{10.1}$$

The predictor of the vector $\widetilde{\mathbf{X}}\left(m,n\right)$ is given by:

$$\overline{\mathbf{X}}\left(m,n\right) = \sum_{k=0}^{K}\sum_{l=0}^{L}\mathbf{A}_{kl}\widetilde{\mathbf{X}}\left(m-k,n-l\right), \tag{10.2}$$

The prediction residual error is defined as follows

$$\mathbf{E}(m,n) = \widetilde{\mathbf{X}}(m,n) - \sum_{k=0}^{K}\sum_{l=0}^{L}\mathbf{A}_{kl}\widetilde{\mathbf{X}}(m-k,n-l), \tag{10.3}$$

where $(k,l) \neq (0,0)$.
As the performance measure of the prediction we propose

$$D_{KL} = \sum_{m}\sum_{n}\mathbf{E}^{T}(m,n)\,\mathbf{E}(m,n) \tag{10.4}$$

The vector predictor of the order $K-L$ will be called the optimal if it minimizes measure (10.4) with respect to the matrix's coefficients $\mathbf{A}_{kl}$. Because of the level of complication in setting the optimal linear predictor in two directions, the problem will be solved first for one direction and then generalized to two directions. For this purpose, we will consider the stationary discrete stochastic process $\left\{\widetilde{\mathbf{X}}(n)\right\}$ with an analogical assumption to (10.1) and with the prediction residual error given by

$$\mathbf{E}(n) = \widetilde{\mathbf{X}}(n) - \sum_{k=1}^{K}\mathbf{A}_{k}\widetilde{\mathbf{X}}(n-k) \tag{10.5}$$

The performance measure will be defined similarly:

$$D_{K} = \sum_{n}\mathbf{E}^{T}(n)\,\mathbf{E}(n) \tag{10.6}$$

The scheme of the linear vector predictor is shown in Fig. 10.1.
We will now derive the equation of the optimal linear vector predictor. We will follow the ideas presented in [51]. To determine the error defined in expression (10.6), we have to determine all the components of the individual vectors $\mathbf{E}(n) = [e_1(n), e_2(n), ..., e_q(n)]^{T}$, i.e.

$$e_i(n) = \widetilde{x}_i(n) - \widetilde{x}_i(n) = \widetilde{x}_i(n) - \sum_{k=1}^{K}\mathbf{a}_{ki}^{T}\,\widetilde{\mathbf{X}}(n-k), \tag{10.7}$$

where $\mathbf{a}_{ki}$ is the i-th row of the k-th prediction coefficient A_K.
Expression (10.7) can be transformed to:

$$e_i(n) = \widetilde{x}_i(n) - \left[\widetilde{\mathbf{X}}^{T}(n-1), \widetilde{\mathbf{X}}^{T}(n-2), ..., \widetilde{\mathbf{X}}^{T}(n-K)\right]\mathbf{f}_i, \tag{10.8}$$

where

$$\mathbf{f}_i = \left[\mathbf{a}_{1i}^T, \mathbf{a}_{2i}^T, ..., \mathbf{a}_{Ki}^T\right]^T \tag{10.9}$$

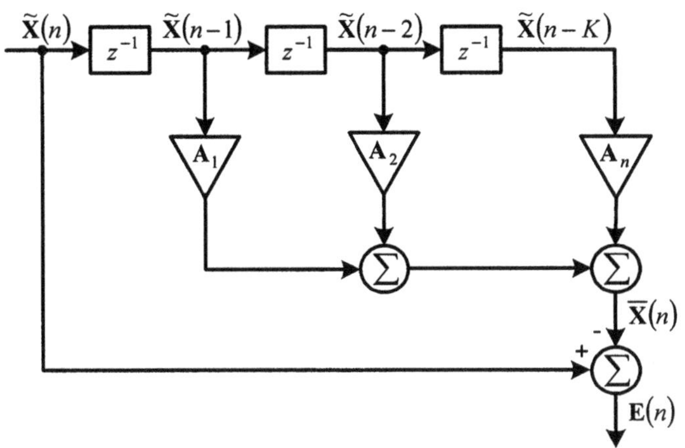

FIGURE 10.1. Scheme of the linear vector predictor

Vector $\mathbf{f}_i$ is created from a composition of the i-th row of matrices $\mathbf{A}_k$, $k = 1, 2, ..., K$, that is, the rows that have an influence on setting the i-th component of vector $\mathbf{E}(n)$. Let N be the length of the accessible sequence of observations. Define

$$\mathbf{H}_i = \left[e_i(1), e_i(2), ..., e_i(N)\right]^T, \tag{10.10}$$

and

$$\mathbf{u}_i = \left[\widetilde{x}_i(1), \widetilde{x}_i(2), ..., \widetilde{x}_i(N)\right]^T \tag{10.11}$$

Consider the equation

$$\mathbf{H}_i = \mathbf{u}_i - \mathbf{Z}\mathbf{f}_i, \tag{10.12}$$

where

$$\mathbf{Z} = \begin{bmatrix} \widetilde{\mathbf{X}}^T(0) & \widetilde{\mathbf{X}}^T(-1) & \cdots & \widetilde{\mathbf{X}}^T(1-K) \\ \widetilde{\mathbf{X}}^T(1) & \widetilde{\mathbf{X}}^T(0) & \cdots & \widetilde{\mathbf{X}}^T(2-K) \\ \vdots & \vdots & \ddots & \vdots \\ \widetilde{\mathbf{X}}^T(N-1) & \widetilde{\mathbf{X}}^T(N-2) & \cdots & \widetilde{\mathbf{X}}^T(N-K) \end{bmatrix} \in \Re^{N\times Kq} \tag{10.13}$$

Based on the above considerations, performance measure (10.6) can be defined as follows:

$$\begin{aligned} D_K &= \sum \|\mathbf{E}(n)\|^2 = \sum_{i=1}^{q} \|\mathbf{H}_i\|^2 \\ &= \sum_{i=1}^{q} \|\mathbf{u}_i - \mathbf{Zf}_i\|^2 = \sum_{i=1}^{q} D_{K_i} \end{aligned} \tag{10.14}$$

In order to minimize D_K we have to minimize every component $D_{K_i} = \|\mathbf{u}_i - \mathbf{Zf}_i\|^2$ separately for $i = 1, 2, ..., q$. The problem of finding the optimal linear vector predictor was reduced to finding q optimal linear scalar predictors that reproduce q components of vector $\widetilde{\mathbf{X}}(n)$. To minimize D_{K_i}, we have to find among all combinations such an $\mathbf{f}_i$, so that $\mathbf{Zf}_i$ would be as close as possible to $\mathbf{u}_i$ in the q-dimensional Euclidean vector space. It is well known from the orthogonality principle that vector $\mathbf{u}_i - \mathbf{Zf}_i$ in this case should be orthogonal to all observations, i.e. to all columns of matrix $\mathbf{Z}$. Thus

$$\mathbf{Z}^T(\mathbf{u}_i - \mathbf{Zf}_i) = \mathbf{0}, \tag{10.15}$$

where $\mathbf{0}$ is a zero – vector.
Therefore we get the equation

$$\mathbf{Z}^T\mathbf{Zf}_i = \mathbf{Z}^T\mathbf{u}_i \tag{10.16}$$

Observe that equation (10.16) expresses the condition defining optimal coefficients of the i-th scalar predictor. Of course, this condition has to be met for all $\mathbf{f}_i$, $i = 1, 2, ..., q$. In order to derive an optimality condition for all $\mathbf{f}_i$, the following vectors will be defined:

$$\mathbf{A} = [\mathbf{f}_1, \mathbf{f}_2, ..., \mathbf{f}_q] = [\mathbf{A}_1, \mathbf{A}_2, ..., \mathbf{A}_K]^T \in \Re^{Kq\times q}, \tag{10.17}$$

$$\widetilde{\mathbf{X}} = [\mathbf{u}_1, \mathbf{u}_2, ..., \mathbf{u}_q] = \left[\widetilde{\mathbf{X}}(1), \widetilde{\mathbf{X}}(2), ..., \widetilde{\mathbf{X}}(N)\right]^T \in \Re^{Nq\times q}, \tag{10.18}$$

Having equations (10.17) and (10.18), equation (10.16) for $i = 1, 2, ..., q$, can be written in a shortened form as:

$$\mathbf{Z}^T\mathbf{ZA} = \mathbf{Z}^T\widetilde{\mathbf{X}} \tag{10.19}$$

For determining the matrix coefficients $\mathbf{A}_k$, $k = 1, 2, ..., K$ of the optimal vector predictor, we should first solve equation (10.19) with respect to matrix $\mathbf{A}$.

Before formulating the final expression defining the optimal linear vector predictor of the K-th order, it would be convenient to define the $K \times K$-dimensional autocorrelation matrix $\mathbf{R}$, whose elements are matrices $\mathbf{R}_{kl}$ such that:

$$\mathbf{R}_{kl} = \mathrm{E}\left[\widetilde{\mathbf{X}}(n-k)\,\widetilde{\mathbf{X}}^T(n-l)\right], \quad k,l = 1,2,...,K, \tag{10.20}$$

$$\mathbf{R}_{k0} = \mathrm{E}\left[\widetilde{\mathbf{X}}(n-k)\,\widetilde{\mathbf{X}}^T(n)\right], \quad k = 1,2,...,K, \tag{10.21}$$

and $\mathbf{R}_{kl} = \mathbf{R}_{lk}^T$. Equation (10.19) can be transformed through multiplications $\mathbf{Z}^T\mathbf{Z}$ and $\mathbf{Z}^T\widetilde{\mathbf{X}}$. After calculating the expected value of each matrix element created as a result of the above steps, we obtain a normal equation of the optimal linear vector predictor in the form:

$$\begin{bmatrix} \mathbf{R}_{11} & \mathbf{R}_{12} & \cdots & \mathbf{R}_{1K} \\ \mathbf{R}_{21} & \mathbf{R}_{22} & \cdots & \mathbf{R}_{2K} \\ \vdots & \vdots & \ddots & \vdots \\ \mathbf{R}_{K1} & \mathbf{R}_{K2} & \cdots & \mathbf{R}_{KK} \end{bmatrix} \begin{bmatrix} \mathbf{A}_1^T \\ \mathbf{A}_2^T \\ \vdots \\ \mathbf{A}_K^T \end{bmatrix} = \begin{bmatrix} \mathbf{R}_{10} \\ \mathbf{R}_{20} \\ \vdots \\ \mathbf{R}_{K0} \end{bmatrix} \tag{10.22}$$

or

$$\sum_{k=1}^{K} \mathbf{A}_k \mathbf{R}_{kl} = \mathbf{R}_{0l} \quad \text{for} \quad l = 1,2,...,K \tag{10.23}$$

Matrix equation (10.22) can be expressed as

$$\mathbf{R}\mathbf{A} = \mathbf{r}, \tag{10.24}$$

where

$$\mathbf{R} = \begin{bmatrix} \mathbf{R}_{11} & \mathbf{R}_{12} & \cdots & \mathbf{R}_{1K} \\ \mathbf{R}_{21} & \mathbf{R}_{22} & \cdots & \mathbf{R}_{2K} \\ \vdots & \vdots & \ddots & \vdots \\ \mathbf{R}_{K1} & \mathbf{R}_{K2} & \cdots & \mathbf{R}_{KK} \end{bmatrix}, \tag{10.25}$$

and

$$\mathbf{A} = \begin{bmatrix} \mathbf{A}_1^T \\ \mathbf{A}_2^T \\ \vdots \\ \mathbf{A}_K^T \end{bmatrix}, \quad \mathbf{r} = \begin{bmatrix} \mathbf{R}_{10} \\ \mathbf{R}_{20} \\ \vdots \\ \mathbf{R}_{K0} \end{bmatrix}, \tag{10.26}$$

The autocorrelation matrix $\mathbf{R}$ includes elements $\mathbf{R}_{kl}$, which are $q \times q$-dimensional correlation matrices, and is a positive semi-definite matrix. Matrix $\mathbf{R}$ is positive definite and hence invertible if the random vector $\widetilde{\mathbf{X}}(n)$ is nondeterministic. We recall that a random process whose k-th order autocorrelation matrices are nonsingular for each $k > 0$, is said to be a nondeterministic process. In this case the prediction problem can be solved by the inversion of matrix $\mathbf{R}$. This is a way that is often quite complicated and may lead to numerical errors, but in the instance of low-level predictors it is acceptable. In more complicated cases, it is advisable to use a generalized version of Levinson's algorithm [132], [161].

For a one-direction vector predictor of the first order, we have a simple matrix equation specifying the coefficient $\mathbf{A}_1$:

$$\mathbf{A}_1 = \mathbf{R}_{01}\mathbf{R}_{11}^{-1} \tag{10.27}$$

Having solved the optimal linear vector predictor problem for the one direction case, we attempt to generalize this equation for two directions. Taking into consideration expressions (10.20), (10.21) and (10.23) for the one direction case, we can analogically write a normal equation for the two-direction case:

$$\sum_{k=0}^{K}\sum_{l=0}^{L}\mathbf{A}_{kl}\mathbf{R}_{kl,ij} = \mathbf{R}_{00,ij}, \tag{10.28}$$

for $i = 0, 1, ..., K,\ j = 0, 1, ..., L,\ (l, k) \neq (0, 0),\ (i, j) \neq (0, 0)$, where

$$\mathbf{R}_{kl,ij} = \mathrm{E}\left[\mathbf{X}(m-k, n-l)\,\mathbf{X}^T(m-i, n-j)\right], \tag{10.29}$$

$$\mathbf{R}_{00,ij} = \mathbf{E}\left[\mathbf{X}(m, n)\,\mathbf{X}^T(m-i, n-j)\right] \tag{10.30}$$

For the first order predictor in the horizontal and vertical direction ($K = 1,\ L = 1$), equation (10.28) can be written as:

$$\begin{cases} \mathbf{A}_{10}\mathbf{R}_{10,10} + \mathbf{A}_{01}\mathbf{R}_{01,10} + \mathbf{A}_{11}\mathbf{R}_{11,10} = \mathbf{R}_{00,10} \\ \mathbf{A}_{10}\mathbf{R}_{10,01} + \mathbf{A}_{01}\mathbf{R}_{01,01} + \mathbf{A}_{11}\mathbf{R}_{11,01} = \mathbf{R}_{00,01} \\ \mathbf{A}_{10}\mathbf{R}_{10,11} + \mathbf{A}_{01}\mathbf{R}_{01,11} + \mathbf{A}_{11}\mathbf{R}_{11,11} = \mathbf{R}_{00,11} \end{cases} \tag{10.31}$$

Thanks to the substitution of parameters (k,l) and (i,j) in accordance with dependencies

$$\begin{aligned}(k,l) &= (1,0) \Longleftrightarrow \tau = 1,\\ (k,l) &= (0,1) \Longleftrightarrow \tau = 2,\\ (k,l) &= (1,1) \Longleftrightarrow \tau = 3,\\ (i,j) &= (1,0) \Longleftrightarrow \vartheta = 1,\\ (i,j) &= (0,1) \Longleftrightarrow \vartheta = 2,\\ (i,j) &= (1,1) \Longleftrightarrow \vartheta = 3,\end{aligned} \tag{10.32}$$

in place of matrix equation system (10.31) we can obtain the following system of matrix equations:

$$\sum_{\tau=1}^{3} \mathbf{A}_\tau \mathbf{R}_{\tau\vartheta} = \mathbf{R}_{0\vartheta} \quad \text{for} \quad \vartheta = 1,2,3 \tag{10.33}$$

or alternatively

$$\begin{cases} \mathbf{A}_1\mathbf{R}_{11} + \mathbf{A}_2\mathbf{R}_{21} + \mathbf{A}_3\mathbf{R}_{31} = \mathbf{R}_{01} \\ \mathbf{A}_1\mathbf{R}_{12} + \mathbf{A}_2\mathbf{R}_{22} + \mathbf{A}_3\mathbf{R}_{32} = \mathbf{R}_{02} \\ \mathbf{A}_1\mathbf{R}_{31} + \mathbf{A}_2\mathbf{R}_{32} + \mathbf{A}_3\mathbf{R}_{33} = \mathbf{R}_{03} \end{cases} \tag{10.34}$$

Equation (10.34) can be rewritten to take form of matrix equation (10.22) and it can be solved in an identical way, having only in mind substitutes (10.32).

10.3 Linear predictor design from empirical data

The solution of equation (10.24) requires the knowledge of $K \cdot K$ elements of matrix $\mathbf{R}$ and K elements of matrix $\mathbf{r}$, that is the autocorrelation matrices $\mathbf{R}_{kl}$ and $\mathbf{R}_{k0}$, $k,l = 1,...,K$, of the discrete stationary stochastic process $\left\{\widetilde{\mathbf{X}}(n)\right\}$. These matrices are not known, but we have a sequence of observations $\widetilde{\mathbf{X}}(1), \widetilde{\mathbf{X}}(2), ..., \widetilde{\mathbf{X}}(N)$ of the process $\left\{\widetilde{\mathbf{X}}(n)\right\}$. Based on the knowledge of this sequence we can estimate the correlation matrix and then calculate the coefficients of the linear vector predictor. Different estimator constructions arise by defining the variability of index n in a different manner.

a) Autocorrelation method

We assume that the input signal is equal to zero outside range $1 \leq n \leq N$, thus

$$\widetilde{\mathbf{X}}(n) = \mathbf{0} \quad \text{for} \quad n \notin \wp = \{1, 2, ..., N\} \tag{10.35}$$

and $\widetilde{\mathbf{X}} = \mathbf{V}(n)$ for $n = 1, 2, ..., M$, where $M = \frac{N_1 \cdot N_2}{q}$.

Under condition (10.35), matrix $\mathbf{Z}$ in expression (10.13) takes the form

$$\mathbf{Z} = \begin{bmatrix} \mathbf{0}^T & \mathbf{0}^T & \cdots & \mathbf{0}^T & \mathbf{0}^T \\ \widetilde{\mathbf{X}}^T(1) & \mathbf{0}^T & \cdots & \mathbf{0}^T & \mathbf{0}^T \\ \widetilde{\mathbf{X}}^T(2) & \widetilde{\mathbf{X}}^T(1) & \cdots & \mathbf{0}^T & \mathbf{0}^T \\ \vdots & \widetilde{\mathbf{X}}^T(2) & \cdots & \mathbf{0}^T & \mathbf{0}^T \\ \vdots & \vdots & \cdots & \widetilde{\mathbf{X}}^T(1) & \mathbf{0}^T \\ \vdots & \vdots & \cdots & \widetilde{\mathbf{X}}^T(2) & \widetilde{\mathbf{X}}^T(1) \\ \widetilde{\mathbf{X}}^T(N) & \vdots & \cdots & \vdots & \widetilde{\mathbf{X}}^T(2) \\ \mathbf{0}^T & \widetilde{\mathbf{X}}^T(N) & \cdots & \vdots & \vdots \\ \mathbf{0}^T & \mathbf{0}^T & \cdots & \vdots & \vdots \\ \mathbf{0}^T & \mathbf{0}^T & \cdots & \widetilde{\mathbf{X}}^T(N) & \vdots \\ \mathbf{0}^T & \mathbf{0}^T & \cdots & \mathbf{0}^T & \widetilde{\mathbf{X}}^T(N) \end{bmatrix} \tag{10.36}$$

Observe that $Z \in \Re^{(N+K) \times Kq}$. If we perform the matrix multiplication $\mathbf{Z}^T\mathbf{Z}$ and $\mathbf{Z}^T\mathbf{X}$ then we get the following matrix equation:

$$\begin{bmatrix} \widetilde{\mathbf{R}}_0 & \widetilde{\mathbf{R}}_1^T & \cdots & \widetilde{\mathbf{R}}_{K-1}^T \\ \widetilde{\mathbf{R}}_1 & \widetilde{\mathbf{R}}_0 & \cdots & \widetilde{\mathbf{R}}_{K-2}^T \\ \vdots & \vdots & \ddots & \vdots \\ \widetilde{\mathbf{R}}_{K-1} & \widetilde{\mathbf{R}}_{K-2} & \cdots & \widetilde{\mathbf{R}}_0 \end{bmatrix} \begin{bmatrix} \mathbf{A}_1^T \\ \mathbf{A}_2^T \\ \vdots \\ \mathbf{A}_K^T \end{bmatrix} = \begin{bmatrix} \widetilde{\mathbf{R}}_1 \\ \widetilde{\mathbf{R}}_2 \\ \vdots \\ \widetilde{\mathbf{R}}_K \end{bmatrix}, \tag{10.37}$$

where

$$\widetilde{\mathbf{R}}_k = \frac{\sum_{n=1}^{N-k} \widetilde{\mathbf{X}}(n)\, \widetilde{\mathbf{X}}^T(n+k)}{N} \in \Re^{q \times q}, \;\; k = 0, 1, ..., K \tag{10.38}$$

Observe that matrix

$$\widetilde{\mathbf{R}} = \begin{bmatrix} \widetilde{\mathbf{R}}_0 & \widetilde{\mathbf{R}}_1^T & \cdots & \widetilde{\mathbf{R}}_{K-1}^T \\ \widetilde{\mathbf{R}}_1 & \widetilde{\mathbf{R}}_0 & \cdots & \widetilde{\mathbf{R}}_{K-2}^T \\ \vdots & \vdots & \ddots & \vdots \\ \widetilde{\mathbf{R}}_{K-1} & \widetilde{\mathbf{R}}_{K-2} & \cdots & \widetilde{\mathbf{R}}_0 \end{bmatrix} \tag{10.39}$$

is block Toeplitz matrix and equation (10.37) can be solved by making use of a generalized Levinson's algorithm [132], [161]. It is easily seen that

$$\begin{aligned}\mathbf{E}\left[\widetilde{\mathbf{R}}_k\right] &= \mathbf{E}\left[\frac{1}{N}\sum_{n=1}^{N-k}\widetilde{\mathbf{X}}(n)\,\widetilde{\mathbf{X}}^T(n+k)\right] \\ &= \frac{1}{N}\sum_{n=1}^{N-k}\mathbf{E}\left[\widetilde{\mathbf{X}}(n)\,\widetilde{\mathbf{X}}^T(n+k)\right] \\ &= \frac{1}{N}\sum_{n=1}^{N-k}\mathbf{R}_k = \frac{N-k}{N}\mathbf{R}_k,\end{aligned} \tag{10.40}$$

for $k = 0, 1, \ldots, K$. Therefore estimator (10.38) is biased. If the number of samples N is much bigger than the order K of the predictor, then the error connected with the estimation will be smaller.

b) Covariance method

In the covariance method we do not make any assumptions about the value of the input signal outside of $K \leq n \leq N$ range. In this case:

$$n \in \wp = \{K, K+1, K+2, ..., N\} \tag{10.41}$$

Simultaneously, we assume that the realizations from the range $n = 0, 1, ..., K-1$ are non-zero. For the covariance method we use the following estimator

$$\widetilde{\mathbf{R}}_{kl} = \frac{\sum_{n=K}^{N-1}\widetilde{\mathbf{X}}(n-k)\,\widetilde{\mathbf{X}}^T(n-l)}{N-K} \tag{10.42}$$

We can determine the expected value of $\tilde{\mathbf{R}}_{kl}$:

$$\begin{aligned}\mathbf{E}\left[\widetilde{\mathbf{R}}_{kl}\right] &= \mathrm{E}\left[\frac{1}{N-K}\sum_{n=K}^{N-1}\widetilde{\mathbf{X}}(n-k)\,\widetilde{\mathbf{X}}^T(n-l)\right] \\ &= \frac{1}{N-K}\sum_{n=K}^{N-1}\mathbf{E}\left[\widetilde{\mathbf{X}}(n-k)\,\widetilde{\mathbf{X}}^T(n-l)\right] \\ &= \frac{1}{N-K}\sum_{n=K}^{N-1}\mathbf{R}_{kl} = \mathbf{R}_{kl},\end{aligned} \tag{10.43}$$

for $k = 1, 2, \ldots K, l = 0, 1, 2, \ldots, K$. Therefore the covariance method is based on unbiased estimators and yields the best results for short data sequences [132], [161]. The supermatrix which contains matrix elements (10.42) is not block Toeplitz and the normal equations cannot be solved by making use of the Levinson's algorithm. However, the normal equations can be solved in a numerically stable way using the Cholesky decomposition.

10.4 Predictor based on neural networks

In this section we propose a multidimensional non-linear predictor based on the feed-forward neural network. Referring to notation given in Chapter 7, the neural net architecture is designed as follows:
a) q inputs given by $v_1(t-1), v_2(t-1), ..., v_q(t-1)$,
b) one hidden layer,
c) q desired signals given by $v_1(t), v_2(t), ..., v_q(t)$,
d) q outputs given by $y_1(t), y_2(t), ..., y_q(t)$.

In the learning process we attempt to minimize the following performance measure

$$Q = \sum_{t=1}^{M} \sum_{i=1}^{q} \left(y_i(t) - v_i(t)\right)^2$$

The weights matrix $\mathbf{WBP}(s)$, see Fig. 12.28 in Chapter 12, can be found by the back-propagation method.

10.5 Concluding remarks

In this chapter first we have presented vector linear predictors with coefficients in the form of matrices. The matrix coefficients can be estimated by the autocorrelation method or by the covariance method. Other estimation techniques can be found in the excellent monographs [132], [161]. A non-linear predictor based on the neural network has been proposed and will be implemented along with parametric predictors in Chapter 12.

11

Design of the Code-book

11.1 Introduction

In the past, the most popular algorithm for the code-book design was the Linde-Buzo-Gray algorithm (LBG) and its variants [154]. However, the LBG algorithm is computationally demanding and in the last decade new methods for the code-book design have been proposed. Most of them are based on competitive learning algorithms [73], [83], [86], [157], [170]. In this chapter we study three such algorithms. Moreover, we discuss the problem of selection of the initial code-book which has a considerable influence on the learning process.

11.2 Competitive algorithms

We will design the optimal code-book $\mathbf{G} = [\mathbf{g}_0, \mathbf{g}_1, ..., \mathbf{g}_J]$, $\mathbf{g}_j = [g_{1j}, g_{2j}, ..., g_{qj}]^T \in \Re^q$, $j = 0, 1, ..., J$, where $J+1$ = size of the code-book, such that the performance measure

$$D = \sum_{t=1}^{M} d\left[\mathbf{E}(t), \mathbf{g}_{j^0}\right]^2, \tag{11.1}$$

is minimized, where $M = \frac{N_1 \cdot N_2}{q}$, $\mathbf{E}(t) = [e_1(t), e_2(t), ..., e_q(t)]^T \in \Re^q$ and

$$d\left[\mathbf{E}(t), \mathbf{g}_{j^0}\right] = \min_{0 \leq j \leq J} \{d[\mathbf{E}(t), \mathbf{g}_j]\}, \tag{11.2}$$

As a measure d in expression (11.1) we can take e.g. the Euclidean distance

$$d = \sqrt{\sum_{i=1}^{q} [e_i(t) - g_{ij}]^2}, \tag{11.3}$$

Prototype vector $\mathbf{g}_{j^0}$ will be called the "winner" of the competition for the closest similarity with vector $\mathbf{E}(t)$ in the vector space $\Re^q$. An implementation of the neural network with a proper learning procedure guarantees an optimal distribution of the prototype vectors $\mathbf{g}_{j^0}$ from the code-book $\mathbf{G}$ in space $\Re^q$ to minimize the error given by expression (11.1). In this application, we take advantage of nice properties of the neural network working as a classifier. The process of the vector quantizer construction with the help of the neural network can be divided into two parts:
- stage of network learning,
- stage of network testing.

Because of its properties, the neural network is able to create the code-book $\mathbf{G}$ as a result of learning. For this purpose we can use the neural network shown in Fig. 11.1.

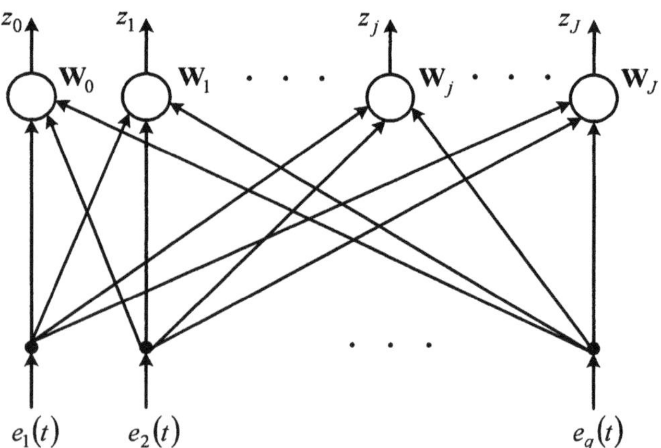

FIGURE 11.1. The code-book based on neural network

In Fig. 11.1. the neural network is shown having $J+1$ neurons, q input nodes on which components of the residual vector $\mathbf{E}(t) = e_1(t), e_2(t), ..., e_q(t)]^T$ are fed for $t = 1, ..., M$. The outputs of neurons are denoted by z_j where $j = 0, 1, ..., J$. Each of the $J+1$ neurons is connected with all the input nodes. The weights of these connections are written as vectors:

$$\mathbf{W}_j = [w_{1j}, w_{2j}, ..., w_{q,j}]^T \in \Re^q, \; j = 0, 1, ..., J, \tag{11.4}$$

which are identified with the prototype vectors $\mathbf{g}_j$ from the code-book $\mathbf{G}$. Therefore

$$\mathbf{g}_j = \mathbf{W}_j \tag{11.5}$$

as we mentioned in Section 7.3. To create the optimal code-book $\mathbf{G}$, we can use one of the three algorithms:
a) competitive learning algorithm (CL),
b) frequency-sensitive competitive learning algorithm (FSCL),
c) Kohonen self-organizing feature maps (KSFM).
All of them are based on one principle. They pick up weights vectors $\mathbf{W}_j = [w_{1j}, w_{2j}, ..., w_{q,j}]^T$ by competition and then make them similar to the network input vector $\mathbf{E}(t) = [e_1(t), e_2(t), ..., e_q(t)]^T$.

This process is shown schematically (during one learning step) in Fig. 11.2, where α is the coefficient of the speed of learning.

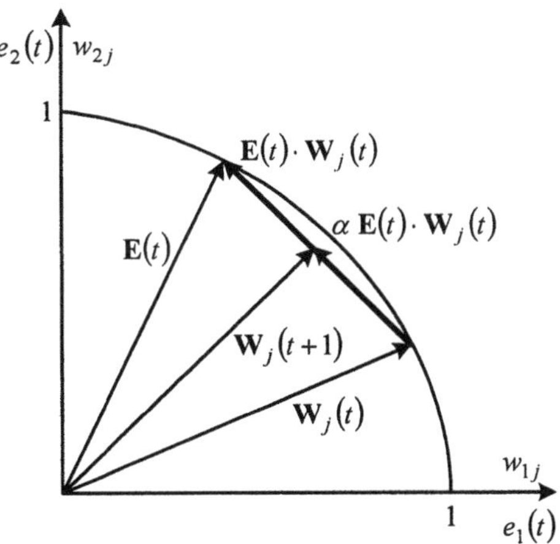

FIGURE 11.2. The process of making vector $\mathbf{W}_j$ similar to vector $\mathbf{E}(t)$ in one learning step

Competitive learning algorithms of the neural network differ as regards setting the "winner neuron", by choosing vectors $\mathbf{W}_j$ intended for learning and in the way they choose the value of the learning coefficient.
Below, we will present network learning algorithms according to consecutive steps allowing their easy implementation by means of a computer program.

a) CL algorithm

Step 1. We define the distance between the input vector $\mathbf{E}(t)$ and weights vector of the j-th neuron $\mathbf{W}_j$ as the Euclidean distance

$$d[\mathbf{E}(t), \mathbf{W}_j] = \|\mathbf{E}(t) - \mathbf{W}_j\| = \sqrt{\sum_{i=1}^{q} [e_i(t) - w_{ij}]^2}$$

Step 2. We present vectors $\mathbf{E}(t)$ on input nodes.
Step 3. We choose the "winner" of the competition j^0 as the one among all neurons for which:

$$d\left[\mathbf{E}(t), \mathbf{W}_{j^0}\right] = \max_{0 \leq j \leq J} \{d[\mathbf{E}(t), \mathbf{W}_j]\}$$

Step 4. We set outputs of neurons in accordance with:

$$z_j = \begin{cases} 1 & \text{for} \quad j = j^0 \\ 0 & \text{for} \quad j \neq j^0 \end{cases}$$

Step 5. We make the weight vector $\mathbf{W}_{j^0}$ similar to the input vector $\mathbf{E}(t)$ according to the recursive rule: $w_{ij}(t+1) = w_{ij}(t) + \alpha(t)[e_i(t) - w_{ij}(t)]\ z_j$, $i = 1, 2, ..., q$, for $t = 1, 2, ...$,where $\alpha(t)$ – is the chosen learning coefficient decreasing to zero as learning progresses, e.g. $\alpha(t) = c_1 e^{-t/c_2}$ where c_1 and c_2 are positive coefficients. The learning process terminates when the error $e_i - w_{ij}$ becomes small enough or a number of learning steps t is bigger than $t_{\max}$ assumed a priori.
Step 6. If the learning set has not been exhausted yet, we go back to step 2 with the next input vector $\mathbf{E}(t+1)$.

b) FSCL algorithm

Step 1. We define the distance between the input vector $\mathbf{E}(t)$ and weights vector $\mathbf{W}_j$ as the Euclidean distance. Additionally, we take into consideration the number of victories f_j of a given neuron:

$$d\left[\mathbf{E}(t), \mathbf{W}_j\right] = F(f_j)\left\|\mathbf{E}(t) - \mathbf{W}_j\right\| = F(f_j)\sqrt{\sum_{i=1}^{q}\left[e_i(t) - w_{ij}\right]^2},$$

where

$F(f_j)$ – function of frequency of victories of the j-th neuron, e.g.:
$F(f_j) = cf_j$, c – positive coefficient,
$F(f_j) = c_1 e^{f_j/c_2}, c_1, c_2$ – positive coefficients,
$F(f_j) = 1 - e^{-(f_j/c_3)+c_4}, c_3, c_4$ – positive coefficients.

Step 2. We present vectors $\mathbf{E}(t)$ on input nodes.
Step 3. We choose the "winner" of the competition j^0 as the one for which:

$$d\left[\mathbf{E}(t), \mathbf{W}_{j^0}\right] = \max_{0 \le j \le J}\left\{d\left[\mathbf{E}(t), \mathbf{W}_j\right]\right\}$$

Step 4. We set outputs of neurons according to:

$$z_j = \begin{cases} 1 & \text{for} \quad j = j^0 \\ 0 & \text{for} \quad j \neq j^0 \end{cases}$$

Step 5. We make weight vector $\mathbf{W}_{j^0}$ similar to the input vector $\mathbf{E}(t)$ in accordance with the recursive rule:

$$\begin{aligned} w_{ij}(t+1) &= w_{ij}(t) + H(f_j)\,\alpha(t)\left[e_i(t) - w_{ij}(t)\right] z_j, \\ i &= 1, 2, ..., q, \quad \text{for} \quad t = 1, 2..., \end{aligned}$$

where

$\alpha(t)$ – is the properly chosen learning coefficient reducing its own value as learning progresses,
$H(f_j)$ – is the properly chosen function of frequency of victories of neuron, e.g.
$H(f_j) = c_5/f_{j^0}, c_5$ – positive coefficient,
$H(f_j) = c_6 e^{-f_{j^0}/c_7}, c_6, c_7$ – positive coefficients.

The learning process terminates when the error $e_i - w_{ij}$ becomes small enough or the number of learning steps t exceeds $t_{\max}$ assumed a priori.
Step 6. If the learning set has not been exhausted yet, we go back to step 2 with the next input vector $\mathbf{E}(t+1)$.

c) KSFM algorithm

Step 1. We change the notation transforming the neurons' order from one-dimensional to two-dimensional. We assume that $J+1=\widehat{\imath}_{\max}\cdot\widehat{\jmath}_{\max}$, $\widehat{\imath}=1,2,...,\widehat{\imath}_{\max}$, $\widehat{\jmath}=1,2,...,\widehat{\jmath}_{\max}$. In this notation a place of a given neuron is determined by pair $\left(\widehat{\imath},\widehat{\jmath}\right)$.

Step 2. We set the distance between the input vector $\mathbf{E}(t)$ and weight vector $\mathbf{W}_{\widehat{\imath},\widehat{\jmath}}$ as follows:

$$d\left[\mathbf{E}(t),\mathbf{W}_{\widehat{\imath},\widehat{\jmath}}\right]=\left\|\mathbf{E}(t)-\mathbf{W}_{\widehat{\imath},\widehat{\jmath}}\right\|=\sqrt{\sum_{i=1}^{q}\left[e_i(t)-w_{i,\left(\widehat{\imath},\widehat{\jmath}\right)}\right]^2}$$

Step 3. We present vectors $\mathbf{E}(t)$ on input nodes.

Step 4. We set the "winner" of the competition $\left(\widehat{\imath}^0,\widehat{\jmath}^0\right)$ as the one for which

$$d\left[\mathbf{E}(t),\mathbf{W}_{\widehat{\imath}^0,\widehat{\jmath}^0}\right]=\min_{\substack{1\le\widehat{\imath}\le\widehat{\imath}_{\max}\\1\le\widehat{\jmath}\le\widehat{\jmath}_{\max}}}\left\{d\left[\mathbf{E}(t),\mathbf{W}_{\widehat{\imath},\widehat{\jmath}}\right]\right\}$$

Step 5. We set outputs of the neurons according to:

$$z_{\widehat{\imath},\widehat{\jmath}}=\begin{cases}1 & \text{for} \quad (\imath,\jmath)=\left(\widehat{\imath}^0,\widehat{\jmath}^0\right)\\0 & \text{for} \quad \left(\widehat{\imath},\widehat{\jmath}\right)\neq\left(\widehat{\imath}^0,\widehat{\jmath}^0\right)\end{cases}$$

Step 6. We define a neighborhood of the neuron $\left(\widehat{\imath}^0,\widehat{\jmath}^0\right)$ in which neighboring neurons will be simultaneously learned, e.g. in the following way:

$$d\left[\left(\widehat{\imath}^0,\widehat{\jmath}^0\right),\left(\widehat{\imath},\widehat{\jmath}\right)\right]=\max\left\{\left|\widehat{\imath}^0,\widehat{\imath}\right|,\left|\widehat{\jmath}^0,\widehat{\jmath}\right|\right\}$$

Step 7. We make weight vectors $\mathbf{W}_{\widehat{\imath},\widehat{\jmath}}$ similar to the input vector $\mathbf{E}(t)$ in accordance with the recursive rule:

$$w_{i,\left(\widehat{\imath},\widehat{\jmath}\right)}(t+1)=w_{i,\left(\widehat{\imath},\widehat{\jmath}\right)}(t)+\alpha(t)$$

$$\cdot\frac{1}{1+d\left[\left(\widehat{\imath},\widehat{\jmath}\right),\left(\widehat{\imath}^0,\widehat{\jmath}^0\right)\right]}\left[e_i(t)-w_{i,\left(\widehat{\imath},\widehat{\jmath}\right)}(t)\right],$$

$$i=1,2,...,q,$$

for $t = 1, 2, ...,$ and for all neurons $\left(\widehat{i}, \widehat{j}\right)$ for which $d\left[\left(\widehat{i^0}, \widehat{j^0}\right), \left(\widehat{i}, \widehat{j}\right)\right] \leq d_{\max}$, where $\alpha(t)$ is the properly chosen learning coefficient.
The learning process terminates when error $e_i - w_{i,\widehat{j}}$ becomes small enough or the number of learning steps t exceeds $t_{\max}$ assumed a priori. If the space around the "winner" neuron goes beyond $\widehat{i} = 1, 2, ..., \widehat{i}_{\max}$, $\widehat{j} = 1, 2, ..., \widehat{j}_{\max}$, then the neurons for the learning process are chosen from the opposite side of the rectangle, which is defined by maximum values of parameters $\widehat{i}$ and $\widehat{j}$. In this way, a closed space without a border is created.
Step 8. If the learning set has not been exhausted yet, we go back to Step 3 with the next input vector $\mathbf{E}(t+1)$.

11.3 Preprocessing

In this section, scaling methods of input vectors $\mathbf{E}(t)$ and weight vectors $\mathbf{W}_j$ will be described. The process of scaling of these vectors should be applied if for setting a distance between those vectors in the neural network's learning algorithms instead of

$$d\left[\mathbf{E}(t), \mathbf{W}_j\right] = \left\|\mathbf{E}(t) - \mathbf{W}_j\right\| = \sqrt{\sum_{i=1}^{q}\left[e_i(t) - w_{ij}\right]^2}, \tag{11.6}$$

we will use:

$$\begin{aligned} d\left[\mathbf{E}(t), \mathbf{W}_j\right] &= 1 - \mathbf{E}^T(t) \circ \mathbf{W}_j \\ &= 1 - \left\|\mathbf{E}(t)\right\| \left\|\mathbf{W}_j\right\| \cos\angle\left(\mathbf{E}(t), \mathbf{W}_j\right) \\ &= 1 - \sqrt{\sum_{i=1}^{q}\left[e_i(t)\right]^2}\sqrt{\sum_{i=1}^{q}\left(w_{ij}\right)^2} \cos\angle\left(\mathbf{E}(t), \mathbf{W}_j\right) \end{aligned} \tag{11.7}$$

The scaling of input vectors $\mathbf{E}(t)$ has an influence on improving the procedure which chooses the "winner" from among all $J+1$ of the neurons of the network. If there is no scaling, it can happen that an influence on the scalar product of vectors $\mathbf{E}(t)$ and $\mathbf{W}_j$ in the space $\Re^q$ will have lengths of these vectors and not only the angle contained between those vectors.

In Fig. 11.3 we present an example of distribution of weight vectors and difference vectors in a $\Re^3$ vector space. Assuming the normalization of vectors $\mathbf{E}(t)$ and $\mathbf{W}_j$, equation (11.7) takes the for

$$d[\mathbf{E}(t), \mathbf{W}_j] = \|\mathbf{W}_j\| \, \|\mathbf{E}(t)\| \cos\theta_j = \cos\theta_j \tag{11.8}$$

where θ_j is the angle between vectors $\mathbf{E}(t)$ and $\mathbf{W}_j$, and $\|\mathbf{W}_j\|$ and $\|\mathbf{E}(t)\|$are lengths of these vectors. In order to eliminate the influence of the length of the vectors on the value of expression (11.8), a normalization of vectors $\mathbf{E}(t)$ and $\mathbf{W}_j$ is carried out. After the normalization, the difference vectors and weights vectors are located on the surface of the hyper sphere with radius equal 1 in space $\Re^q$. In Fig. 11.4, we illustrate space $\Re^3$ along with a unit circle and vectors $\mathbf{E}(t)$ and $\mathbf{W}_j$ put there as a result of normalization.

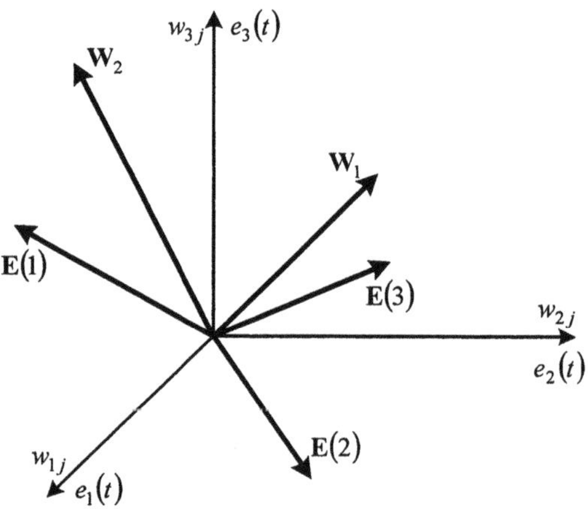

FIGURE 11.3. An example of distribution of weight vectors and difference vectors in $\Re^3$ vector space

Expressions (11.9) and (11.10) after the vector normalization of vectors $\mathbf{E}(t)$ and $\mathbf{W}_j$ will take the following values:

$$\|\mathbf{E}(t)\| = 1 \quad \text{and} \quad \|\mathbf{W}_j\| = 1$$

Simultaneously, expression (11.9) is reduced to the following form:

$$d[\mathbf{E}(t), \mathbf{W}_j] = \cos\theta_j \tag{11.9}$$

In points a) and b) two methods of normalizing vectors $\mathbf{E}(t)$ and $\mathbf{W}_j$ will be presented. They are suitable for the computer realization.

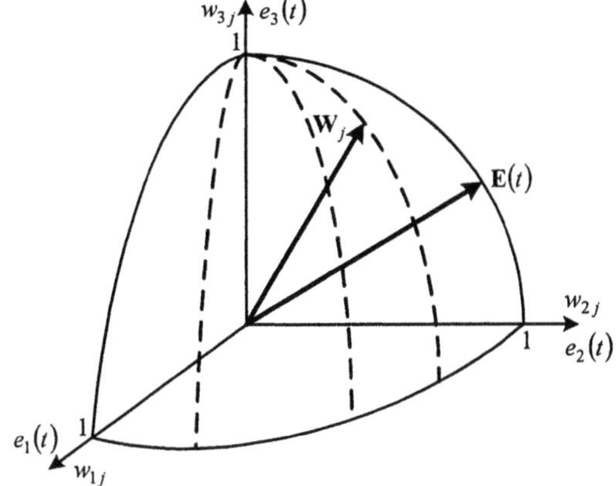

FIGURE 11.4. Vectors $\mathbf{E}(t)$ and $\mathbf{W}_j$ after normalization

The normalization process will be discussed in the context of the difference vector $\mathbf{E}(t)$, but the weight vectors normalization is the same.

a) Normalization – method I

In method I, the normalization of vector $\mathbf{E}'(t)$ is done on the basis of the expression:

$$\mathbf{E}(t) = \frac{\mathbf{E}'(t)}{\|\mathbf{E}'(t)\|}, \tag{11.10}$$

where

$\mathbf{E}'(t) = \left[e'_1(t), e'_2(t), ..., e'_q(t)\right]^T \in \Re^q$ – q-dimensional non-standard difference vector,

$\mathbf{E}(t) = \left[e_1(t), e_2(t), ..., e_q(t)\right]^T \in \Re^q$ – q-dimensional normalized difference vector,

$\|\mathbf{E}'(t)\| = \sqrt{\sum_{i=1}^{q} e'_i(t)^2}$ – length of vector $\mathbf{E}'(t)$.

In other words, the components of the difference vector $\mathbf{E}'(t)$ are divided by the length of vector $\mathbf{E}'(m,n)$, which was schematically shown in Fig. 11.5. for a two-dimensional vector space. After this operation, the length of vector $\mathbf{E}'(t)$ equals 1.

b) Normalization – method II

This method is more complicated than method I and is based on sequences of operations done on the difference vector $\mathbf{E}''(t)$:

1. To the nonstandard vector

$$\mathbf{E}''(t) = \left[e_1''(t), e_2''(t), ..., e_q''(t)\right]^T \in \Re^q$$

one more dimension s is added, so a new vector is created as follows

$$\mathbf{E}'(t) = \left[s, e_1'(t), e_2'(t), ..., e_q'(t)\right]^T \in \Re^{q+1}$$

with

$$s = \sqrt{S^2 - \|\mathbf{E}''(t)\|^2}, \tag{11.11}$$

where
S – number imperceptibly greater than the length of the longest difference vector $\mathbf{E}''(t)$,

$\|\mathbf{E}''(t)\| = \sqrt{\sum_{i=1}^{q} e_i''(t)^2}$ – length of vector $\mathbf{E}''(t)$.

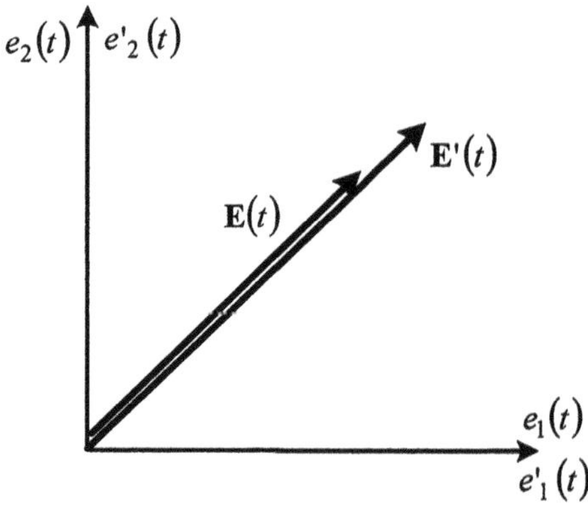

FIGURE 11.5. Process of normalization of difference vector with application of method no 1

In this way, vector $\mathbf{E}''(t)$ is on a hyper sphere with radius S as is shown in the three-dimensional vector space in Fig. 11.6.
2. Difference vector $\mathbf{E}'(t)$ is divided by its length S, i.e.

$$\mathbf{E}(t) = \frac{\mathbf{E}'(t)}{S} \tag{11.12}$$

The created normalized vector $\mathbf{E}(t)$ with length equal to 1 has an additional dimension s.

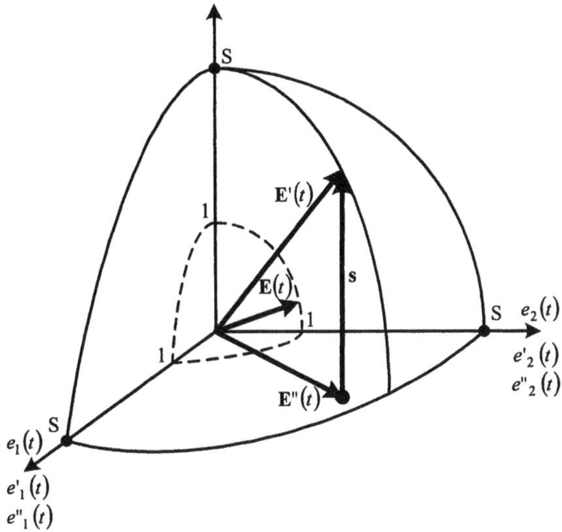

FIGURE 11.6. Distribution of difference vectors in each phase of normalization by means of method II

11.4 Selection of initial code-book

The selection of initial values for the weight vectors $\mathbf{W}_j(0) \in \Re^q$ i.e. a suitable distribution of these vectors in space $\Re^q$ has a considerable influence on the speed and quality of the learning process. To avoid negative occurrences during the starting phase of the learning process, we should avoid:

1. Placing the initial weight vectors $\mathbf{W}_j(0)$ in space $\Re^q$ far from the difference vectors $\mathbf{E}(t)$, because they can be skipped in the learning phase and, therefore, lost.

2. Grouping too few initial weight vectors $\mathbf{W}_j(0)$ in space $\Re^q$ next to a large group of difference vectors $\mathbf{E}(t)$ from the learning set because to different classes of residual vectors the same weight vectors can be assigned.

3. Placing too many weight vectors $\mathbf{W}_j(0)$ in space $\Re^q$ next to the difference vectors which belong to one class because such an arrangement can break one class of difference vectors into a few different classes which means losing a few weight vectors.

There are several ways to initialize the initial weights of the network, which allows minimizing the influence of the above mentioned occurrences on the learning process:

a) Random distribution of weight vectors $\mathbf{W}_j(0) \in \Re^q$ in space $\Re^q$ which is shown in Fig. 11.7. The weight vectors are spread on the surface of the hyper sphere with radius $= 1$. This would happen after the normalization of the weight vectors. This method has its weak points because random distribution of weight vectors in the vector space makes them equally distributed and difference vectors usually group together in a small part of the space. In this way, many weight vectors may never be chosen as "winners" and lost forever. Besides, the number of weight vectors lying next to a group of difference vectors $\mathbf{E}(t)$ from the learning set can be too small to properly divide individual classes of difference vectors (one weight vector is assigned only to one class of difference vectors). The best solution in this situation would be such a distribution of initial weight vectors $\mathbf{W}_j(0) \in \Re^q$ so that their density in different regions of space $\Re^q$ corresponds to the density of difference vectors in those regions of the vector space.

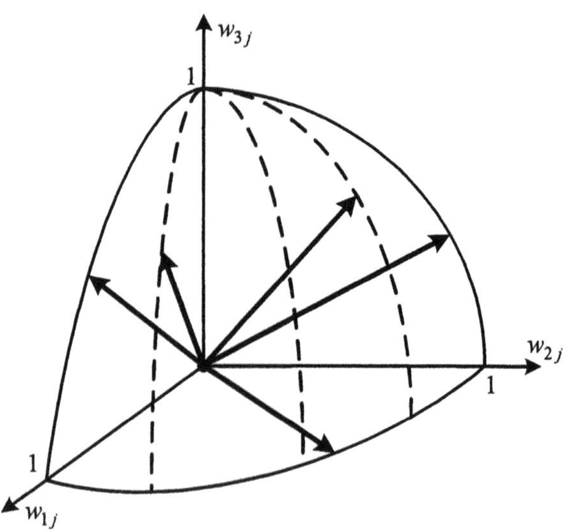

FIGURE 11.7. Random distribution of weight vectors $\mathbf{W}_j(0) \in \Re^q$

b) The second method of weight vectors $\mathbf{W}_j(0) \in \Re^q$ initialization is the method of convex combination. It proposes at the beginning of the learning process to assign value $1/\sqrt{q}$ to each component of all the weight vectors where q is the number of input nodes. It leads to a normalization of the weight vectors and their coincidental distribution in space $\Re^q$. Difference vectors are transformed so that their components would satisfy the following expression:

$$e_i' = \alpha(t)\, e_i + (1/\sqrt{q})\,(1 - \alpha(t)), \tag{11.13}$$

where q – number of input nodes of network, $\alpha(t)$ – learning coefficient. The difference vectors, at first very similar to each other, slowly break the group of coincidental weight vectors. During the learning process, weight vectors follow the difference vectors until one weight vector is assigned to each category of the difference vectors. This slows down the learning process because of the slow motion of weight vectors following a "moving target" in the form of difference vectors from the learning set.

c) The third method of setting the first value of weight vectors recommends adding noise to the difference vectors. The difference vectors in space $\Re^q$ catch randomly placed weight vectors and "pull" them towards one another, i.e. make them similar. This method is unfortunately slower than the convex combination method.

d) The fourth method of weight vector initialization is based on the random spreading of weight vectors in space $\Re^q$ but the pulling of those vectors to difference vectors is different. At the beginning of the learning process, the learning of all weight vectors is enforced, not only of the "winner". Therefore, all the weight vectors gather in the area where difference vectors are present. During learning, the space of weight vectors which are subject to learning becomes smaller until only the "winner" is learned. This process improves the learning speed because the next vector from the learning set will have a shorter route to the weight vectors, which are grouped more closely to the difference vectors.

e) This method, known as the Desieno method, gives every winner a "conscience" in the form of a number of learning steps within which it has to get close enough to a difference vector. If it does not get close enough within the limit of learning steps, then the requirements in the "contest" for the winner are raised for this neuron, which makes winning easier for other neurons.

11.5 Concluding remarks

In this chapter we have presented three algorithms for the code-book design. The CL algorithm is the simplest one. The main problem with the CL algorithm is that some of the neurons may be underutilized. The FSCL algorithm keeps a count of how frequently each neuron is the winner. Therefore, during the learning process, all neurons are modified about an equal number of times. The Kohonen self-organizing feature map (KSFM) algorithm requires a proper choice of the neighborhood of the wining neuron.

12
Design of the PVQ Schemes

12.1 Introduction

In this chapter we present four methods to design the PVQ scheme:

a) with "open-loop",

b) with "closed-loop" and parametric predictor,

c) with "modified closed-loop" and parametric predictor,

d) with "modified closed-loop" and neural predictor.

In each case we explain how to determine the predictor and vector quantizer. The design with the "modified closed-loop" is studied in two versions: with the parametric predictor and with the neural predictor.

We also compare the PVQ schemes given in this chapter and present the results of the simulations. The tested image is a standard picture "Lena" (see Fig. 12.1) characterized by

(a) $N_1 \times N_2 = 512 \times 512$ frame of size,

(b) 256 grey levels for each pixel,

(c) $n_1 \times n_2 = 4 \times 4$ pixels blocks of image.

We will illustrate the performance of design schemes by making use of the MSE and SNR measures.

FIGURE 12.1. "Lena" original image

12.2 Open-loop design

The scheme of determination of parameters in an "open-loop" is described in Fig. 12.2.

As we can see in Fig. 12.2, determining the parameters in an "open-loop" is a simplified scheme of design. It starts from the calculation of the predictor coefficients based on the statistics of $\mathbf{V}(m,n)$. Details of this are described in Chapter 10. On the input of the predictor there are given delayed vectors $\mathbf{V}(m-k,n-l)$, that is original input vectors. The learning sequence for the code-book design is generated basing on the equation

$$\mathbf{E}(m,n) = \mathbf{V}(m,n) - \sum_{k=0}^{K}\sum_{l=0}^{L}\mathbf{A}_{kl}\mathbf{V}(m-k,n-l), \qquad (12.1)$$

where:
$m = 1,2,...,N_1/n_1,\ n = 1,2,...,N_2/n,\ (k,l) \neq (0,0)$.

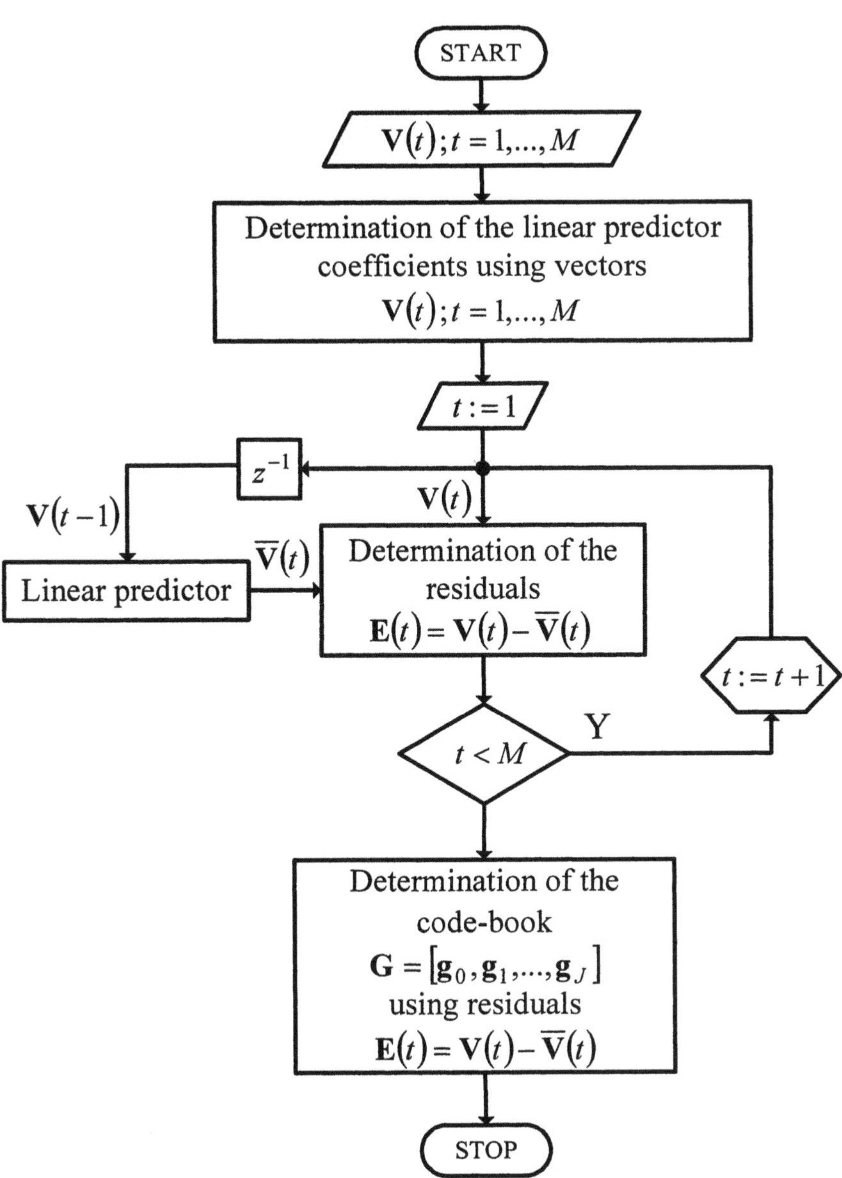

FIGURE 12.2. "Open-loop" PVQ design

After generating the learning sequence (12.1), we determine the vector quantizer parameters, which means the prototype vectors in the code-book $\mathbf{G} = [\mathbf{g}_0, \mathbf{g}_1, ..., \mathbf{g}_J]$. Algorithms for setting the code-book will be presented in Chapter 12. At this moment, we can close the loop because during a normal operation of the PVQ system, the difference vectors $\mathbf{E}(m, n)$ will be determined on the basis of formulas (9.1) and (9.4), that is, with regard to the vector quantizer operation. The scheme of the calculation of the parameters of the PVQ system with an "open-loop" is the simplest one. We can only have doubts whether residual vectors $\mathbf{E}(m, n)$ calculated in the "open-loop" scheme will properly reflect the statistics of these vectors that occur later during a normal system operation.

In Fig. 12.3 and Fig. 12.4 we present the results of simulations for a varying code-book size (MSE and SNR versus code-book size). Figures 12.5 and 12.6 illustrate the dependence of MSE and SNR on the number of epochs (determination of the code-book in Fig. 12.2) for the code-book size = 512.

In Fig. 12.7 we show the reconstructed "Lena" image for the code-book size = 512, two epochs and the CL algorithm. In this case MSE = 360 and SNR = 16.91.

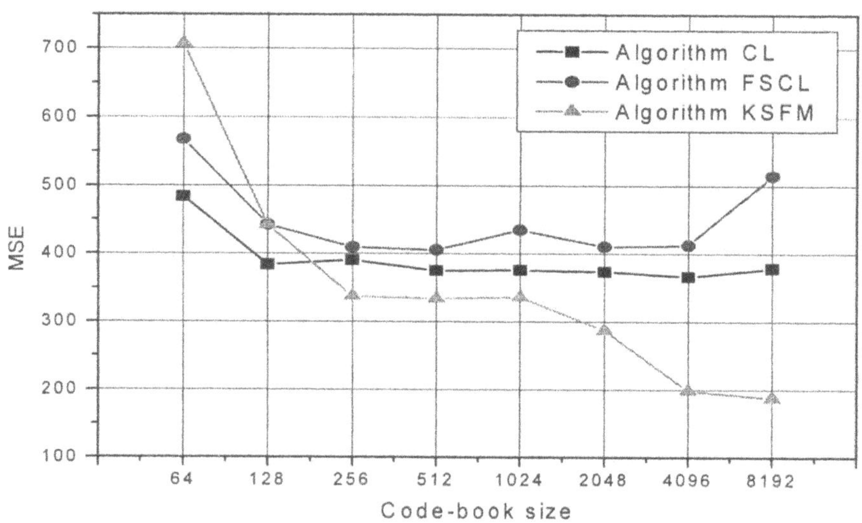

FIGURE 12.3. "Open-loop" design for the code-book size = 512 (MSE versus code-book size)

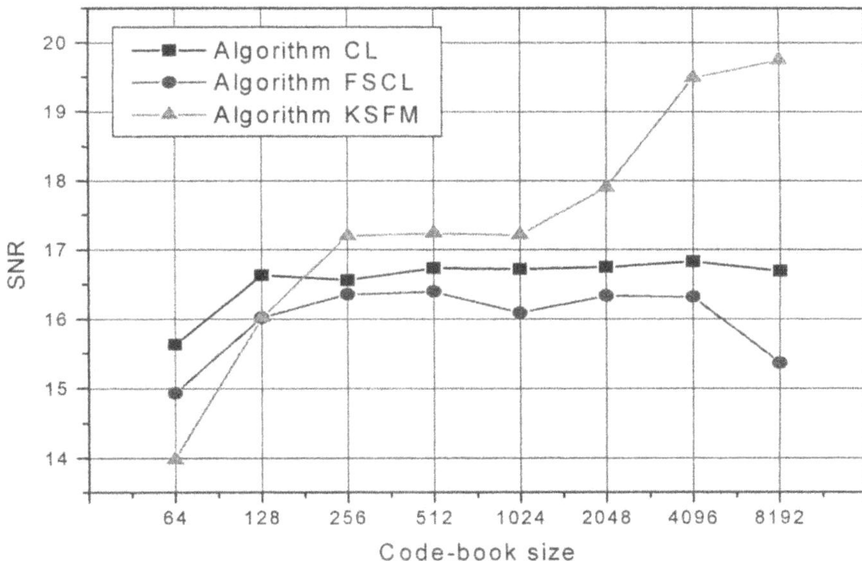

FIGURE 12.4. "Open-loop" design (SNR versus code-book size)

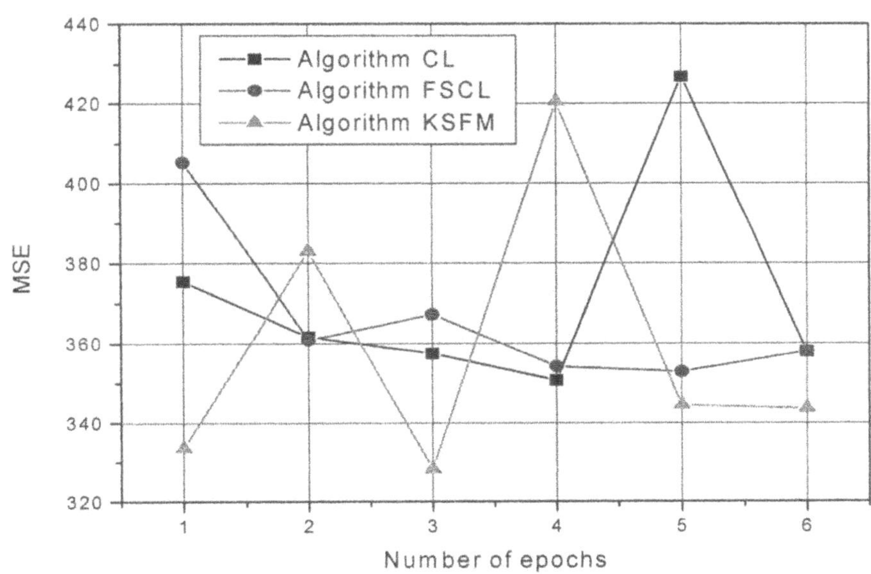

FIGURE 12.5. "Open-loop" design for the code-book size = 512 (MSE versus number of epochs)

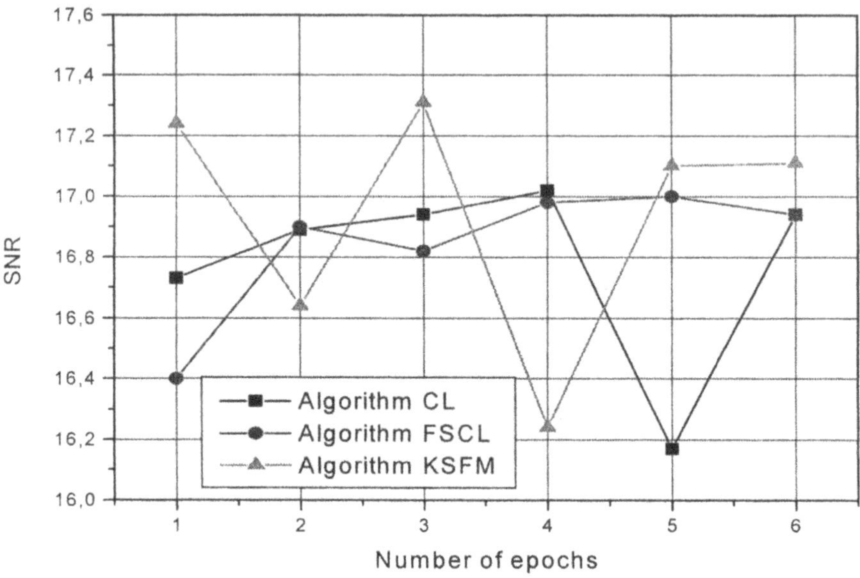

FIGURE 12.6. "Open-loop" design for the code-book size = 512 (SNR versus number of epochs)

FIGURE 12.7. The reconstructed "Lena" image-open loop design

Fig. 12.8 presents the difference between the reconstructed and original images.

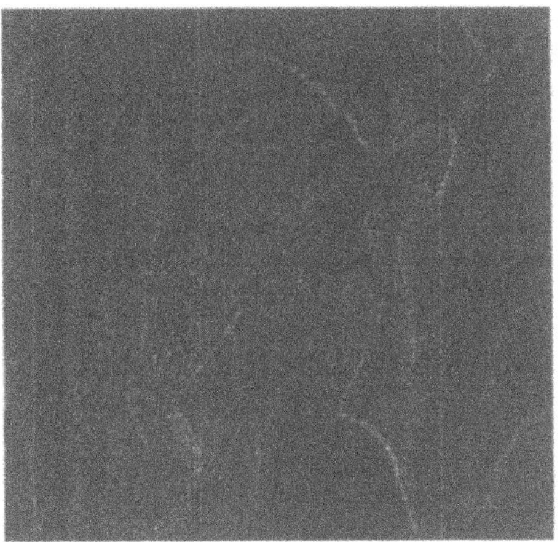

FIGURE 12.8. The difference between the reconstructed and original images: "open- loop" design

12.3 Closed-loop design

As previously, first the predictor coefficients (matrices) are determined in the open-loop scheme. Then the starting code-book $\mathbf{G}(0) = [\mathbf{g}_0(0), \mathbf{g}_1(0), ..., \mathbf{g}_j(0), ..., \mathbf{g}_J(0)]$, $j = 0, 1, ..., J$ is set. In this moment, the process of the learning sequence generation starts according to the following equation:

$$\mathbf{E}(m, n; s) = \mathbf{V}(m, n) - \sum_{k=0}^{K}\sum_{l=0}^{L} \mathbf{A}_{kl}\widetilde{\mathbf{V}}(m-k, n-l; s), \qquad (12.2)$$

where:
$m = 1, 2, ..., N_1/n_1$, $n = 1, 2, ..., N_2/n_2$, $(k, l) \neq (0, 0)$,
$s = 0, 1, 2, ...$

Comparing with the "open-loop" scheme, the PVQ in the "closed-loop" is enriched with a feedback loop, whose counter is variable (index) s. The block-diagram of the "closed-loop" PVQ design is shown in Fig. 12.9.

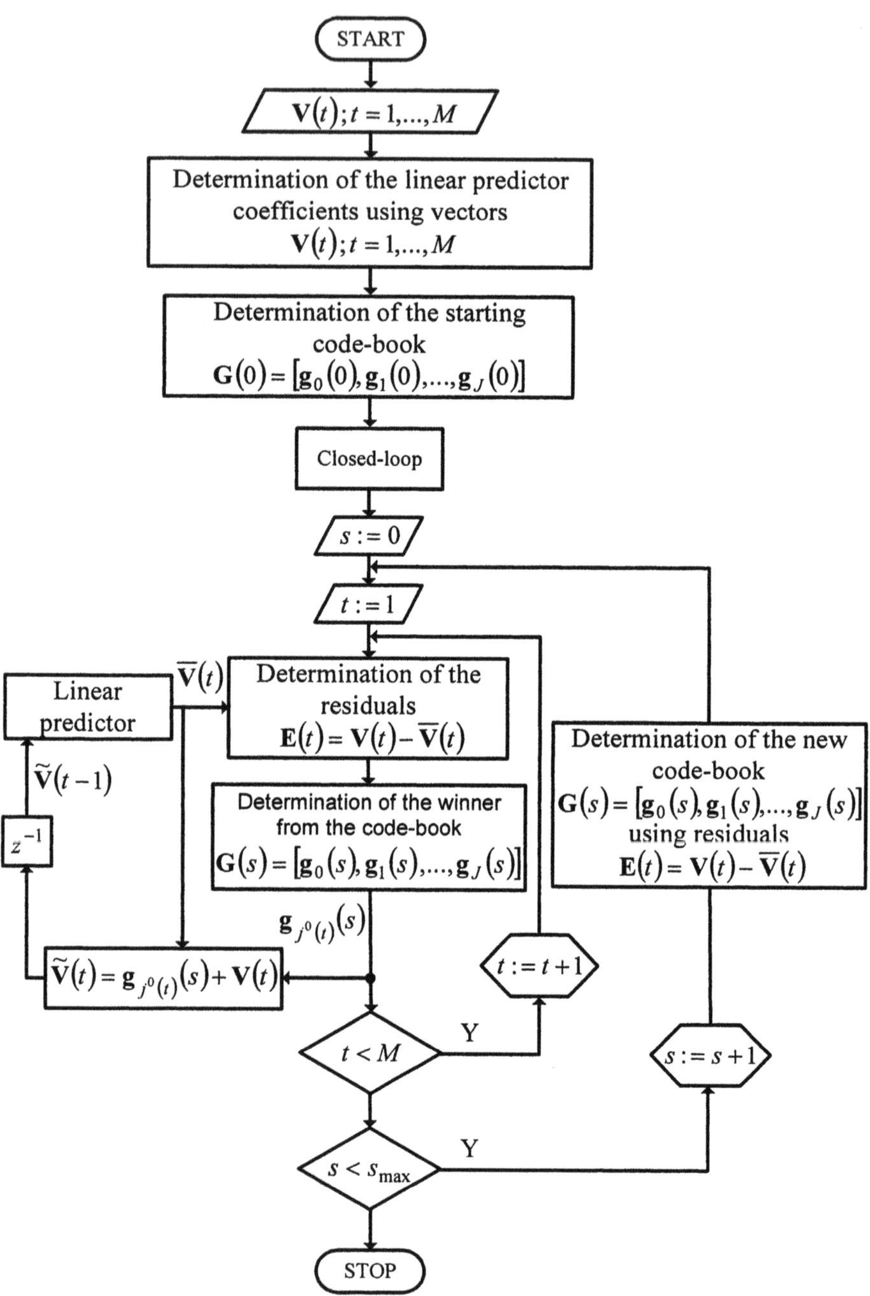

FIGURE 12.9. "Closed-loop" PVQ design

On the basis of the learning set $\mathbf{E}(m, n; s)$, the code-book $\mathbf{G}(s) = [\mathbf{g}_0(s), \mathbf{g}_1(s), ..., \mathbf{g}_J(s)]$, is set. To leave the loop, that is, to finish the process of designing the system parameters, the satisfaction of the inequality $D(s) \leq D_{\max}$ is necessary where $D(s)$ is calculated at every pass through the loop. The value of this error indicator can be calculated in a few possible variants, e.g. with a version of expression for the average square error that was adapted to this case as follows:

$$D(s) = \frac{\sum_{m=1}^{N_1/n_1} \sum_{n=1}^{N_2/n_2} \sum_{i=1}^{q} \left[v_i(m,n) - \widetilde{v}_i(m,n,s)\right]^2}{N_1 \cdot N_2} \tag{12.3}$$

Alternatively, the design process is terminated when $s < s_{\max}$, as it is shown in Fig. 12.9. The method of designing the parameters with the "closed-loop" is characterized by a partial elimination of the influence of the starting code-book $\mathbf{G}(0)$ on the accuracy of the determination of prototype vectors $\mathbf{g}_j(s)$ because this process is continued until the distribution of the prototype vectors in the space of vectors $\mathbf{E}(m, n)$ reaches optimum. It is hard to say whether this means achieving minimal measure D in a local or global sense but it will be an improvement in comparision with the "closed-loop" design.

In Fig. 12.10 and Fig. 12.11 we present the results of simulations for a varying code-book size (MSE and SNR versus code-book size). Figures 12.12 and 12.13 illustrate the dependence of MSE and SNR on the number of epochs (determination of the new code-book in Fig. 12.9) for the code-book size = 512 and the CL algorithm. In Fig. 12.14 and Fig. 12.15 we illustrate the dependence of MSE and SNR on the number of epochs for the code-book size = 512 and the FSCL algorithm.

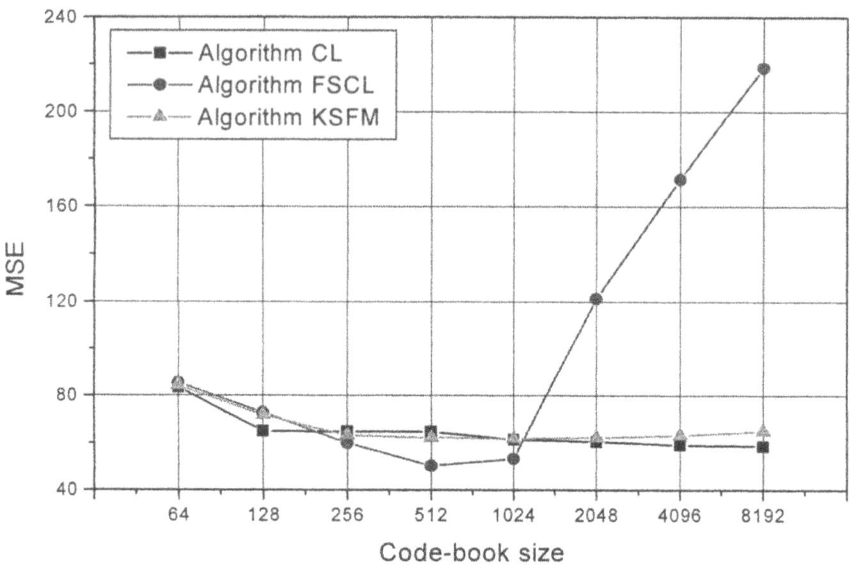

FIGURE 12.10. "Closed-loop" design (MSE versus code-book size)

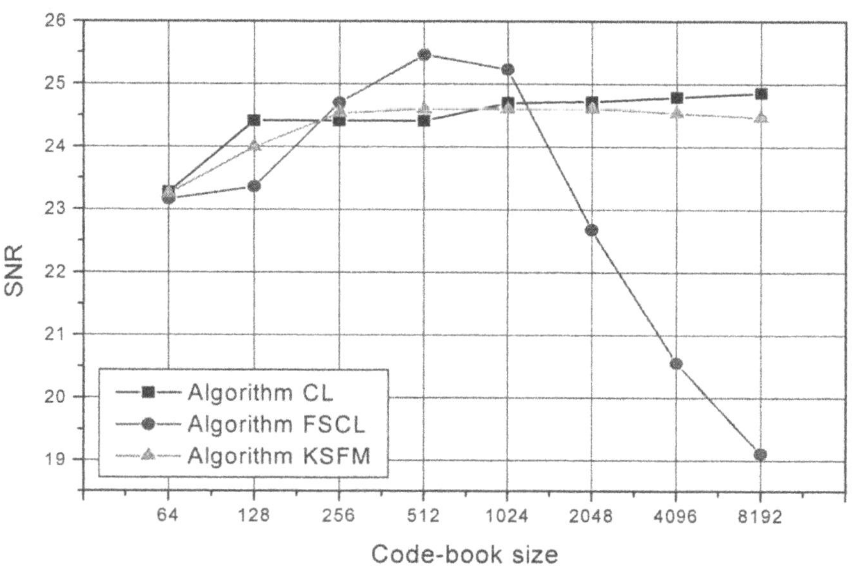

FIGURE 12.11. "Closed-loop" design (SNR versus code-book size)

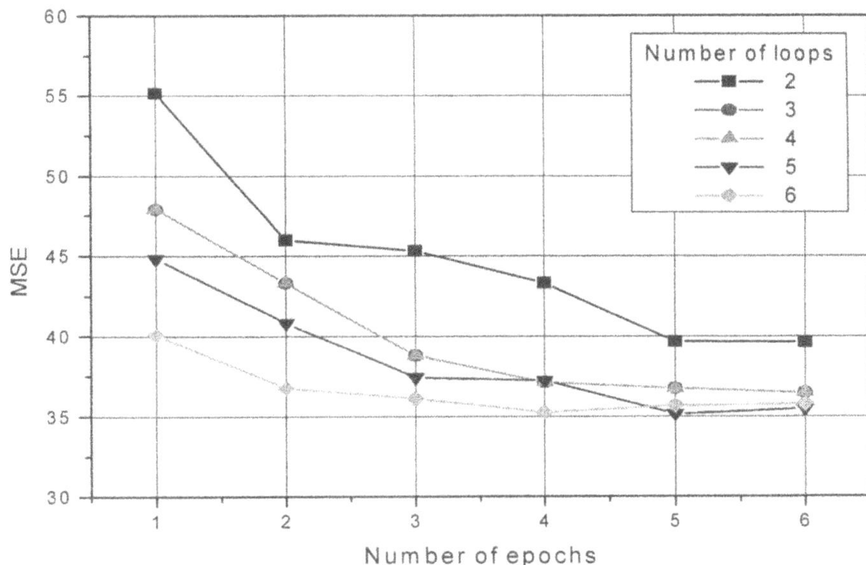

FIGURE 12.12. "Closed-loop" design (MSE versus number of epochs) with the CL algorithm and code-book size = 512

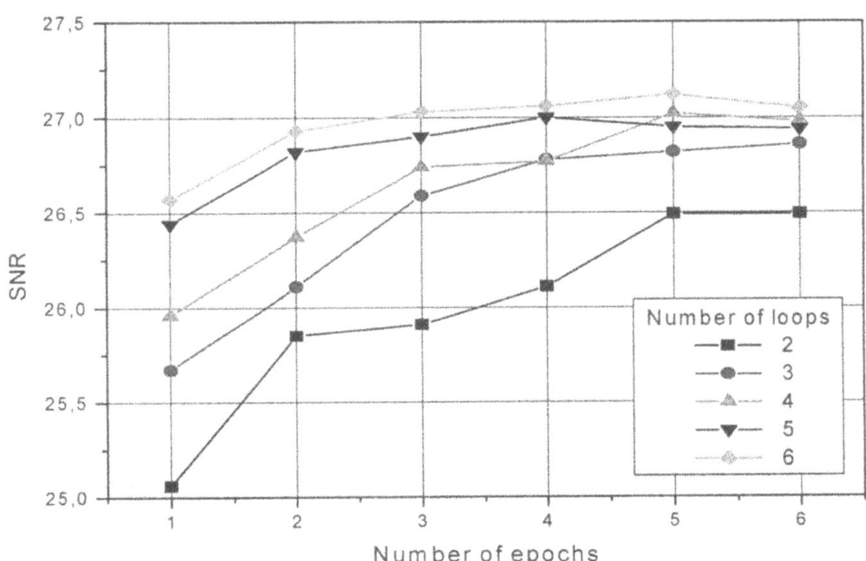

FIGURE 12.13. "Closed-loop" design (SNR versus number of epochs) with the CL algorithm and code-book size = 512

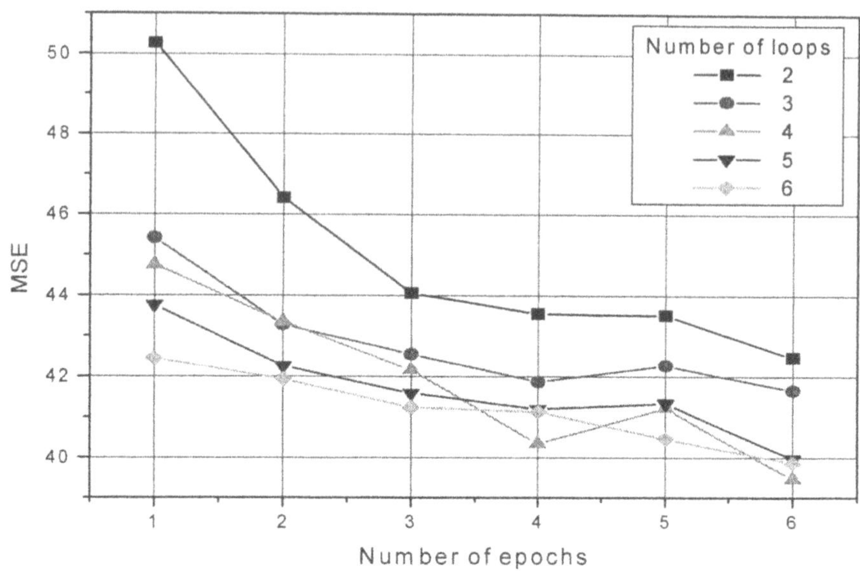

FIGURE 12.14. "Closed-loop" design (MSE versus number of epochs) with the FSCL algorithm and code-book size = 512

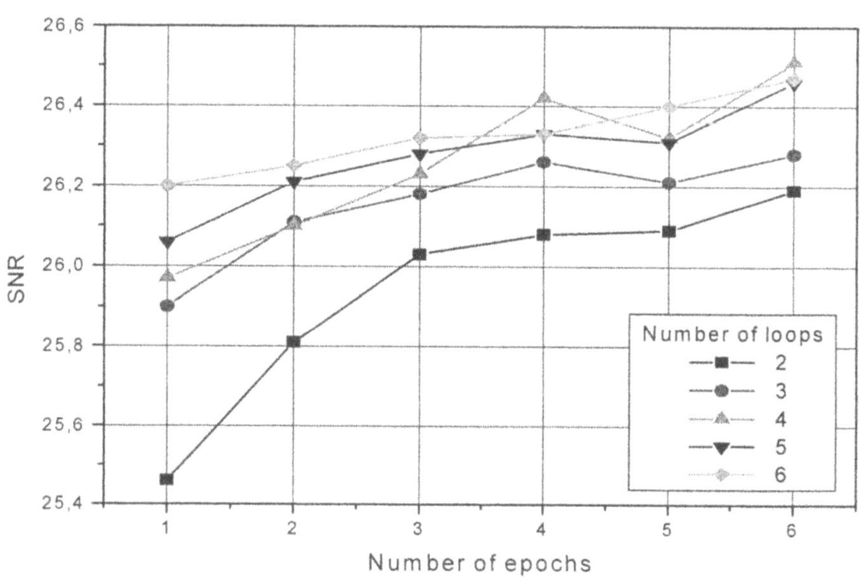

FIGURE 12.15. "Closed-loop" design (SNR versus number of epochs) with the FSCL algorithm and code-book size = 512

In Fig. 12.16 we show the reconstructed "Lena" image for the code-book size = 2048, one epoch, two loops and the FSCL algorithm. In this case MSE = 121.4 and SNR = 22.68. Figure 12.17 presents the difference between the reconstructed and original images.

FIGURE 12.16. The reconstructed "Lena" image: closed loop design

FIGURE 12.17. The difference between the reconstructed and original images: closed- loop design

In Fig. 12.18 and Fig. 12.19 we illustrate the dependence of MSE and SNR on the number of epochs for the code-book size = 512 and the KSFM algorithm.

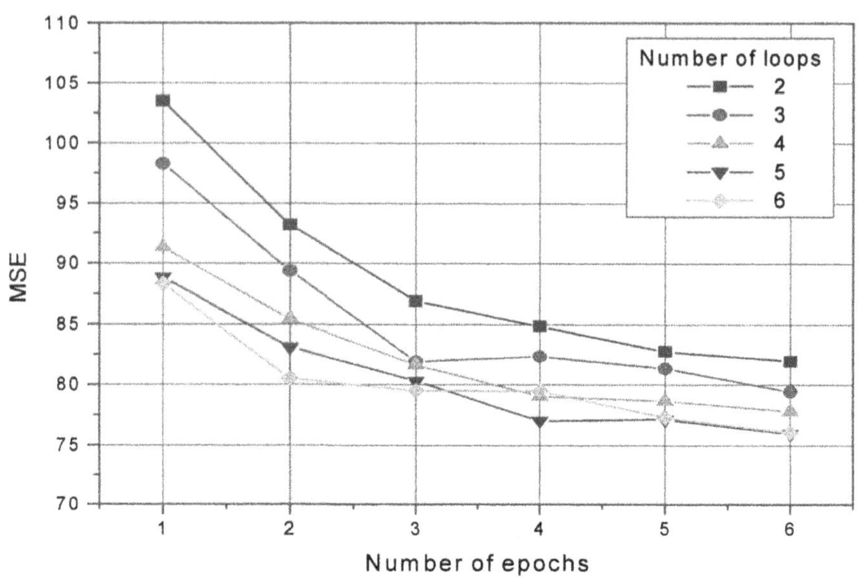

FIGURE 12.18. "Closed-loop" design (MSE versus number of epochs) with the KSFM algorithm and code-book size = 512

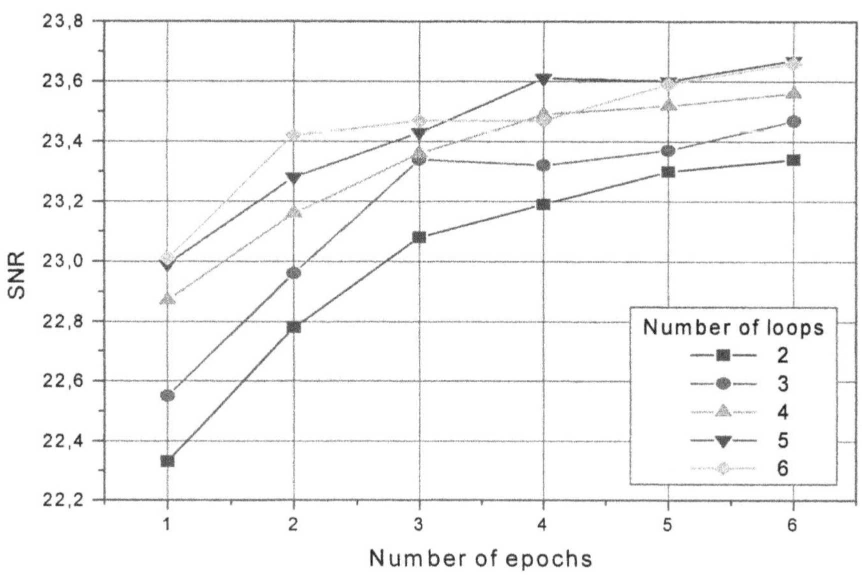

FIGURE 12.19. "Closed-loop" design (SNR versus number of epochs) with the KSFM algorithm and code-book size = 512

12.4 Modified closed-loop design

The algorithm for designing parameters in the VQ DPCM system with the "modified closed-loop" is shown in Fig. 12.20. The only difference between this algorithm and the system with the "closed-loop" is that the code-book is modified every time when the learning difference $\mathbf{E}(m, n; s)$ is generated.

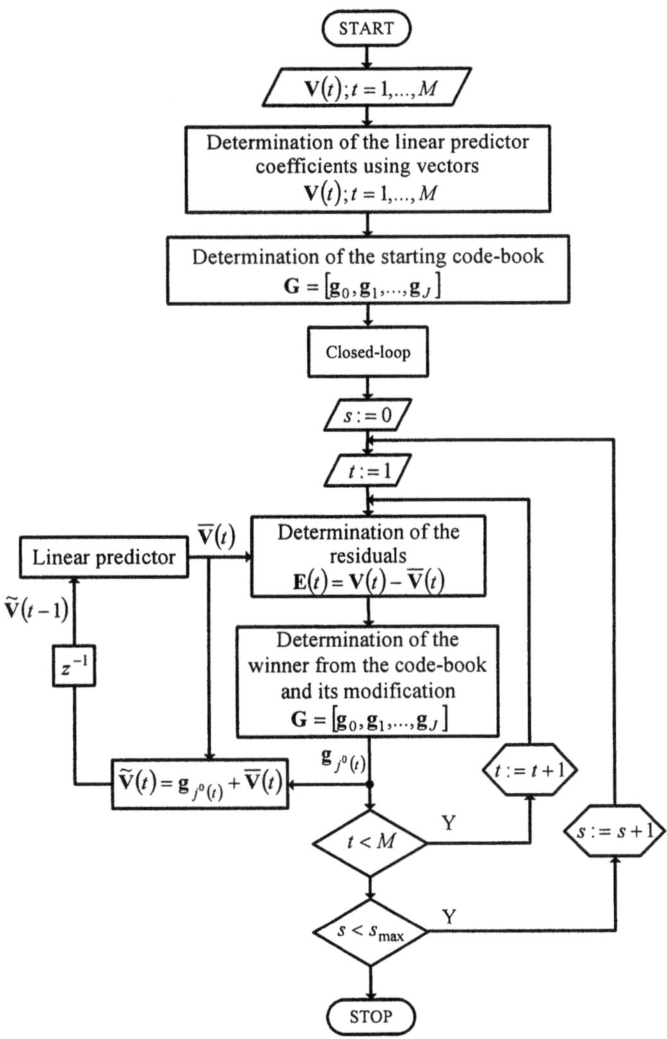

FIGURE 12.20. "Modified closed-loop" PVQ design

One advantage of the system with the "modified closed-loop" over the standard scheme with the "closed-loop" is based on the fact that the code-book is modified more often which leads to the conclusion that at a given moment the code-book is better adjusted to the input vectors. The outer loop, whose activity depends on the degree of minimizing of the reproduction error in the whole image, is stopped after achieving a satisfactory error level or after it has passed a fixed number of iterations $s_{\max}$, as it was the case in the system with the "closed-loop".

In Fig. 12.21 and Fig. 12.22 we present the results of simulations for a varying code-book size (MSE and SNR versus code-book size).

Figures 12.23 and 12.24 illustrate the dependence of MSE and SNR on the number of loops for the code-book size = 512.

In Fig. 12.25 we show the reconstructed "Lena" image for the code-book size = 512, twenty loops and the KFSM algorithm. I this case MSE = 33.29 and SNR = 27.25. Figure 12.26 presents the difference between the reconstructed and original images.

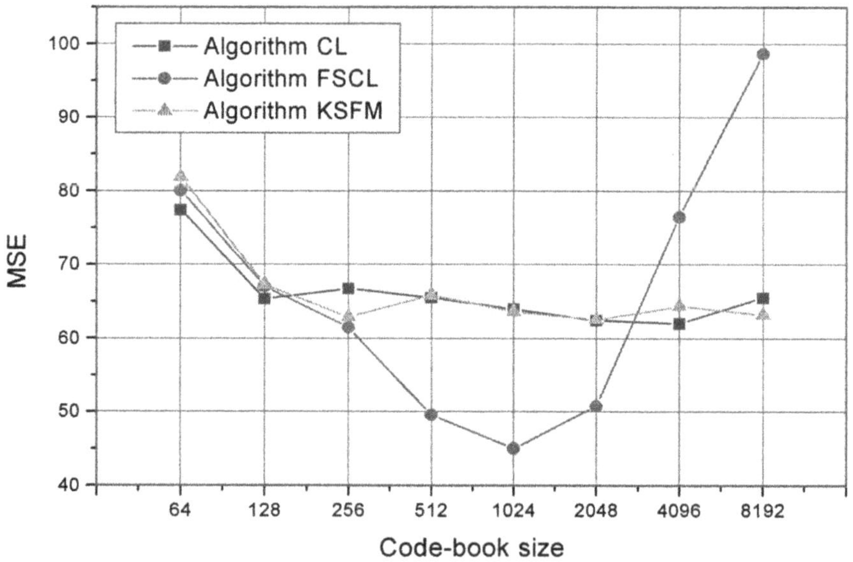

FIGURE 12.21. "Modified closed-loop" design (MSE versus code-book size)

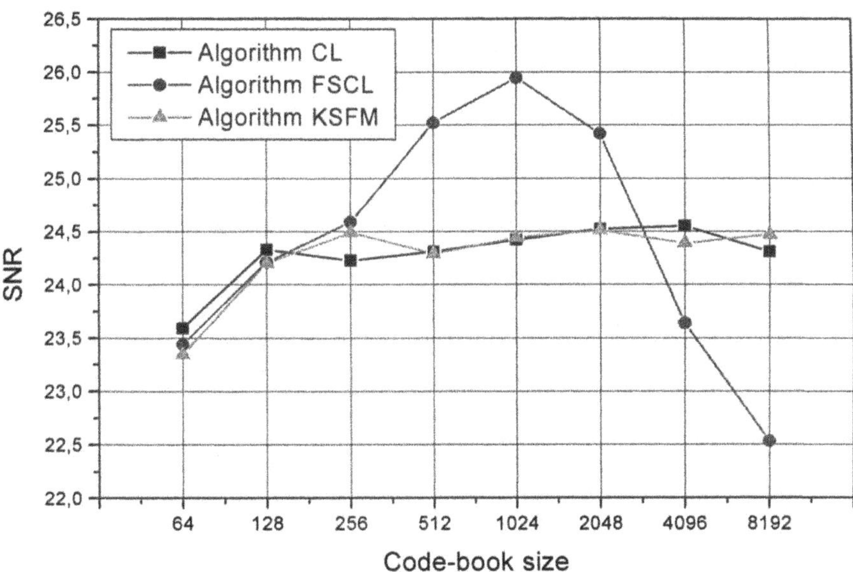

FIGURE 12.22. "Modified closed-loop" design (SNR versus code-book size)

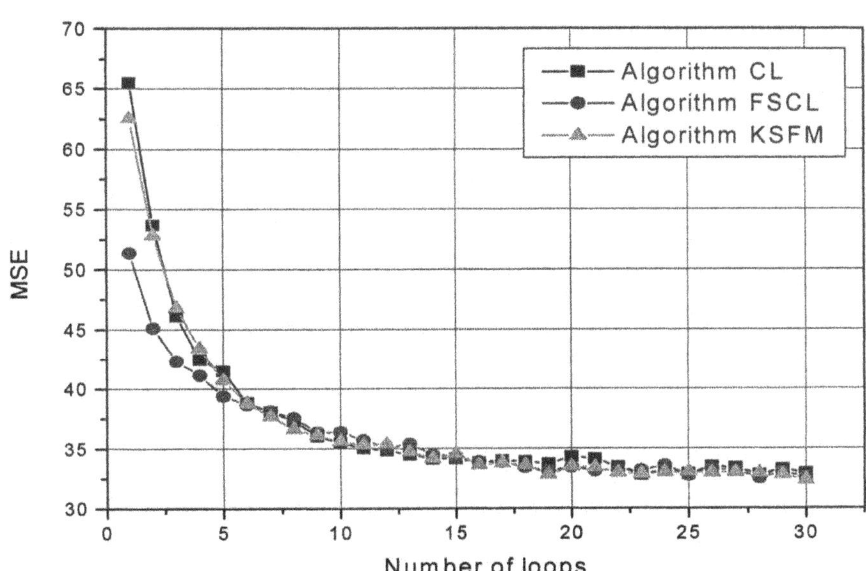

FIGURE 12.23. "Modified closed-loop" design (SNR versus number of loops) for the code-book size = 512

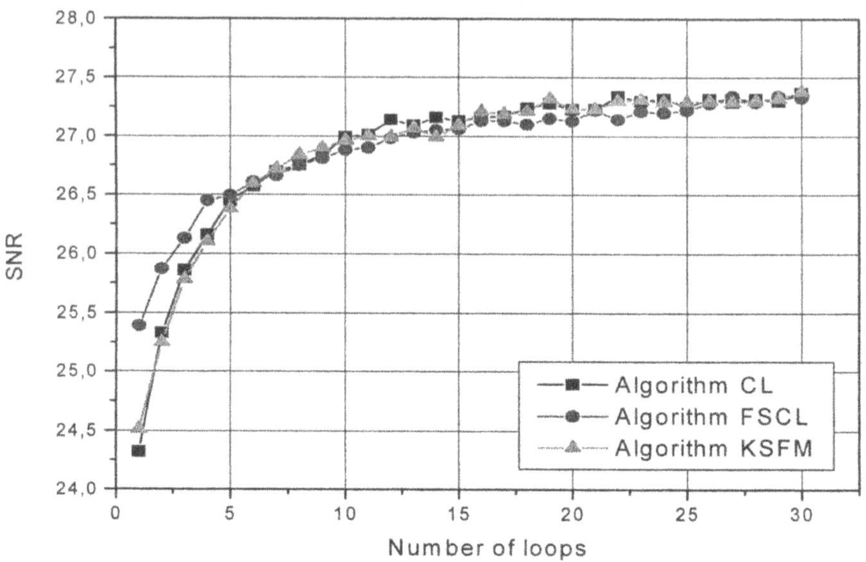

FIGURE 12.24. "Modified closed-loop" design (SNR versus number of loops) for the code-book size = 512

FIGURE 12.25. The reconstructed "Lena" image: modified closed-loop design

FIGURE 12.26. The difference between the reconstructed and original images: "modified closed-loop" design

12.5 Neural PVQ design

In Fig. 12.27 we present the PVQ block diagram with a neural predictor. Comparing with the classical PVQ scheme shown in Fig. 9.1, a linear vector predictor has been replaced by a neuro-predictor in the form of the feed-forward neural network trained by the back-propagation method. The predictor is described in detail in Section 10.4. In Fig. 12.28 we show the "modified closed-loop" PVQ design with the neural predictor. Observe that like in scheme depicted in Fig. 12.20, the code-book is modified every time when the learning difference $\mathbf{E}(m, n; s)$ is generated. Moreover, the neuro-predictor is trained after every epoch of the lenght M.

In Fig. 12.29 and Fig. 12.30 we show the dependence of MSE and SNR on the number of loops for the code-book size = 512

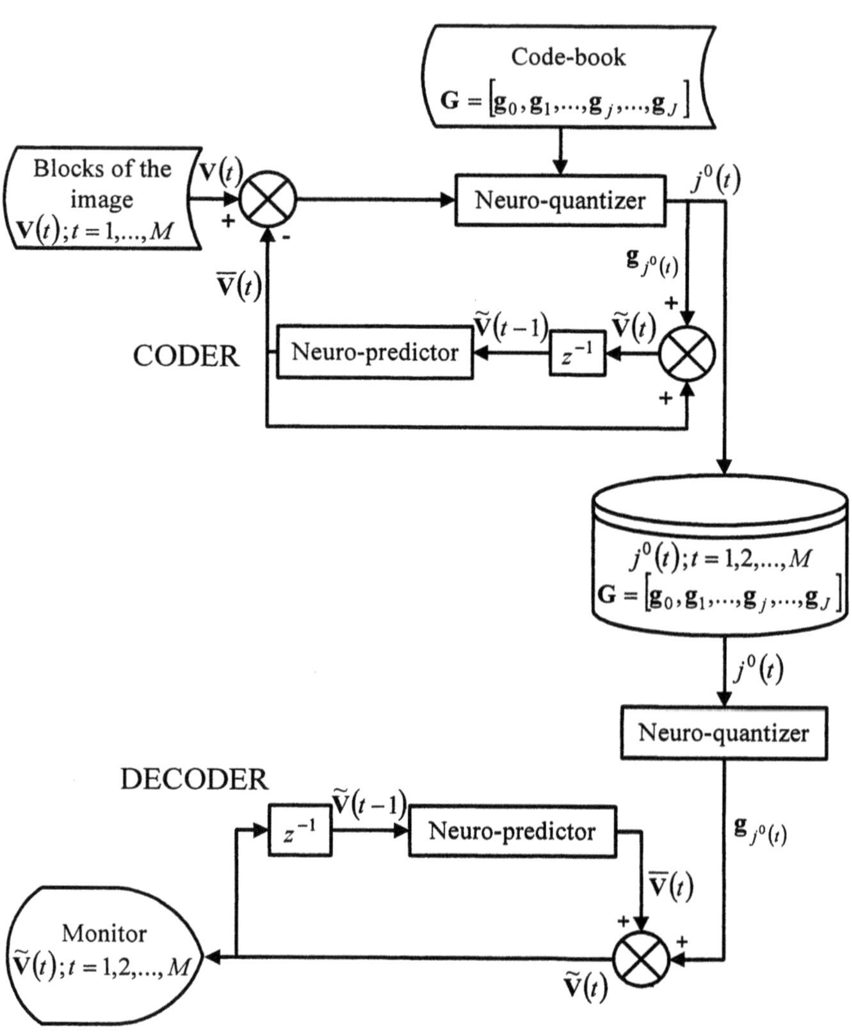

FIGURE 12.27. The PVQ block diagram with neural predictor

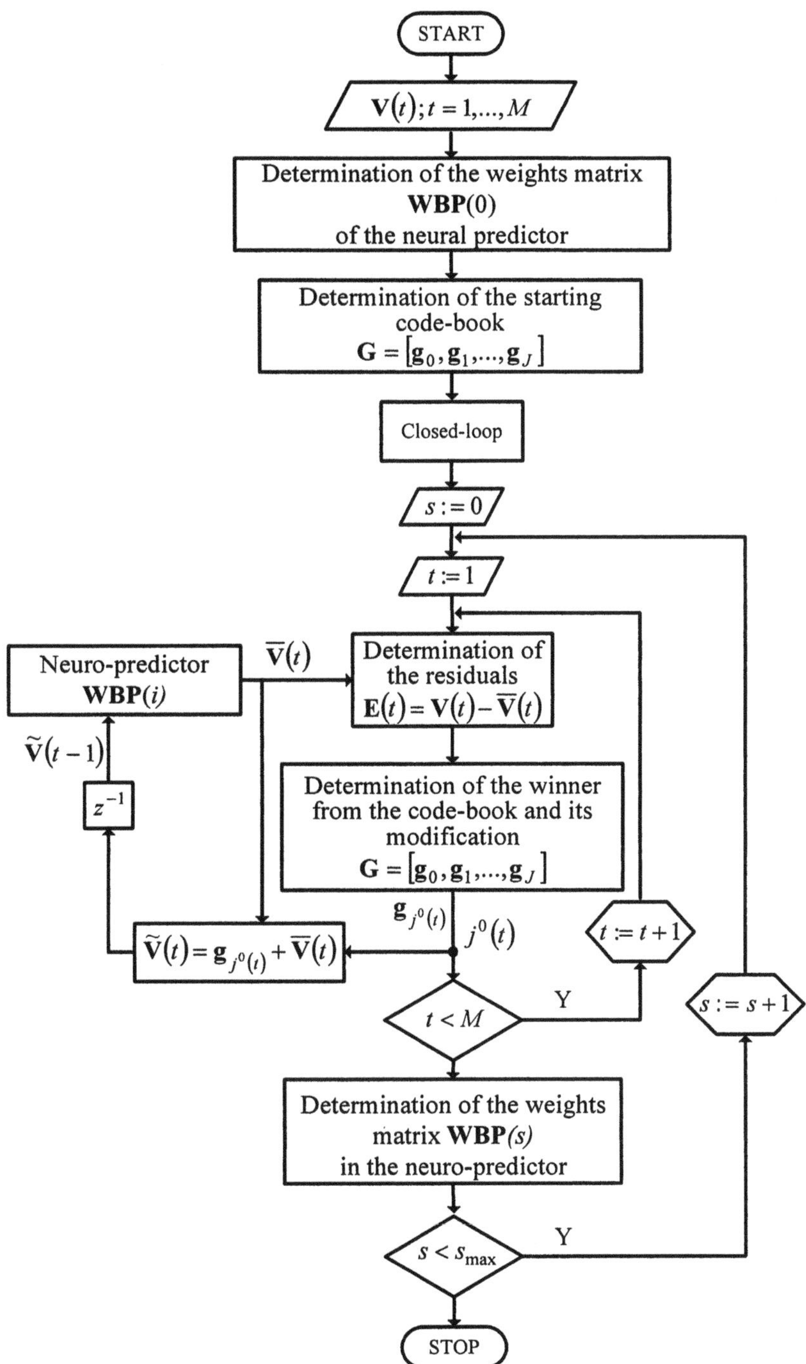

FIGURE 12.28. "Modified closed-loop" PVQ design with neural predictor

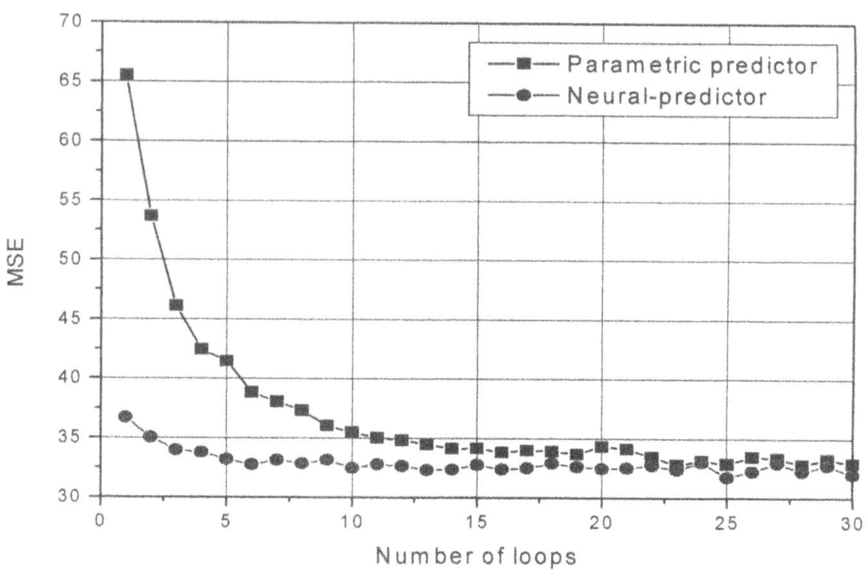

FIGURE 12.29. Neural PVQ design (MSE versus number of loops) for the code-book size = 512

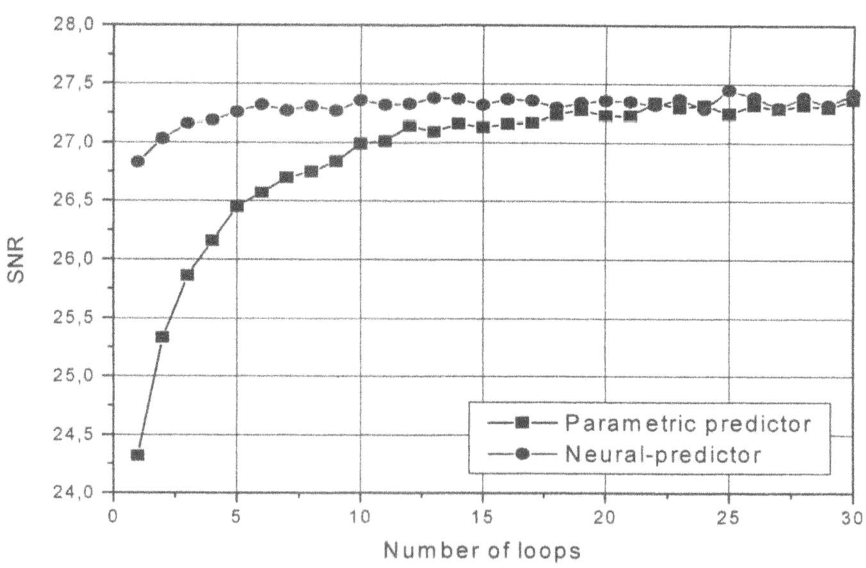

FIGURE 12.30. Neural PVQ design (SNR versus number of loops) for the code-book size = 512

12.6 Concluding remarks

In this chapter we have tested various schemes for the PVQ design. The simplest method is the open-loop design. It leads to the worse performance (MSE and SNR measures) of the image compression algorithm. The closed-loop design gives a significant improvement over the open-loop design. The modified closed-loop design works slightly better than the closed-loop design. The neural design gives almost the same results as modified closed-loop design if the number of loops in both PVQ schemes is sufficiently large.

Part III

Recursive Least Squares Methods for Neural Network Learning and their Systolic Implementations

13
A Family of the RLS Learning Algorithms

13.1 Introduction

In the last decade various methods have been developed for the feed-forward neural network training. They include the following schemes:

(i) classical back – propagation learning algorithms, its momentum version and some modifications [1], [49], [78], [82], [115], [117], [124], [128], [142], [174], [176], [185], [238], [244], [245], [273], [281], [282], [294], [332], [333],

(ii) heuristic algorithms designed to accelerate the convergence of the back-propagation algorithm [198], [269],

(iii) conjugate gradient-based algorithms [35], [165],

(iv) second ordered-algorithms including a powerful-Marquardt algorithm [24], [36], [106], [107], [151],

(v) combination of the BP with the layer-wise optimization method [205],

(vi) recursive least squares methods [26], [33], [34],

(vii) the extended Kalman filter (EKF) techniques [123], [152], [188], [239], [242], [266], [267], [330].

Other interesting results in this area can be found in [5], [27], [28], [29], [54], [60], [74], [97], [118], [119], [130], [131], [137], [143], [150], [167], [199], [240], [251], [259], [273], [298]. Despite of so many techniques further improvement is highly desirable with regard to numerical stability, accuracy, generalization capability and computational complexity.

In this chapter we develop a family of neural networks learning algorithms based on the recursive least square procedures adopted from the filters theory [265]. The main results of the chapter are summarized as follows:

1. We study the classical RLS algorithms and their robust counterparts, called QQ^T– RLS and UD – RLS algorithms, which are less liable to round-up error accumulation within the classical RLS algorithm.

2. In the paper each algorithm is derived for two different cases: the error is determined in the linear part of the neurons (Error Transferred Back – ETB) or "as usually" in the back propagation neural networks. The motivation to study the ETB-type neural networks stems from the fact that in several papers [239] authors indicated an improved convergence rate.

3. We assume that the net is completely connected in the pyramid structure: each neuron in any layer receives inputs from neurons in all previous layers (not only from the directly preceding layer). We will investigate and demonstrate increased computational abilities of the pyramid network. A new insight into the Vapnic-Chervonenkis dimension in the case of the classical back-propagation (see [120]) gives a good recommendation for completely connected nets trained by the RLS algorithms.

13.2 Notation

In this section we describe a classical feed-forward structure (not fully connected) and introduce a notation for the i-th neuron in the k-th layer. The following terminology will be used:

L – number of layers in the network

N_k –number of neurons in the k-th layer, $k = 1, ..., L$

N_0 –number of inputs of the neural networks

$\mathbf{u} = [u_1, ..., u_{N_0}]^T$– vector of input signals of the neural network

$y_i^{(k)}$ – the output signal of the i-th neuron, $i = 1, ..., N_k$, in the k-th layer, $k = 1, ..., L$,

$y_i^{(k)}(n) = f\left(s_i^{(k)}(n)\right)$

$\mathbf{y}^{(k)} = \left[y_i^{(k)}, ..., y_{N_k}^{(k)}\right]^T$ – vector of output signals of the k-th layer, $k = 1, ..., L$,

$x_i^{(k)}$ – the i-th input, $i = 0, ..., N_{k-1}$, for the k-th layer, $k = 1, ..., L$, where

$$x_i^{(k)} = \begin{cases} u_i & \text{for} \quad k = 1 \\ y_i^{(k-1)} & \text{for} \quad k = 2, ..., L \\ 1 & \text{for} \quad i = 0, \ k = 1, ..., L \end{cases}$$

$\mathbf{x}^{(k)} = \left[x_0^{(k)}, ..., x_{N_{k-1}}^{(k)}\right]^T$ – vector of input signals of the k-th layer, $k = 1, ..., L$

$w_{ij}^{(k)}$ – weight of the i-th neuron, $i = 1, ..., N_k$, of the k-th layer, $k = 1, ..., L$, connecting this neuron with the j-th input $x_j^{(k)}, j = 0, ..., N_{k-1}$

$\mathbf{w}_i^{(k)} = \left[w_{i0}^{(k)}, ..., w_{iN_{k-1}}^{(k)}\right]^T$ – vector of weights of the i-th neuron, $i = 1, ..., N_k$, in the k-th layer, $k = 1, ..., L$

$s_i^{(k)}(n) = \sum_{j=0}^{N_{k-1}} w_{ij}^{(k)}(n-1)\, x_j^{(k)}(n)$ – the linear output of the i-th neuron, $i = 1, ..., N_k$, in the k-th layer, $k = 1, ..., L$

$\mathbf{s}^{(k)} = \left[s_1^{(k)}, ..., s_{N_k}^{(k)}\right]^T$ – vector of linear outputs in the k-th layer, $k = 1, ..., L$

$d_i^{(k)}$ – the desired output of the i-th neuron, $i = 1, ..., N_k$, in the k-th layer, $k = 1, ..., L$

$\mathbf{d}^{(k)} = \left[d_i^{(k)}, ..., d_{N_k}^{(k)}\right]^T$ – vector of desired outputs in layer k, $k = 1, ..., L$

$b_i^{(k)} = f^{-1}\left(d_i^{(k)}\right)$ – the desired linear output of the i-th neuron,

$i = 1, \ldots, N_k$, in the k-th layer, $k = 1, \ldots, L$

$\mathbf{b}^{(k)} = \left[b_i^{(k)}, \ldots, b_{N_k}^{(k)}\right]^T$ – vector of desired linear outputs in layer k, $k = 1, \ldots, L$

$\varepsilon_i^{(k)}(n) = d_i^{(k)}(n) - y_i^{(k)}(n)$ – the error of the i-th neuron, $i = 1, \ldots, N_k$, in the k-th layer, $k = 1, \ldots, L$

$\boldsymbol{\varepsilon}^{(k)} = \left[\varepsilon_i^{(k)}, \ldots, \varepsilon_{N_k}^{(k)}\right]^T$ – vector of the error signals in the k-th layer, $k = 1, \ldots, L$

$e_i^{(k)}(n) = b_i^{(k)}(n) - f'\left(y_i^{(k)}(n)\right)$ – the error of the linear part of the i-th neuron, $i = 1, \ldots, N_k$, in the k-th layer, $k = 1, \ldots, L$

$\mathbf{e}^{(k)} = \left[e_1^{(k)}, \ldots, e_{N_k}^{(k)}\right]^T$ – vector of the error signals in the k-th layer, $k = 1, \ldots, L$

λ – forgetting factor in the RLS algorithm

μ – learning coefficient of the back-propagation (BP) algorithm

α – momentum coefficient of the momentum back-propagation (MBP) algorithm

In Fig. 13.1 we show a model of the i-th neuron in the k-th layer.

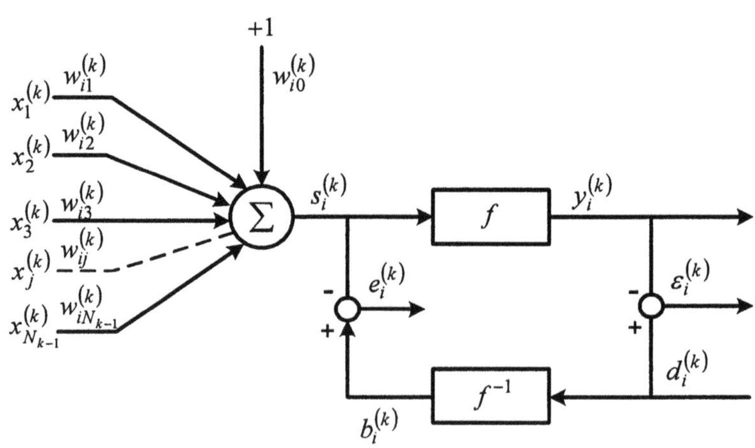

FIGURE 13.1. Model of the i-th neuron in the k-th layer

In descriptions of single layer networks the index k will be omitted. The above notation will be supplemented in the next section to describe the pyramid neural network structure.

13.3 Problem description

For the clarity of presentations we derive each algorithm starting from a single-layer neural network having a linear transfer function. Derivations for a single-layer network having a non-linear transfer function and multi-layer networks are a consequence of that single layer linear case. To be more specific we formalize our investigations as follows (the notation is given in Section 13.2):

a) For the single layer NN we derive RLS algorithms based on the performance measure:

$$\begin{aligned} J_1(n) &= \sum_{t=1}^{n} \lambda^{n-t} \sum_{j=1}^{N} \varepsilon_j^2(t) \\ &= \sum_{t=1}^{n} \lambda^{n-t} \sum_{j=1}^{N} \left[d_j(t) - \mathbf{x}^T(t)\, \mathbf{w}_j(n)\right]^2, \end{aligned} \tag{13.1}$$

$$\begin{aligned} J_2(n) &= \sum_{t=1}^{n} \lambda^{n-t} \sum_{i=1}^{N} e_i^2(t) \\ &= \sum_{t=1}^{n} \lambda^{n-t} \sum_{i=1}^{N} \left[b_i(t) - \mathbf{x}^T(t)\, \mathbf{w}_i(n)\right]^2, \end{aligned} \tag{13.2}$$

where

$$b_i(t) = f^{-1}(d_i(t)) = \begin{cases} \ln \frac{d_i(t)}{1-d_i(t)} \text{ for the sigmoidal transfer function} \\ \frac{1}{2} \ln \frac{1+d_i(t)}{1-d_i(t)} \text{ for the tangent hyperbolic transfer function} \end{cases} \tag{13.3}$$

$$\begin{aligned} J_3(n) &= \sum_{t=1}^{n} \lambda^{n-t} \sum_{j=1}^{N} \varepsilon_j^2(t) \\ &= \sum_{t=1}^{n} \lambda^{n-t} \sum_{j=1}^{N} \left[d_j(t) - f\left(\mathbf{x}^T(t)\, \mathbf{w}_j(n)\right)\right]^2, \end{aligned} \tag{13.4}$$

Obviously J_1 relates to the linear neuron, J_2 to the nonlinear neuron with the error defined in the linear part whereas J_3 is the standard case of the sigmoidal neuron.

b) We derive learning algorithms minimizing performance measures J_1, J_2, J_3 assuming that the autocorrelation matrix is of the form

$$\mathbf{R}(n) = \sum_{t=1}^{n} \lambda^{n-t} \mathbf{x}(t)\, \mathbf{x}^T(t) \tag{13.5}$$

and next we study factorized forms on this matrix given by

$$\mathbf{R}(n) = \mathbf{Q}(n)\, \mathbf{Q}^T(n) \tag{13.6}$$

or

$$\mathbf{P}(n) = \mathbf{R}^{-1}(n) = \mathbf{U}(n)\, \mathbf{D}(n)\, \mathbf{U}^T(n) \tag{13.7}$$

Note that $\mathbf{Q}$ is a certain matrix [265] which is not related to the orthogonal matrix of the $\mathbf{QR}$ decomposition, $\mathbf{U}$ is an upper triangular matrix and $\mathbf{D}$ is a diagonal matrix.

c) We extend the results described in a) and b) to the case of a multi-layer network having a pyramid structure and derive updated equations minimizing the following objective functions:

$$\begin{aligned} J_5(n) &= \sum_{t=1}^{n} \lambda^{n-t} \sum_{j=1}^{N_L} \varepsilon_j^{(L)^2}(t) = \sum_{t=1}^{n} \lambda^{n-t} \sum_{j=1}^{N_L} \\ &\left[d_j^{(L)}(t) - \mathbf{x}^{(L)^T}(t)\, \mathbf{w}_j^{(L)}(n)\right]^2, \end{aligned} \tag{13.8}$$

$$\begin{aligned} J_4(n) &= \sum_{t=1}^{n} \lambda^{n-t} \sum_{j=1}^{N_L} e_j^{(L)^2}(t) = \sum_{t=1}^{n} \lambda^{n-t} \sum_{j=1}^{N_L} \\ &\left[b_j^{(L)}(t) - \mathbf{x}^{(L)^T}(t)\, \mathbf{w}_j^{(L)}(n)\right]^2, \end{aligned} \tag{13.9}$$

$$\begin{aligned} J_6(n) &= \sum_{t=1}^{n} \lambda^{n-t} \sum_{j=1}^{N_L} \varepsilon_j^{(L)^2}(t) = \sum_{t=1}^{n} \lambda^{n-t} \sum_{j=1}^{N_L} \\ &\left[d_j^{(L)}(t) - f\left(\mathbf{x}^{(L)^T}(t)\, \mathbf{w}_j^{(L)}(n)\right)\right]^2, \end{aligned} \tag{13.10}$$

We assume that each neuron in a pyramid structure has its own internal weights as it is shown in Fig.13.1

$$\left[w_{i0}^{(k)}, \ldots, w_{iN_{k-1}}^{(k)}\right]^T$$

and moreover external weights

$$\left[w_{iN_{k-1}+1}^{(k)}(n)\ldots w_{i(N_{k-1}+N_{k-2})}^{(k)}(n)\ldots w_{i(N_{k-1}+\ldots+N_0)}^{(k)}(n)\right]^T$$

connecting that neuron with all previous layers and neurons (including input signals). Combining internal and external weights we get vector $\mathbf{w}_i^{(k)}$ of weights $w_{ij}^{(k)}$ of the i-th neuron, $i = 1, ..., N_k$, in the k-th layer, $k = 1, \ldots, L$, given by

$$\mathbf{w}_i^{(k)}(n) = \begin{bmatrix} w_{i0}^{(k)}(n) \\ w_{i1}^{(k)}(n) \\ \vdots \\ w_{iN_{k-1}}^{(k)}(n) \\ w_{iN_{k-1}+1}^{(k)}(n) \\ \vdots \\ w_{i(N_{k-1}+N_{k-2})}^{(k)}(n) \\ \vdots \\ w_{i(N_{k-1}+\ldots+N_0)}^{(k)}(n) \end{bmatrix} \tag{13.11}$$

Observe that vector $\mathbf{w}_i^{(k)}$ consists of: bias $w_{i0}^{(k)}$, weights connecting the corresponding neuron with directly preceding layer and with all previous layers, and finally weights connecting that neuron with the inputs of the net.

In notation (13.11) weights $\mathbf{w}_{ij}^{(k)}$connect the i-th neuron in the k-th layer with the j-th input $x_j^{(k)}$, $j = 0, ..., \mathrm{N}_{k-1}, \mathrm{N}_{k-1} + 1, ..., \mathrm{N}_{k-1} + \mathrm{N}_{k-2}, ..., \mathrm{N}_{k-1} + ... \mathrm{N}_0$.We may redefine all weights in notation (13.11) as follows:

$$\mathbf{w}_i^{(k)}(n) = \begin{bmatrix} w_{i0}^{(k)}(n) \\ w_{i1}^{(k,k-1)}(n) \\ \vdots \\ w_{iN_{k-1}}^{(k,k-1)}(n) \\ w_{i1}^{(k,k-2)}(n) \\ \vdots \\ w_{iN_{k-2}}^{(k,k-2)}(n) \\ \vdots \\ w_{iN_0}^{(k,0)}(n) \end{bmatrix} \tag{13.12}$$

In the equivalent notation (13.12) weight $w_{im}^{(k,l)}$ connects the i-th neuron of the k-th layer with the m-th output of the l-th layer. It will be assumed that initial values $\mathbf{w}_i^{(k)}(0)$ of the weights are small random numbers. The input of the k-th layer takes the form:

$$\mathbf{x}_i^{(k)}(n) = \begin{bmatrix} 1 \\ x_1^{(k)}(n) \\ \vdots \\ x_{N_{k-1}}^{(k)}(n) \\ x_{N_{k-1}+1}^{(k)}(n) \\ \vdots \\ x_{(N_{k-1}+N_{k-2})}^{(k)}(n) \\ \vdots \\ x_{(N_{k-1}+\ldots+N_0)}^{(k)}(n) \end{bmatrix} = \begin{bmatrix} 1 \\ y_1^{(k-1)}(n) \\ \vdots \\ y_{N_{k-1}}^{(k-1)}(n) \\ y_1^{(k-2)}(n) \\ \vdots \\ y_{N_{k-2}}^{(k-2)}(n) \\ \vdots \\ x_{N_0}^{(0)}(n) \end{bmatrix} \tag{13.13}$$

This chapter is organized as it is shown in Table 13.1 (SL-stands for single layer NN, ML- stands for multi-layer NN).

TABLE 13.1. Organization of Chapter 13

Algorithm	Architecture	Linear Neural Network	Nonlinear ETB Neural Network	Nonlinear Neural Network
RLS	SL	13.4.1a-J_1	13.4.1b-J_2	13.4.1c-J_3
	ML	13.4.2a-J_4	13.4.2b-J_5	13.4.2c-J_6
QQ^T-RLS	SL	13.5.1a-J_1	13.5.1.b-J_2	13.5.1c-J_3
	ML	13.5.2a-J_4	13.5.2b-J_5	13.5.2c-J_6
UD-RLS	SL	13.6.1a-J_1	13.6.1b-J_2	13.6.1c-J_3
	ML	13.6.2a-J_4	13.6.2b-J_5	13.6.2c-J_6

13.4 RLS learning algorithms

In this section we derive RLS learning algorithms for performance measures J_1, J_2, J_3 (single layer NN) and next J_4, J_5, J_6 (multi-layer NN). We do not assume factorization of autocorrelation matrices.

13.4.1 Single layer neural network

a) Objective function J_1

The performance index (13.1) is minimized by taking the partial derivative of $J_1(n)$ with respect to $\mathbf{w}_i = \left[w_{i0}, ..., w_{iN_{k-1}}\right]^T$ and setting it equal to zero, i.e.

$$\begin{aligned}\frac{\partial J(n)}{\partial \mathbf{w}_i(n)} &= \frac{\partial \sum\limits_{t=1}^{n} \lambda^{n-t} \sum\limits_{j=1}^{N} \varepsilon_j^2(j)}{\partial \mathbf{w}_i(n)} \\ &= \frac{\partial \sum\limits_{t=1}^{n} \lambda^{n-t} \sum\limits_{j=1}^{N} \left[d_j(t) - \mathbf{x}^T(t)\mathbf{w}_j(n)\right]^2}{\partial \mathbf{w}_i(n)} \\ &= -2\sum_{t=1}^{n} \lambda^{n-t} \left[d_i(t) - \mathbf{x}^T(t)\mathbf{w}_i(n)\right]\mathbf{x}(t) = \mathbf{0}\end{aligned} \tag{13.14}$$

where $\mathbf{0}$ is $\mathrm{N}_0 + 1$ dimensional zero vector.

We get the normal equation

$$\sum_{t=1}^{n} \lambda^{n-t} \left[d_i(t) - \mathbf{x}^T(t)\mathbf{w}_i(n)\right]\mathbf{x}(t) = \mathbf{0} \tag{13.15}$$

which is presented in the vector form

$$\mathbf{r}_i(n) = \mathbf{R}(n)\,\mathbf{w}_i(n) \tag{13.16}$$

where

$$\mathbf{R}(n) = \sum_{t=1}^{n} \lambda^{n-t}\mathbf{x}(t)\,\mathbf{x}^T(t) \tag{13.17}$$

$$\mathbf{r}_i(n) = \sum_{t=1}^{n} \lambda^{n-t} d_i(t)\,\mathbf{x}(t) \tag{13.18}$$

Obviously $\mathbf{R}$ is the autocorrelation matrix of the input signals and $\mathbf{r}$ is the correlation vector between the input signal and the desired

signal.
The normal equation can be solved taking the inverse

$$\mathbf{w}_i(n) = \mathbf{R}^{-1}(n)\,\mathbf{r}_i(n) \tag{13.19}$$

if $\det \mathbf{R}(n) \neq 0$.

We will apply the RLS algorithm in order to avoid the inverse operation in (13.19) and solve equation (13.16) recursively.
Observe that matrix $\mathbf{R}_i^{(k)}(n)$ and vector $\mathbf{r}_i^{(k)}(n)$ can be presented in the form

$$\mathbf{R}(n) = \lambda\mathbf{R}(n-1) + \mathbf{x}(n)\,\mathbf{x}^T(n) \tag{13.20}$$

$$\mathbf{r}_i(n) = \lambda\mathbf{r}_i(n-1) + \mathbf{x}(n)\,d_i(n) \tag{13.21}$$

We will use the matrix inversion lemma (also known as the Sherman-Morrison identity).

Theorem 13.1
If positive definite matrices **A** and **B** (N×N dimensional) satisfy

$$\mathbf{A} = \mathbf{B}^{-1} + \mathbf{C}\mathbf{D}^{-1}\mathbf{C}^T \tag{13.22}$$

where **C** is a N×M dimensional matrix and **D** is positive definite M×M dimensional matrix, then

$$\mathbf{A}^{-1} = \mathbf{B} - \mathbf{B}\mathbf{C}\left(\mathbf{D} + \mathbf{C}^T\mathbf{B}\mathbf{C}\right)^{-1}\mathbf{C}^T\mathbf{B} \tag{13.23}$$

Comparing (13.22) and (13.20) we get

$$\begin{aligned} \mathbf{A} &= \mathbf{R}(n) \\ \mathbf{B}^{-1} &= \lambda\mathbf{R}(n-1) \\ \mathbf{C} &= \mathbf{x}(n) \\ \mathbf{D} &= 1 \end{aligned} \tag{13.24}$$

Therefore

$$\mathbf{P}(n) = \lambda^{-1}\left[\mathbf{I} - \mathbf{g}(n)\,\mathbf{x}^T(n)\right]\mathbf{P}(n-1) \tag{13.25}$$

where

$$\mathbf{P}(n) = \mathbf{R}^{-1}(n) \tag{13.26}$$

and

$$\mathbf{g}(n) = \frac{\mathbf{P}(n-1)\,\mathbf{x}(n)}{\lambda + \mathbf{x}^T(n)\,\mathbf{P}(n-1)\,\mathbf{x}(n)} \tag{13.27}$$

Moreover

$$\mathbf{g}(n) = \mathbf{P}(n)\,\mathbf{x}(n) \tag{13.28}$$

Equation (13.28) is a consequence of a simple algebra:

$$\mathbf{g}(n) = \frac{\mathbf{P}(n-1)\,\mathbf{x}(n)}{\lambda + \mathbf{x}^T(n)\,\mathbf{P}(n-1)\,\mathbf{x}(n)}$$

$$= \frac{\lambda^{-1}\left[\lambda\mathbf{P}(n-1)\,\mathbf{x}(n) + \mathbf{P}(n-1)\,\mathbf{x}(n)\,\mathbf{x}^T(n)\,\mathbf{P}(n-1)\,\mathbf{x}(n)\right]}{\lambda + \mathbf{x}^T(n)\,\mathbf{P}(n-1)\,\mathbf{x}(n)}$$

$$- \frac{\lambda^{-1}\left[\mathbf{P}(n-1)\,\mathbf{x}(n)\,\mathbf{x}^T(n)\,\mathbf{P}(n-1)\,x(n)\right]}{\lambda + \mathbf{x}^T(n)\,\mathbf{P}(n-1)\,\mathbf{x}(n)}$$

$$= \frac{\lambda^{-1}\left[\left(\lambda + \mathbf{x}^T(n)\,\mathbf{P}(n-1)\,\mathbf{x}(n)\right)\mathbf{I}\right]\mathbf{P}(n-1)\,\mathbf{x}(n)}{\lambda + \mathbf{x}^T(n)\,\mathbf{P}(n-1)\,\mathbf{x}(n)} \tag{13.29}$$

$$- \frac{\lambda^{-1}\left[\mathbf{P}(n-1)\,\mathbf{x}(n)\,\mathbf{x}^T(n)\right]\mathbf{P}(n-1)\,\mathbf{x}(n)}{\lambda + \mathbf{x}^T(n)\,\mathbf{P}(n-1)\,\mathbf{x}(n)}$$

$$= \lambda^{-1}\left[\mathbf{I} - \frac{\mathbf{P}(n-1)\,\mathbf{x}(n)\,\mathbf{x}^T(n)}{\lambda + \mathbf{x}^T(n)\,\mathbf{P}(n-1)\,\mathbf{x}(n)}\right]\mathbf{P}(n-1)\,\mathbf{x}(n)$$

$$= \lambda^{-1}\left[\mathbf{I} - \mathbf{g}(n)\,\mathbf{x}^T(n)\right]\mathbf{P}(n-1)\,\mathbf{x}(n) = \mathbf{P}(n)\,\mathbf{x}(n)$$

From (13.19) and (13.21) it follows that

$$\begin{aligned}\mathbf{w}_i(n) &= \mathbf{R}^{-1}(n)\,\mathbf{r}_i(n) = \lambda\mathbf{P}(n)\,\mathbf{r}_i(n-1) \\ &\quad + \mathbf{P}(n)\,\mathbf{x}(n)\,d_i(n)\end{aligned} \tag{13.30}$$

The combination of (13.25) and (13.30) gives

$$\begin{aligned}\mathbf{w}_i(n) &= \left[\mathbf{I} - \mathbf{g}(n)\,\mathbf{x}^T(n)\right]\mathbf{P}(n-1)\,\mathbf{r}_i(n-1) \\ &\quad + \mathbf{P}(n)\,\mathbf{x}(n)\,d_i(n)\end{aligned} \tag{13.31}$$

From (13.31) and (13.19) we get

$$\begin{aligned}\mathbf{w}_i(n) &= \mathbf{w}_i(n-1) - \mathbf{g}(n)\,\mathbf{x}^T(n)\,\mathbf{w}_i(n-1) \\ &\quad + \mathbf{P}(n)\,\mathbf{x}(n)\,d_i(n)\end{aligned} \tag{13.32}$$

The substitution of equation (13.28) in recursion (13.32) gives

$$\mathbf{w}_i(n) = \mathbf{w}_i(n-1) + \mathbf{g}(n)\left[d_i(n) - \mathbf{x}^T(n)\,\mathbf{w}_i(n-1)\right] \tag{13.33}$$

and the updated equations of the RLS algorithm are derived as follows:

$$\varepsilon_i(n) = d_i(n) - \mathbf{x}^T(n)\,\mathbf{w}_i(n-1) = d_i(n) - y_i(n) \tag{13.34}$$

$$\mathbf{g}(n) = \frac{\mathbf{P}(n-1)\,\mathbf{x}(n)}{\lambda + \mathbf{x}^T(n)\,\mathbf{P}(n-1)\,\mathbf{x}(n)} \tag{13.35}$$

$$\mathbf{P}(n) = \lambda^{-1}\left[\mathbf{I} - \mathbf{g}(n)\,\mathbf{x}^T(n)\right]\mathbf{P}(n-1) \tag{13.36}$$

$$\mathbf{w}_i(n) = \mathbf{w}_i(n-1) + \mathbf{g}(n)\,\varepsilon_i(n) \tag{13.37}$$

where δ is a constant and $\mathbf{I}$ is the identity matrix:

The initial conditions are given by:

$$\mathbf{P}(0) = \delta\mathbf{I}, \quad \delta > 0 \tag{13.38}$$

b) Objective function J_2

Analogously to case (a), the normal equation takes the form

$$\frac{J_2(n)}{\partial \mathbf{w}_i(n)} = -2\sum_{t=1}^{n} \lambda^{n-t}\left[b_i(t) - \mathbf{x}^T(t)\,\mathbf{w}_i(n)\right]\mathbf{x}^T(t) \tag{13.39}$$

or

$$\mathbf{r}_i(n) = \mathbf{R}(n)\,\mathbf{w}_i(n) \tag{13.40}$$

where

$$\mathbf{R}(n) = \sum_{t=1}^{n} \lambda^{n-t}\mathbf{x}(t)\,\mathbf{x}^T(n) \tag{13.41}$$

and

$$\mathbf{r}_i(n) = \sum \lambda^{n-t} b_i(t)\,\mathbf{x}(t) \tag{13.42}$$

Observe that equations (13.41) and (13.42) are analogous to equations (13.17) and (13.18). Therefore the update formulas correspond to (13.34) – (13.37) as follows:

$$e_i(n) = b_i(n) - \mathbf{x}^T(n)\,\mathbf{w}_i(n-1) = b_i(n) - s_i(n) \tag{13.43}$$

$$\mathbf{g}(n) = \frac{\mathbf{P}(n-1)\,\mathbf{x}(x)}{\lambda + \mathbf{x}^T(n)\,\mathbf{P}(n-1)\,\mathbf{x}(n)} \tag{13.44}$$

$$\mathbf{P}(n) = \lambda^{-1}\left[\mathbf{I} - \mathbf{g}(n)\,\mathbf{x}^T(n)\right]\mathbf{P}(n-1) \tag{13.45}$$

$$\mathbf{w}_i(n) = \mathbf{w}_i(n-1) + \mathbf{g}(n)\,e_i(n) \tag{13.46}$$

The initial conditions are given by (13.38).

c) Objective function J_3

Taking the partial derivative of $J_3(n)$ with respect to $\mathbf{w}_i(n)$ and setting it equal to zero, we have

$$\frac{\partial J_3(n)}{\partial \mathbf{w}_i(n)} = 2\sum_{t=1}^{n} \lambda^{n-t} \sum_{j=1}^{N} \frac{\partial \varepsilon_j(t)}{\partial \mathbf{w}_i(n)} \varepsilon_j(t) \tag{13.47}$$

$$= -2\sum_{t=1}^{n} \lambda^{n-t} \sum_{j=1}^{N} \frac{\partial y_j(t)\, \partial s_j(t)}{\partial s_j(t)\, \partial \mathbf{w}_i(n)} \varepsilon_j(t) = \mathbf{0}$$

Simple algebra converts (13.47) to:

$$\sum_{t=1}^{n} \lambda^{n-t} \frac{\partial y_i(t)}{\partial s_i(t)} \frac{\partial s_i(t)}{\partial \mathbf{w}_i(n)} [d_i(t) - y_i(t)]$$

$$= \sum_{t=1}^{n} \lambda^{n-t} \frac{\partial y_i(t)}{\partial s_i(t)} \mathbf{x}^T(t) [d_i(t) - y_i(t)] \tag{13.48}$$

$$= \sum_{t=1}^{n} \lambda^{n-t} \frac{\partial y_i(t)}{\partial s_i(t)} \mathbf{x}^T(t) [f(b_i(t) - f(s_i(t)))] = \mathbf{0}$$

Applying Taylor's formula to the expression in the brackets of (13.48), we get

$$f(b_i(t)) \approx f(s_i(t)) + f'(s_i(t))(b_i(t) - s_i(t)) \tag{13.49}$$

where

$$b_i(n) = f^{-1}(d_i(n)) \tag{13.50}$$

The combination of (13.49) with (13.50) leads to the equation

$$\sum_{t=1}^{n} \lambda^{n-t} f'^2(s_i(t)) \left[b_i(t) - \mathbf{x}^T(t)\mathbf{w}_i(n)\right] x(t) = \mathbf{0} \tag{13.51}$$

which can be rewritten in the vector form

$$\mathbf{r}_i(n) = \mathbf{R}_i(n)\mathbf{w}_i(n) \tag{13.52}$$

where

$$\mathbf{R}_i(n) = \sum_{t=1}^{n} \lambda^{n-t} f'^2(s_i(t))\, \mathbf{x}(t)\mathbf{x}^T(t) \tag{13.53}$$

and

$$\mathbf{r}_i(n) = \sum_{t=1}^{n} \lambda^{n-t} f'^2(s_i(t))\, b_i(t)\, \mathbf{x}(t) \tag{13.54}$$

Note that the substitution of $\mathbf{x}(t)$ and $b_i(t)$ in (13.43) – (13.46) by:

$$\mathbf{x}(t) \to f'(s_i(t))\,\mathbf{x}(t) \tag{13.55}$$

$$b_i(t) \to f'(s_i(t))\,b_i(t) \tag{13.56}$$

gives the following recursive learning algorithm:

$$\begin{aligned}\varepsilon_i(n) &= f'(s_i(n))\left[b_i(n) - \mathbf{x}^T(n)\,\mathbf{w}_i(n-1)\right] \\ &\approx d_i(n) - y_i(n)\end{aligned} \tag{13.57}$$

$$\mathbf{g}_i(n) = \frac{f'(s_i(n))\,\mathbf{P}_i(n-1)\,\mathbf{x}(n)}{\lambda + f'^2(s_i(n))\,\mathbf{x}^T(n)\,\mathbf{P}_i(n-1)\,\mathbf{x}(n)} \tag{13.58}$$

$$\mathbf{P}_i(n) = \lambda^{-1}\left[\mathbf{I} - f'(s_i(n))\,\mathbf{g}_i(n)\,\mathbf{x}^T(n)\right]\mathbf{P}_i(n-1) \tag{13.59}$$

$$\mathbf{w}_i(n) = \mathbf{w}_i(n-1) + \mathbf{g}_i(n)\,\varepsilon_i(n) \tag{13.60}$$

The initial conditions corresponding to (13.38) are given by

$$\mathbf{P}_i(0) = \delta\mathbf{I}, \quad \delta > 0.$$

13.4.2 Multi-layer neural networks

a) Objective function J_4

We minimize the performance measure by taking the partial derivative with respect to $\mathbf{w}_i^{(k)}$ and setting it equal to zero:

$$\begin{aligned}\frac{\partial J_4(t)}{\partial \mathbf{w}_i^{(k)}(n)} &= \sum_{t=1}^{n}\lambda^{n-t}\sum_{j=1}^{N_L}\frac{\partial \varepsilon_j^{(L)}(t)}{\partial \mathbf{w}_i^{(k)}(n)} \\ &= \sum_{t=1}^{n}\lambda^{n-t}\sum_{j=1}^{N_L}\frac{\partial \varepsilon_j^{(L)^2}(t)}{\partial \mathbf{s}_i^{(k)}(n)}\frac{\partial s_i^{(k)}(t)}{\partial \mathbf{w}_i^{(k)}(n)} \\ &= -2\sum_{t=1}^{n}\lambda^{n-t}\left(-\frac{1}{2}\right)\sum_{j=1}^{N_L}\frac{\partial \varepsilon_j^{(L)^2}(t)}{\partial \mathbf{s}_i^{(k)}(n)}\mathbf{x}^{(k)}(n) \\ &= -2\sum_{t=1}^{n}\lambda^{n-t}\varepsilon_i^{(k)}\mathbf{x}^{(k)}(n) = \mathbf{0}\end{aligned} \tag{13.61}$$

where

$$\varepsilon_i^{(k)}(n) = -\frac{1}{2}\frac{\partial \sum_{j=1}^{N_L}\varepsilon_j^{(L)^2}(n)}{\partial s_i^{(k)}(n)} \tag{13.62}$$

It is easily seen that

$$\begin{aligned} \varepsilon_i^{(k)}(n) &= -\frac{1}{2}\sum_{l=k+1}^{L}\sum_{m=1}^{N_l}\sum_{j=1}^{N_L}\frac{\partial\varepsilon_j^{(L)^2}(t)}{\partial s_m^{(l)}(n)}\frac{\partial s_m^{(l)}(n)}{\partial s_i^{(k)}(n)} \\ &= \sum_{l=k+1}^{L}\sum_{m=1}^{N_l}\varepsilon_m^{(l)}(n)\,w_{mi}^{(lk)}(n) \end{aligned} \tag{13.63}$$

Observe that equation (13.61) takes the form

$$\sum_{t=1}^{n}\lambda^{n-t}\varepsilon_i^{(k)}\mathbf{x}^{(k)}(n) = \mathbf{0} \tag{13.64}$$

or

$$\sum_{t=1}^{n}\lambda^{n-t}\left[d_i^{(k)}(t) - x^{(k)^T}(t)\,\mathbf{w}_i^{(k)}(n)\right]\mathbf{x}^{(k)}(t) = \mathbf{0} \tag{13.65}$$

where $d_i^{(k)}$ and $\varepsilon_i^{(k)}$ are defined by

$$d_i^{(k)}(n) = \begin{cases} d_i^{(L)}(n) & \text{for} \quad k = L \\ y_i^{(k)}(n) + \varepsilon_i^{(k)}(n) & \text{for} \quad k = 1, ..., L-1 \end{cases} \tag{13.66}$$

$$\varepsilon_i^{(k)}(n) = \sum_{l=k+1}^{L}\sum_{m=1}^{N_l}\varepsilon_m^{(l)}(n)\,w_{mi}^{(lk)}(n) \ \text{ for } k = 1, ..., L-1 \tag{13.67}$$

It is easily seen that formula (13.67) corresponds to the back propagation algorithm. The normal equation is given by:

$$\mathbf{r}_i^{(k)}(n) = \mathbf{R}^{(k)}(n)\,\mathbf{w}_i^{(k)}(n) \tag{13.68}$$

where

$$\mathbf{R}^{(k)}(n) = \sum_{t=1}^{n}\lambda^{n-t}\mathbf{x}^{(k)}(t)\,\mathbf{x}^{(k)^T}(t) \tag{13.69}$$

and

$$r_i^{(k)}(n) = \sum_{t=1}^{n}\lambda^{n-t}d_i^{(k)}(t)\,\mathbf{x}^{(k)}(t) \tag{13.70}$$

Observing that the normal equation (13.68) for a fixed layer k is the same as that in (13.16), we derive the following recursive algorithm:

$$\begin{aligned} \varepsilon_i^{(k)}(n) &= d_i^{(k)}(n) - \mathbf{x}^{(k)^T}(n)\,\mathbf{w}_i^{(k)}(n-1) \\ &= d_i^{(k)}(n) - y_i^{(k)}(n) \end{aligned} \tag{13.71}$$

$$\mathbf{g}^{(k)}(n) = \frac{\mathbf{P}^{(k)}(n-1)\,\mathbf{x}^{(k)}(n)}{\lambda + \mathbf{x}^{(k)^T}(n)\,\mathbf{P}^{(k)}(n-1)\,\mathbf{x}^{(k)}(n)} \tag{13.72}$$

$$\mathbf{P}^{(k)}(n) = \lambda^{-1}\left[\mathbf{I} - \mathbf{g}^{(k)}(n)\,\mathbf{x}^{(k)^T}(n)\right]\mathbf{P}^{(k)}(n-1) \tag{13.73}$$

$$\mathbf{w}_i^{(k)}(n) = \mathbf{w}_i^{(k)}(n-1) + \mathbf{g}^{(k)}(n)\,\varepsilon_i^{(k)}(n) \tag{13.74}$$

where $i = 1, ..., N_k$, $k = 1, ..., L$. The initial conditions are given by

$$\mathbf{P}^{(k)}(0) = \delta\mathbf{I}, \quad \delta > 0 \tag{13.75}$$

The errors given by (13.71) are propagated by making use of formulas (13.66) and (13.67).

b) Objective function J_5

In this case we start with the gradient

$$\begin{aligned}
\frac{\partial J_5(n)}{\partial \mathbf{w}_i^{(k)}(n)} &= \sum_{t=1}^{n} \lambda^{n-t} \sum_{j=1}^{N_L} \frac{\partial e_j^{(L)^2}(t)}{\partial \mathbf{w}_i^{(k)}(n)} \\
&= \sum_{t=1}^{n} \lambda^{n-t} \sum_{j=1}^{N_L} \frac{\partial e_j^{(L)^2}(t)}{\partial \mathbf{s}_i^{(k)}(n)} \frac{\partial s_i^{(k)}(n)}{\partial \mathbf{w}_i^{(k)}(n)} \\
&= -2 \sum_{t=1}^{n} \lambda^{n-t} \left(-\frac{1}{2}\right) \sum_{j=1}^{N_L} \frac{\partial e_j^{(L)^2}(t)}{\partial \mathbf{s}_i^{(k)}(n)} \mathbf{x}^{(k)}(n) \\
&= -2 \sum_{t=1}^{n} \lambda^{n-t} e_i^{(k)} \mathbf{x}^{(k)}(n) = \mathbf{0}
\end{aligned} \tag{13.76}$$

where

$$e_i^{(k)}(n) \mathrel{\hat{=}} \left(-\frac{1}{2}\right) \frac{\partial \sum_{i=1}^{N_L} e_i^{(L)^2}(n)}{\partial \mathbf{s}_i^{(k)}(n)} \tag{13.77}$$

Observe that

$$\begin{aligned}
e_i^{(k)}(n) - \frac{1}{2} \sum_{l=k+1}^{L} \sum_{m=1}^{N_l} \frac{\partial \sum_{i=1}^{N_L} e_i^{(L)^2}(n)}{\partial \mathbf{s}_m^{(l)}(n)} \frac{\partial s_m^{(l)}(n)}{\partial s_i^{(k)}(n)} \\
= \sum_{l=k+1}^{L} \sum_{m=1}^{N_l} e_m^{(l)}(n)\, w_{mi}^{(lk)}(n)\, f'\left(s_i^{(k)}(n)\right)
\end{aligned} \tag{13.78}$$

$$= f'\left(s_i^{(k)}(n)\right) \sum_{l=k+1}^{L} \sum_{m=1}^{N_l} e_m^{(l)}(n)\, w_{mi}^{(lk)}(n)$$

Formula (13.78) describes how to propagate errors of linear parts of neurons. The normal equation is of the form

$$\sum_{t=1}^{n} \lambda^{n-t} \left[b_i^{(k)}(t) - \mathbf{x}^{(k)^T}(t)\, \mathbf{w}_i^{(k)}(n)\right] \mathbf{x}^{(k)^T}(t) = \mathbf{0} \tag{13.79}$$

or in the vector notation

$$\mathbf{r}_i^{(k)}(n) = \mathbf{R}^{(k)}(n)\, \mathbf{w}_i^{(k)}(n) \tag{13.80}$$

where

$$\mathbf{R}^{(k)}(n) = \sum_{t=1}^{n} \lambda^{n-t} \mathbf{x}^{(k)}(t)\, \mathbf{x}^{(k)^T}(t) \tag{13.81}$$

and

$$\mathbf{r}_i^{(k)}(n) = \sum_{t=1}^{n} \lambda^{n-t} b_i^{(k)}(t)\, \mathbf{x}^{(k)}(t) \tag{13.82}$$

Due to the similarity of normal equations (13.80) and (13.68) we immediately get the RLS algorithm analogous to equations (13.71) – (13.74).

$$\begin{aligned} e_i^{(k)}(n) &= b_i^{(k)}(n) - \mathbf{x}^{(k)^T}(n)\, \mathbf{w}_i^{(k)}(n-1) \\ &= b_i^{(k)}(n) - s_i^{(k)}(n) \end{aligned} \tag{13.83}$$

$$b_i^{(k)}(n) = \begin{cases} b_i^{(L)}(n) & \text{for } k = L \\ s_i^{(k)}(n) + e_i^{(k)}(n) & \text{for } k = 1, ..., L-1 \end{cases} \tag{13.84}$$

$$e_i^{(k)}(n) = \begin{cases} b_i^{(L)}(n) - s_i^{(L)}(n) \quad \text{for } k = L \\ f'\left(s_i^{(k)}(n)\right) \sum_{l=k+1}^{L} \sum_{m=1}^{N_l} e_m^{(l)}(n)\, w_{mi}^{(lk)}(n) \\ \text{for } k = 1, ..., L-1 \end{cases} \tag{13.85}$$

$$\mathbf{g}^{(k)}(n) = \frac{\mathbf{P}^{(k)}(n-1)\, \mathbf{x}^{(k)}(n)}{\lambda + \mathbf{x}^{(k)^T}(n)\, \mathbf{P}^{(k)}(n-1)\, \mathbf{x}^{(k)}(n)} \tag{13.86}$$

$$\mathbf{P}^{(k)}(n) = \lambda^{-1}\left[\mathbf{I} - \mathbf{g}^{(k)}(n)\, \mathbf{x}^{(k)^T}(n)\right] \mathbf{P}^{(k)}(n-1) \tag{13.87}$$

$$\mathbf{w}_i^{(k)}(n) = \mathbf{w}_i^{(k)}(n-1) + \mathbf{g}^{(k)}(n)\, e_i^{(k)}(n) \tag{13.88}$$

Obviously, the actual computations of the new weights are based on formulas (13.85) – (13.88). The initial conditions are given by (13.75).

c) Objective function J_6

The appropriate gradient is calculated and set to zero as follows

$$\frac{\partial J_6(n)}{\partial \mathbf{w}_i^{(k)}(n)} = \sum_{t=1}^{n} \lambda^{n-t} \sum_{j=1}^{N_L} \frac{\partial e_j^{(L)^2}(t)}{\partial \mathbf{w}_i^{(k)}(n)}$$

$$= \sum_{t=1}^{n} \lambda^{n-t} \sum_{j=1}^{N_L} \frac{\partial \varepsilon_j^{(L)^2}(t)}{\partial y_i^{(k)}(n)} \frac{\partial y_i^{(k)}(n)}{\partial s_i^{(k)}(n)} \frac{\partial s_i^{(k)}(n)}{\partial \mathbf{w}_i^{(k)}(n)} \tag{13.89}$$

$$= -2\sum_{t=1}^{n} \lambda^{n-t} \left(-\frac{1}{2}\right) \sum_{j=1}^{N_L} \frac{\partial \varepsilon_j^{(L)^2}(t)}{\partial y_i^{(k)}(n)} \frac{\partial y_i^{(k)}(n)}{\partial s_i^{(k)}(n)} \mathbf{x}^{(k)}(n)$$

$$= -2\sum_{t=1}^{n} \lambda^{n-t} \varepsilon_i^{(k)} f'\left(s_i^{(k)}(n)\right) \mathbf{x}^{(k)}(n) = \mathbf{0}$$

where

$$\varepsilon_i^{(k)}(n) \mathrel{\hat{=}} -\frac{1}{2} \frac{\partial \sum_{j=1}^{N_L} e_j^{(L)^2}(n)}{\partial y_i^{(k)}(n)} \tag{13.90}$$

After some calculations we get

$$\begin{aligned} \varepsilon_i^{(k)}(n) &= -\frac{1}{2} \sum_{l=k+1}^{L} \sum_{m=1}^{N_l} \sum_{j=1}^{N_L} \frac{\partial \varepsilon_j^{(L)^2}(t)}{\partial s_m^{(l)}(n)} \frac{\partial y_m^{(l)}(n)}{\partial s_m^{(l)}(n)} \\ \cdot \frac{\partial s_m^{(l)}(n)}{\partial y_i^{(k)}(n)} &= \sum_{l=k+1}^{L} \sum_{m=1}^{N_l} \varepsilon_m^{(l)}(n) f'\left(s_m^{(l)}(n)\right) w_{mi}^{(lk)}(n) \end{aligned} \tag{13.91}$$

The right-hand side of equation (13.89) can be presented in the form

$$\sum_{t=1}^{n} \lambda^{n-t} \left[d_i^{(k)}(t) - y_i^{(k)}(t)\right] f'\left(s_i^{(k)}(t)\right) x^{(k)}(t)$$

$$= \sum_{t=1}^{n} \lambda^{n-t} \left[f\left(b_i^{(k)}(t)\right) - f\left(s_i^{(k)}(t)\right)\right] \tag{13.92}$$

$$\cdot f'\left(s_i^{(k)}(t)\right) x^{(k)}(t) = \mathbf{0}$$

Applying Taylor's formula to the expression in the brackets of (13.92), we have

$$f\left(b_i^{(k)}(t)\right) \approx f\left(s_i^{(k)}(t)\right) + f'\left(s_i^{(k)}(t)\right)\left(b_i^{(k)}(t) - s_i^{(k)}(t)\right) \quad (13.93)$$

From (13.92) and (13.93) it follows that the normal equation is of the form

$$\sum_{t=1}^{n} \lambda^{n-t} f'^2\left(s_i^{(k)}(t)\right)\left[b_i^{(k)}(t) - \mathbf{x}^{(k)^T} \cdot (t)\,\mathbf{w}_i^{(k)}(n)\right]\mathbf{x}^{(k)^T}(t) = \mathbf{0} \quad (13.94)$$

or

$$\mathbf{r}_i^{(k)}(n) = \mathbf{R}_i^{(k)}(n)\,\mathbf{w}_i^{(k)}(n) \quad (13.95)$$

where

$$\mathbf{R}_i^{(k)}(n) = \sum_{t=1}^{n} \lambda^{n-t} f'^2\left(s_i^{(k)}(t)\right)\mathbf{x}^{(k)}(t)\,\mathbf{x}^{(k)^T}(t) \quad (13.96)$$

$$\mathbf{r}_i^{(k)}(n) = \sum_{t=1}^{n} \lambda^{n-t} f'^2\left(s_i^{(k)}(t)\right) b_i^{(k)}(t)\,\mathbf{x}^{(k)}(t) \quad (13.97)$$

We take advantage of the analogy between formulas (13.95) – (13.97) and formulas (13.52) – (13.54). The update equations are copied from (13.57) – (13.60) as follows:

$$\varepsilon_i^{(k)}(n) = f'\left(s_i^{(k)}(n)\right)\left[b_i^{(k)}(n) - \mathbf{x}^{(k)^T}(n) \cdot \mathbf{w}_i^{(k)}(n-1)\right] \approx d_i^{(k)}(n) - y_i^{(k)}(n) \quad (13.98)$$

$$b_i^{(k)}(n) = f^{-1}\left(y_i^{(k)}(n) + \varepsilon_i^{(k)}(n)\right) \quad (13.99)$$

$$\varepsilon_i^{(k)}(n) = \begin{cases} d_i^{(L)}(n) - y_i^{(L)}(n) \quad \text{for} \quad k = L \\ \sum_{l=k+1}^{L} \sum_{m=1}^{N_l} \varepsilon_m^{(l)}(n)\, f'\left(s_m^{(l)}(n)\right) w_{mi}^{(lk)}(n) \\ \text{for} \quad k = 1, ..., L-1 \end{cases} \quad (13.100)$$

$$\mathbf{g}_i^{(k)}(n) = \frac{f'\left(s_i^{(k)}(n)\right)\mathbf{P}_i^{(k)}(n-1)\,\mathbf{x}^{(k)}(n)}{\lambda + f'^2\left(s_i^{(k)}(n)\right)\mathbf{x}^{(k)^T}(n)\,\mathbf{P}_i^{(k)}(n-1)\,\mathbf{x}^{(k)}(n)} \quad (13.101)$$

$$\mathbf{P}_i^{(k)}(n) = \lambda^{-1}\left[\mathbf{I} - f'\left(s_i^{(k)}(n)\right)\mathbf{g}_i^{(k)}\right. \quad (13.102)$$

$$\cdot (n)\, \mathbf{x}^{(k)^T}(n)\Big]\, \mathbf{P}_i^{(k)}(n-1)$$

$$\mathbf{w}_i^{(k)}(n) = \mathbf{w}_i^{(k)}(n-1) + \mathbf{g}_i^{(k)}(n)\, \varepsilon_i^{(k)}(n) \tag{13.103}$$

The initial conditions corresponding to (13.75) are given by

$$\mathbf{P}_i^{(k)}(0) = \delta I, \quad \delta > 0 \tag{13.104}$$

Obviously, the actual computations of new weights are based on formulas (13.100) – (13.103).

13.5 QQ–RLS learning algorithms

We derive RLS learning algorithms for performance measures J_1, J_2, J_3 (single layer NN) and next J_4, J_5, J_6 (multi-layer NN) assuming that autocorrelation matrices are factorized as the product of a certain matrix and its transposition: the so-called Potter's square-root factorization [265], [295].

13.5.1 Single layer

a) Objective function J_1

The normal equation is given by formulas (13.16) – (13.18). Combining the factorized form of $\mathbf{R}(n)$ given by (13.6) with the update equation (13.20), we get

$$\mathbf{Q}(n)\, \mathbf{Q}^T(n) = \lambda \mathbf{Q}(n-1)\, \mathbf{Q}^T(n-1) + \mathbf{x}(n)\, \mathbf{x}^T(n) \tag{13.105}$$

Let as introduce the normalized data vector $\mathbf{z}(n)$ defined as follows:

$$\mathbf{x}(n) = \mathbf{Q}(n-1)\, \mathbf{z}(n) \tag{13.106}$$

Consequently, equation (13.105) takes the form

$$\mathbf{Q}(n)\, \mathbf{Q}^T(n) = \mathbf{Q}(n-1)\left[\lambda \mathbf{I} + \mathbf{z}(n)\, \mathbf{z}^T(n)\right] \mathbf{Q}^T(n-1) \tag{13.107}$$

Let as define the projection operators

$$\mathbf{K}(n) = \mathbf{z}(n)\left(\mathbf{z}^T(n)\, \mathbf{z}(n)\right)^{-1} \mathbf{z}^T(n) \tag{13.108}$$

and $\mathbf{K}^{\perp}$

$$\mathbf{K}^{\perp}(n) = \mathbf{I} - \mathbf{K}(n) \tag{13.109}$$

Formula (13.108) can be rewritten as follows

$$\mathbf{z}(n)\,\mathbf{z}^T(n) = \mathbf{K}(n)\left(\mathbf{z}^T(n)\,\mathbf{z}(n)\right) \tag{13.110}$$

If equation (13.110) is substituted into equation (13.107) then

$$\mathbf{Q}(n)\,\mathbf{Q}^T(n) = \mathbf{Q}(n-1) \tag{13.111}$$
$$\cdot\left[\lambda\mathbf{I} + \mathbf{K}(n)\left(\mathbf{z}^T(n)\,\mathbf{z}(n)\right)\right]\mathbf{Q}^T(n-1)$$

From (13.111) and (13.109) it follows that

$$\mathbf{Q}(n)\,\mathbf{Q}^T(n) = \mathbf{Q}(n-1)\left[\lambda\mathbf{K}^{\perp}(n)\right. \tag{13.112}$$
$$\left.+\left(\lambda + \mathbf{z}^T(n)\,\mathbf{z}(n)\right)\mathbf{K}(n)\right]\mathbf{Q}^T(n-1)$$

It can be easily verified that the projection operator $\mathbf{K}$ has the following properties

$$\begin{aligned} \mathbf{K}^T(n) &= \mathbf{K}(n) \\ \mathbf{K}(n)\,\mathbf{x}(n) &= \mathbf{K}(n)\,\mathbf{K}(n)\ldots\mathbf{K}(n)\,\mathbf{x}(n) \\ \mathbf{K}^{\perp}(n)\,\mathbf{K}(n) &= 0 \end{aligned} \tag{13.113}$$

From properties (13.113) it follows that formula (13.112) takes the form

$$\begin{aligned} &\mathbf{Q}(n)\,\mathbf{Q}^T(n) = \mathbf{Q}(n-1) \\ &\cdot\left[\lambda\mathbf{K}^{\perp}(n)\,\mathbf{K}^{\perp}(n)\right]\mathbf{Q}^T(n-1) + \mathbf{Q}(n-1) \\ &\cdot\left[2\lambda^{\frac{1}{2}}\mathbf{K}^{\perp}(n)\left(\lambda + \mathbf{z}^T(n)\,\mathbf{z}(n)\right)^{\frac{1}{2}}\mathbf{K}(n)\right]\mathbf{Q}^T(n-1) \\ &+\mathbf{Q}(n-1)\left[\left(\lambda + \mathbf{z}^T(n)\,\mathbf{z}(n)\right)\mathbf{K}(n)\,\mathbf{K}(n)\right]\mathbf{Q}^T(n-1) \\ &= \mathbf{Q}(n-1)\left[\lambda^{\frac{1}{2}}\mathbf{K}^{\perp}(n) + \left(\lambda + \mathbf{z}^T(n)\,\mathbf{z}(n)\right)^{\frac{1}{2}}\mathbf{K}(n)\right] \\ &\cdot\left[\lambda^{\frac{1}{2}}\mathbf{K}^{\perp}(n) + \left(\lambda + \mathbf{z}^T(n)\,\mathbf{z}(n)\right)^{\frac{1}{2}}\mathbf{K}(n)\right]\mathbf{Q}^T(n-1) \end{aligned} \tag{13.114}$$

Consequently

$$\mathbf{Q}(n) = \mathbf{Q}(n-1)\left[\lambda^{\frac{1}{2}}\mathbf{K}^{\perp}(n)\right. \tag{13.115}$$
$$\left.+\left(\lambda + \mathbf{z}^T(n)\,\mathbf{z}(n)\right)^{\frac{1}{2}}\mathbf{K}(n)\right]$$

and

$$\mathbf{Q}^{-1}(n) = \left[\lambda^{-\frac{1}{2}}\mathbf{K}^{\perp}(n)\right. \tag{13.116}$$

$$+\left(\lambda+\mathbf{z}^{T}(n)\,\mathbf{z}(n)\right)^{-\frac{1}{2}}\mathbf{K}(n)\Big]\,\mathbf{Q}^{-1}(n-1)$$

From (13.6) and (13.106) it follows that

$$\mathbf{R}^{-1}(n)=\mathbf{Q}^{-1^{T}}(n)\,\mathbf{Q}^{-1}(n) \tag{13.117}$$

and

$$\mathbf{z}^{T}(n)\,\mathbf{z}(n)=\mathbf{x}^{T}(n)\,\mathbf{R}^{-1}(n)\,\mathbf{x}(n) \tag{13.118}$$

Let as denote the nominator and denominator of equation (13.35) as vector **k** and a scalar, respectively. Combining that equation with (13.6) and (13.106), we get

$$\mathbf{k}(n)=\mathbf{R}^{-1}(n-1)\,\mathbf{x}(n)=\mathbf{Q}^{-1^{T}}(n-1)\,\mathbf{z}(n) \tag{13.119}$$

and

$$\alpha(n)=\lambda+\mathbf{x}^{T}(n)\,\mathbf{R}^{-1}(n-1)\,\mathbf{x}(n)=\lambda+\mathbf{z}^{T}(n)\,\mathbf{z}(n) \tag{13.120}$$

By making use of (13.119) and (13.120) we convert equation (13.116) into

$$\begin{aligned}\mathbf{Q}^{-1}(n)&=\Big[\lambda^{-\frac{1}{2}}\left(\mathbf{I}-\mathbf{K}(n)\right)\\&+\left(\lambda+\mathbf{z}^{T}(n)\,\mathbf{z}(n)\right)^{-\frac{1}{2}}\mathbf{K}(n)\Big]\,\mathbf{Q}^{-1}(n-1)\\&=\lambda^{-\frac{1}{2}}\mathbf{Q}^{-1}(n-1)+\left[\left(\lambda+\mathbf{z}^{T}(n)\,\mathbf{z}(n)\right)^{-\frac{1}{2}}-\lambda^{-\frac{1}{2}}\right]\\&\cdot\mathbf{K}(n)\,\mathbf{Q}^{-1}(n-1)=\lambda^{-\frac{1}{2}}\mathbf{Q}^{-1}(n-1)\\&+\left(\alpha^{-\frac{1}{2}}(n)-\lambda^{-\frac{1}{2}}\right)\left(\alpha(n)-\lambda\right)^{-1}\mathbf{z}(n)\,\mathbf{k}^{T}(n)\end{aligned} \tag{13.121}$$

The QQ^{T} – RLS algorithm is summarized as follows:

$$\mathbf{z}(n)=\mathbf{Q}^{-1}(n-1)\,\mathbf{x}(n) \tag{13.122}$$

$$\mathbf{k}(n)=\mathbf{Q}^{-1^{T}}(n-1)\,\mathbf{z}(n) \tag{13.123}$$

$$\alpha(n)=\lambda+\mathbf{z}^{T}(n)\,\mathbf{z}(n) \tag{13.124}$$

$$\mathbf{g}(n)=\frac{\mathbf{k}(n)}{\alpha(n)} \tag{13.125}$$

$$\mathbf{Q}^{-1}(n)=\lambda^{-\frac{1}{2}}\mathbf{Q}^{-1}(n-1) \tag{13.126}$$

$$+\left(\alpha^{-\frac{1}{2}}(n)-\lambda^{-\frac{1}{2}}\right)\cdot(\alpha(n)-\lambda)^{-1}\mathbf{z}(n)\mathbf{k}^T(n)$$

$$\varepsilon_i(n)=d_i(n)-\mathbf{x}^T(n)\mathbf{w}_i(n-1)=d_i(n)-y_i(n) \tag{13.127}$$

$$\mathbf{w}_i(n)=\mathbf{w}_i(n-1)+\mathbf{g}(n)\varepsilon_i(n) \tag{13.128}$$

The initial conditions are given by

$$\mathbf{Q}^{-1}(0)=\delta\mathbf{I},\ \ \delta>0 \tag{13.129}$$

b) Objective function J_2

Using the same idea as in 13.4.1b and 13.5.1a we get the following algorithm:

$$\mathbf{z}(n)=\mathbf{Q}^{-1}(n-1)\mathbf{x}(n) \tag{13.130}$$

$$\mathbf{k}(n)=\mathbf{Q}^{-1^T}(n-1)\mathbf{z}(n) \tag{13.131}$$

$$\alpha(n)=\lambda+\mathbf{z}^T(n)\mathbf{z}(n) \tag{13.132}$$

$$\mathbf{g}(n)=\frac{\mathbf{k}(n)}{\alpha(n)} \tag{13.133}$$

$$\begin{aligned}\mathbf{Q}^{-1}(n) &= \lambda^{-\frac{1}{2}}\mathbf{Q}^{-1}(n-1)+\left(\alpha^{-\frac{1}{2}}(n)-\lambda^{-\frac{1}{2}}\right)\\ &\quad\cdot(\alpha(n)-\lambda)^{-1}\mathbf{z}(n)\mathbf{k}^T(n)\end{aligned} \tag{13.134}$$

$$\varepsilon_i(n)=b_i(n)-\mathbf{x}^T(n)\mathbf{w}_i(n-1)=b_i(n)-s_i(n) \tag{13.135}$$

$$\mathbf{w}_i(n)=\mathbf{w}_i(n-1)+\mathbf{g}(n)\varepsilon_i(n) \tag{13.136}$$

The initial conditions are given by (13.129).

c) Objective function J_3

Using arguments similar to those in Sections 13.4.1c and 13.5.2b we obtain the following algorithm

$$\mathbf{z}_i(n)=\mathbf{Q}_i^{-1}(n-1)\mathbf{x}(n) \tag{13.137}$$

$$\mathbf{k}_i(n)=f'(s_i(n))\mathbf{Q}_i^{-1^T}(n-1)\mathbf{z}_i(n) \tag{13.138}$$

$$\alpha_i(n)=\lambda+f'^2(s_i(n))\mathbf{z}_i^T(n)\mathbf{z}_i(n) \tag{13.139}$$

$$\mathbf{g}_i(n)=\frac{\mathbf{k}_i(n)}{\alpha_i(n)} \tag{13.140}$$

$$\mathbf{Q}_i^{-1}(n) = \lambda^{-\frac{1}{2}}\mathbf{Q}_i^{-1}(n-1) + \left(\alpha_i^{-\frac{1}{2}}(n) - \lambda^{-\frac{1}{2}}\right) \cdot (\alpha_i(n) - \lambda)^{-1} \mathbf{z}_i(n) f'(s_i(n)) \mathbf{k}_i^T(n) \tag{13.141}$$

$$\begin{aligned} \varepsilon_i(n) &= f'(s_i(n)) \left[b_i(n) - \mathbf{x}^T(n)\mathbf{w}_i(n-1)\right] \\ &\approx d_i(n) - y_i(n) \end{aligned} \tag{13.142}$$

$$\mathbf{w}_i(n) = \mathbf{w}_i(n-1) + \mathbf{g}_i(n)\varepsilon_i(n) \tag{13.143}$$

The initial conditions are given by

$$\mathbf{Q}_i^{-1}(0) = \delta\mathbf{I}, \quad \delta > 0 \tag{13.144}$$

13.5.2 Multi-layer neural network

The learning algorithms for multi-layer neural networks can be easily derived using arguments similar to those in Section 13.2.2. The errors are propagated by making use of formulas (13.66) and (13.67), (13.85) and (13.100) for objective functions J_4, J_5, J_6 respectively. We present the final forms of the update equations.

a) Objective function J_4

$$\mathbf{z}^{(k)}(n) = \mathbf{Q}^{(k)^{-1}}(n-1)\mathbf{x}^{(k)}(n) \tag{13.145}$$

$$\mathbf{k}^{(k)}(n) = \mathbf{Q}^{(k)^{-1^T}}(n-1)\mathbf{z}^{(k)}(n) \tag{13.146}$$

$$\alpha^{(k)}(n) = \lambda + \mathbf{z}^{(k)^T}(n)\mathbf{z}^{(k)}(n) \tag{13.147}$$

$$\mathbf{g}^{(k)}(n) = \frac{\mathbf{k}^{(k)}(n)}{\alpha^{(k)}(n)} \tag{13.148}$$

$$\mathbf{Q}^{(k)^{-1}}(n) = \lambda^{-\frac{1}{2}}\mathbf{Q}^{(k)^{-1}}(n-1) + \left(\alpha^{(k)-\frac{1}{2}}(n) - \lambda^{-\frac{1}{2}}\right)\left(\alpha^{(k)}(n) - \lambda\right)^{-1} \mathbf{z}^{(k)}(n)\mathbf{k}^{(k)^T}(n) \tag{13.149}$$

$$\begin{aligned} \varepsilon_i^{(k)}(n) &= d_i^{(k)}(n) - \mathbf{x}^{(k)^T}(n)\mathbf{w}_i^{(k)}(n-1) \\ &= d_i^{(k)}(n) - y_i^{(k)}(n) \end{aligned} \tag{13.150}$$

$$\mathbf{w}_i^{(k)}(n) = \mathbf{w}_i^{(k)}(n-1) + \mathbf{g}^{(k)}(n)\varepsilon^{(k)}(n) \tag{13.151}$$

The initial conditions are given by

$$\mathbf{Q}^{(k)^{-1}}(0) = \delta \mathbf{I}, \ \ \delta > 0 \tag{13.152}$$

b) Objective function J_5

$$\mathbf{z}^{(k)}(n) = \mathbf{Q}^{(k)^{-1}}(n-1)\,\mathbf{x}^{(k)}(n) \tag{13.153}$$

$$\mathbf{k}^{(k)}(n) = \mathbf{Q}^{(k)^{-1^T}}(n-1)\,\mathbf{z}^{(k)}(n) \tag{13.154}$$

$$\alpha^{(k)}(n) = \lambda + \mathbf{z}^{(l)^T}(n)\,\mathbf{z}^{(k)}(n) \tag{13.155}$$

$$\mathbf{g}^{(k)}(n) = \frac{\mathbf{k}^{(k)}(n)}{\alpha^{(k)}(n)} \tag{13.156}$$

$$\mathbf{Q}^{(k)^{-1}}(n) = \lambda^{-\frac{1}{2}}\mathbf{Q}^{(k)^{-1}}(n-1) + \left(\alpha^{(k)-\frac{1}{2}}(n) - \lambda^{-\frac{1}{2}}\right)\left(\alpha^{(k)}(n) - \lambda\right)^{-1}\mathbf{z}^{(k)}(n)\,\mathbf{k}^{(k)^T}(n) \tag{13.157}$$

$$\begin{aligned} e_i^{(k)}(n) &= b_i^{(k)}(n) - \mathbf{x}^{(k)^T}(n)\,\mathbf{w}_i^{(k)}(n-1) \\ &= b_i^{(k)}(n) - s_i^{(k)}(n) \end{aligned} \tag{13.158}$$

$$\mathbf{w}_i^{(k)}(n) = \mathbf{w}_i^{(k)}(n-1) + \mathbf{g}^{(k)}(n)\,e^{(k)}(n) \tag{13.159}$$

The initial conditions are given by (13.152) and the actual error is calculated by formula (13.85).

c) Objective function J_6

$$\mathbf{z}_i^{(k)}(n) = \mathbf{Q}_i^{(k)^{-1}}(n-1)\,\mathbf{x}^{(k)}(n) \tag{13.160}$$

$$\mathbf{k}_i^{(k)}(n) = f'\left(s_i^{(k)}(n)\right)\mathbf{Q}_i^{(k)^{-1^T}}(n-1)\,\mathbf{z}_i^{(k)}(n) \tag{13.161}$$

$$\alpha_i^{(k)}(n) = \lambda + f'^2\left(s_i^{(k)}(n)\right)\mathbf{z}_i^{(k)^T}(n)\,\mathbf{z}_i^{(k)}(n) \tag{13.162}$$

$$\mathbf{g}_i^{(k)}(n) = \frac{\mathbf{k}_i^{(k)}(n)}{\alpha_i^{(k)}(n)} \tag{13.163}$$

$$\mathbf{Q}_i^{(k)^{-1}}(n) = \lambda^{-\frac{1}{2}}\mathbf{Q}_i^{(k)^{-1}}(n-1)$$

$$+\left(\alpha_i^{(k)-\frac{1}{2}}(n)-\lambda^{-\frac{1}{2}}\right)\cdot\left(\alpha_i^{(k)}(n)-\lambda\right)^{-1} \tag{13.164}$$

$$\cdot\mathbf{z}_i^{(k)}(n)\, f'\left(s_i^{(k)}(n)\right)\mathbf{k}_i^{(k)^T}(n)$$

$$\varepsilon_i^{(k)}(n)=f'\left(s_i^{(k)}(n)\right)$$

$$\cdot\left[b_i^{(k)}(n)-\mathbf{x}^{(k)^T}(n)\cdot\mathbf{w}_i^{(k)}(n-1)\right] \tag{13.165}$$

$$\approx d_i^{(k)}(n)-y_i^{(k)}(n)$$

$$\mathbf{w}_i(n)=\mathbf{w}_i(n-1)+\mathbf{g}_i(n)\,\varepsilon_i(n) \tag{13.166}$$

The initial conditions corresponding to (13.152) are given by

$$\mathbf{Q}_i^{(k)^{-1}}(0)=\delta\mathbf{I},\quad \delta>0 \tag{13.167}$$

and the actual error is calculated by making use of formula (13.100).

13.6 UD-RLS learning algorithms

We derive RLS learning algorithms for performance measures J_1, J_2, J_3 (single layer NN) and next J_4, J_5, J_6 (multi-layer NN) assuming that inverse of the autocorrelation matrix is factorized as the product of an upper triangular matrix, a diagonal matrix and a transposition of an upper triangular matrix.

13.6.1 Single layer

We present some details for performance measure J_1. In a similar way we can derive learning algorithms for J_2 and J_3.

a) Objective function J_1

We will derived a new algorithm by modification of procedures (13.34) – (13.37) in Section 13.4.1. Let us assume that $\mathbf{R}^{-1}(n)=\mathbf{P}(n)=\mathbf{U}(n)\,\mathbf{D}(n)\,\mathbf{U}^T(n)$ where $\mathbf{U}$ is a triangular matrix with zeros in the bottom elements and ones in the main diagonal, $\mathbf{D}$ is a diagonal matrix, i.e.

$$\mathbf{U}(n) = \begin{bmatrix} 1 & u_{01}(n) & u_{02}(n) & \cdots & u_{0N_0}(n) \\ 0 & 1 & u_{12}(n) & \cdots & u_{1N_0}(n) \\ 0 & 0 & 1 & & \vdots \\ \vdots & \vdots & \vdots & \ddots & u_{N_0-1N_0}(n) \\ 0 & 0 & 0 & \cdots & 1 \end{bmatrix} \tag{13.168}$$

and

$$\mathbf{D}(n) = \begin{bmatrix} c_0 & 0 & \cdots & 0 \\ 0 & c_1 & \cdots & 0 \\ \vdots & \vdots & \ddots & \vdots \\ 0 & 0 & \cdots & c_{N_0} \end{bmatrix} \tag{13.169}$$

By defining vectors

$$\mathbf{f} = \mathbf{U}^T(n-1)\,\mathbf{x}(n), \quad \mathbf{f} = [f_0, ..., f_{N_0}]^T \tag{13.170}$$

$$\mathbf{h} = \mathbf{D}(n-1)\,\mathbf{f}, \quad \mathbf{h} = [h_0, ..., h_{N_0}]^T \tag{13.171}$$

and the scalar

$$\begin{aligned} \beta &= \lambda + \mathbf{x}^T(n)\,\mathbf{P}(n-1)\,\mathbf{x}(n) \\ &= \lambda + \mathbf{x}^T(n)\,\mathbf{U}(n-1)\,\mathbf{D}(n-1)\,\mathbf{U}^T \\ &\cdot (n-1)\,\mathbf{x}(n) = \lambda + \mathbf{f}^T\mathbf{D}(n-1)\,\mathbf{f} = \boldsymbol{\lambda} + \mathbf{f}^T\mathbf{h} \end{aligned} \tag{13.172}$$

equation (13.35) takes the form

$$\begin{aligned} \mathbf{g}(n) &= \mathbf{P}(n-1)\,\mathbf{x}(n)\,\beta^{-1} \\ &= \mathbf{U}(n-1)\,\mathbf{D}(n-1)\,\mathbf{U}^T(n-1)\,\mathbf{x}(n)\,\beta^{-1} \\ &= \mathbf{U}(n-1)\,\mathbf{D}(n-1)\,\mathbf{f}\boldsymbol{\beta}^{-1} = \mathbf{U}(n-1)\,\mathbf{h}\boldsymbol{\beta}^{-1} \end{aligned} \tag{13.173}$$

We introduce notations

$$\beta_m = \lambda + \sum_{i=0}^{m} f_i h_i, \quad m = 0, ..., N_0 \tag{13.174}$$

and

$$\beta_{N_0} = \beta$$

Using similar arguments to those in [295] we get the following UD-RLS algorithm for a single layer linear neural network:
Step 1

$$\varepsilon_i(n) = d_i(n) - \mathbf{x}^T(n)\,\mathbf{w}_i(n-1) = d_i(n) - y_i(n) \tag{13.175}$$

$$\mathbf{f} = \mathbf{U}^T(n-1)\,\mathbf{x}(n) \tag{13.176}$$

$$\mathbf{h} = \mathbf{D}(n-1)\,\mathbf{f} \tag{13.177}$$

$$\beta_{-1} = \lambda \tag{13.178}$$

Step 2
For $j = 0$ to N_{k-1}

$$\beta_j = \beta_{j-1} + f_j h_j \tag{13.179}$$

$$c_j(n) = c_j(n-1)\,\frac{\beta_{j-1}}{\beta_j \lambda} \tag{13.180}$$

$$k_j = h_j \tag{13.181}$$

$$\mu_j = \frac{-f_j}{\beta_{j-1}} \tag{13.182}$$

Step 2.1
For $m = 0$ to $j-1 (j > 0)$

$$u_{mj}(n) = u_{mj}(n-1) + \mu_j k_m \tag{13.183}$$

$$k_m = k_m + u_{mj}(n-1)\,k_j \tag{13.184}$$

Step 3

$$\mathbf{g}(n) = \frac{\left[k_0, ..., k_{N_{k-1}}\right]^T}{\beta_{N_{k-1}}} \tag{13.185}$$

$$\mathbf{w}_i(n) = \mathbf{w}_i(n-1) + \mathbf{g}(n)\,\varepsilon_i(n) \tag{13.186}$$

Initial values are set as

$$\begin{aligned} \mathbf{D}(0) &= \delta\mathbf{I}, \;\; \delta > 0 \\ \mathbf{U}(0) &= \mathbf{I} \end{aligned} \tag{13.187}$$

b) Objective function J_2
Step 1

$$e_i(n) = b_i(n) - \mathbf{x}^T(n)\,\mathbf{w}_i(n-1) = b_i(n) - s_i(n) \tag{13.188}$$

$$\mathbf{f} = \mathbf{U}^T(n-1)\,\mathbf{x}(n) \tag{13.189}$$

$$\mathbf{h} = \mathbf{D}(n-1)\,\mathbf{f} \tag{13.190}$$

$$\beta_{-1} = \lambda \tag{13.191}$$

Step 2
For $j = 0$ to N_{k-1}

$$\beta_j = \beta_{j-1} + f_j h_j \tag{13.192}$$

$$c_j(n) = c_j(n-1)\frac{\beta_{j-1}}{\beta_j \lambda} \tag{13.193}$$

$$k_j = h_j \tag{13.194}$$

$$\mu_j = \frac{-f_j}{\beta_{j-1}} \tag{13.195}$$

Step 2.1
For $m = 0$ to $j-1 (j>0)$

$$u_{mj}(n) = u_{mj}(n-1) + \mu_j k_m \tag{13.196}$$

$$k_m = k_m + u_{mj}(n-1)\,k_j \tag{13.197}$$

Step 3

$$\mathbf{g}(n) = \frac{\left[k_0, \ldots, k_{N_{k-1}}\right]^T}{\beta_{N_{k-1}}} \tag{13.198}$$

$$\mathbf{w}_i(n) = \mathbf{w}_i(n-1) + \mathbf{g}(n)\,e_i(n) \tag{13.199}$$

Initial values are given by (13.187).

c) Objective function J_3

Step 1

$$\varepsilon_i(n) = f'(s_i(n))\left[b_i(n) - \mathbf{x}^T(n)\right. \tag{13.200}$$
$$\left.\cdot\mathbf{w}_i(n-1)\right] \approx d_i(n) - y_i(n)$$

$$\mathbf{f}_i = \mathbf{U}_i^T(n-1)\,\mathbf{x}(n) \tag{13.201}$$

$$\mathbf{h}_i = \mathbf{D}_i(n-1)\,\mathbf{f}_i \tag{13.202}$$

$$\beta_{i,-1} = \lambda \tag{13.203}$$

Step 2
For $j = 0$ to N_{k-1}

$$\beta_{ij} = \beta_{ij-1} + f'^2(s_i(n))\,f_{ij}h_{ij} \tag{13.204}$$

$$c_{ij}(n) = c_{ij}(n-1)\frac{\beta_{ij-1}}{\beta_{ij}\lambda} \tag{13.205}$$

$$k_{ij} = h_{ij} \tag{13.206}$$

$$\mu_{ij} = -f'^2(s_i(n))\frac{f_{ij}}{\beta_{ij-1}} \tag{13.207}$$

Step 2.1
For $m = 0$ to $j-1 (j>0)$

$$u_{imj}(n) = u_{imj}(n-1)\mu_{ij}k_{im} \tag{13.208}$$

$$k_{im} = k_{im} + u_{imj}(n-1)k_{ij} \tag{13.209}$$

Step 3

$$\mathbf{g}_i(n) = \frac{\left[k_{i0}, \ldots, k_{iN_{k-1}}\right]^T}{\beta_{iN_{k-1}}} \tag{13.210}$$

$$\mathbf{w}_i(n) = \mathbf{w}_i(n-1) + g_i(n)\varepsilon_i(n) \tag{13.211}$$

The initial conditions corresponding to (13.187) are given by

$$\mathbf{D}_i(0) = \delta\mathbf{I}, \quad \delta > 0 \tag{13.212}$$

$$\mathbf{U}_i(0) = \mathbf{I}.$$

13.6.2 Multi-layer neural networks

The results for multi-layer NN generalize the results for a single-layer NN. The errors are propagated by making use of formulas (13.66) and (13.67), (13.85) and (13.100) for objective functions J_4, J_5, J_6 respectively.

a) Objective function J_4
Step 1

$$\begin{aligned} \varepsilon_i^{(k)}(n) &= d_i^{(k)}(n) - \mathbf{x}^{(k)^T}(n)\mathbf{w}_i^{(k)}(n-1) \\ &= d_i^{(k)}(n) - y_i^{(k)}(n) \end{aligned} \tag{13.213}$$

$$\mathbf{f} = \mathbf{U}^{(k)^T}(n-1)\mathbf{x}^{(k)}(n) \tag{13.214}$$

$$\mathbf{h} = \mathbf{D}^{(k)}(n-1)\mathbf{f} \tag{13.215}$$

$$\beta_{-1} = \lambda \tag{13.216}$$

Step 2
For $j = 0$ it N_{k-1}

$$\beta_j = \beta_{j-1} + f_j h_j \tag{13.217}$$

$$c_j^{(k)}(n) = c_j^{(k)}(n-1)\frac{\beta_{j-1}}{\beta_j \lambda} \tag{13.218}$$

$$k_j = h_j \tag{13.219}$$

$$\mu_j = \frac{-f_j}{\beta_{j-1}} \tag{13.220}$$

Step 2.1
For $m = 0$ it $j-1 (j > 0)$

$$u_{mj}^{(k)}(n) = u_{mj}^{(k)}(n-1) + \mu_j k_m \tag{13.221}$$

$$k_m = k_m + u_{mj}^{(k)}(n-1)\, k_j \tag{13.222}$$

Step 3.

$$\mathbf{g}^{(k)}(n) = \frac{[k_0, ..., k_{N-1}]^T}{\beta_{N_{k-1}}} \tag{13.223}$$

$$\mathbf{w}_i^{(k)}(n) = \mathbf{w}_i^{(k)}(n-1) + \mathbf{g}^{(k)}(n)\, \varepsilon_i^{(k)}(n) \tag{13.224}$$

The initial conditions are given by

$$\begin{aligned} \mathbf{D}^{(k)}(0) &= \delta \mathbf{I}, \ \delta > 0 \\ \mathbf{U}^{(k)}(0) &= \mathbf{I} \end{aligned} \tag{13.225}$$

b) Objective function J_5

Step 1

$$\begin{aligned} e_i^{(k)}(n) &= b_i^{(k)}(n) - \mathbf{x}^{(k)^T}(n)\, \mathbf{w}_i^{(k)}(n-1) \\ &= b_i^{(k)}(n) - s_i^{(k)}(n) \end{aligned} \tag{13.226}$$

$$\mathbf{f} = \mathbf{U}^{(k)^T}(n-1)\, \mathbf{x}^{(k)}(n) \tag{13.227}$$

$$\mathbf{h} = \mathbf{D}^{(k)}(n-1)\, \mathbf{f} \tag{13.228}$$

$$\beta_{-1} = \lambda \tag{13.229}$$

Step 2
For $j = 0$ to N_{k-1}

$$\beta_j = \beta_{j-1} + f_j h_j \tag{13.230}$$

$$c_j^{(k)}(n) = c_j^{(k)}(n-1)\frac{\beta_{j-1}}{\beta_j \lambda} \tag{13.231}$$

$$k_j = h_j \tag{13.232}$$

$$\mu_j = \frac{-f_j}{\beta_{j-1}} \tag{13.233}$$

Step 2.1
For $m = 0$ to $j-1 (j > 0)$

$$u_{mj}^{(k)}(n) = u_{mj}^{(k)}(n-1) + \mu_j k_m \tag{13.234}$$

$$k_m = k_m + u_{mj}^{(k)}(n-1)\, k_j \tag{13.235}$$

Step 3

$$\mathbf{g}^{(k)}(n) = \frac{\left[k_0, ..., k_{N_{k-1}}\right]^T}{\beta_{N_{k-1}}} \tag{13.236}$$

$$\mathbf{w}_i^{(k)}(n) = \mathbf{w}_i^{(k)}(n-1) + \mathbf{g}^{(k)}(n)\, e_i^{(k)}(n) \tag{13.237}$$

The initial conditions are given by (13.225) and the actual error $e_i^{(k)}(n)$ is calculated by making use of formula (13.85).

c) Objective function J_6

Step 1

$$\varepsilon_i^{(k)}(n) = f'\left(s_i^{(k)}(n)\right)\left[b_i^{(k)}(n)\right. \tag{13.238}$$

$$\left. -\mathbf{x}^{(k)^T}(n)\, \mathbf{w}_i^{(k)}(n-1)\right] \approx d_i^{(k)}(n) - y_i^{(k)}(n)$$

$$\mathbf{f}_i^{(k)} = \mathbf{U}_i^{(k)^T}(n-1)\, \mathbf{x}^{(k)}(n) \tag{13.239}$$

$$\mathbf{h}_i^{(k)} = \mathbf{D}_i^{(k)}(n-1)\, \mathbf{f}_i^{(k)} \tag{13.240}$$

$$\beta_{i,-1}^{(k)} = \lambda \tag{13.241}$$

Step 2
For $j = 0$ to N_{k-1}

$$\beta_{ij}^{(k)} = \beta_{i,j-1}^{(k)} = +f'^2\left(s_i^{(k)}(n)\right) f_{ij}^{(k)} h_{ij}^{(k)} \tag{13.242}$$

$$c_{ij}^{(k)}(n) = c_{ij}^{(k)}(n-1)\frac{\beta_{i,j-1}^{(k)}}{\beta_{ij}^{(k)} \lambda} \tag{13.243}$$

$$k_{ij}^{(k)} = h_{ij}^{(k)} \tag{13.244}$$

$$\mu_{ij}^{(k)} = -f'^2\left(s_i^{(k)}(n)\right)\frac{f_{ij}^{(k)}}{\beta_{i,j-1}^{(k)}} \tag{13.245}$$

Step 2.1
For $m = 0$ to $j-1 (j>0)$

$$u_{imj}^{(k)}(n) = u_{imj}^{(k)}(n-1) + \mu_{ij}^{(k)} k_{im}^{(k)} \tag{13.246}$$

$$k_{im}^{(k)} = k_{im}^{(k)} + u_{imj}^{(k)}(n-1)\, k_{ij}^{(k)} \tag{13.247}$$

Step 3

$$\mathbf{g}_i^{(k)}(n) = \frac{\left[k_{i0}^{(k)},\ldots,k_{iN_{k-1}}^{(k)}\right]^T}{\beta_{iN_{k-1}}} \tag{13.248}$$

$$\mathbf{w}_i^{(k)}(n) = \mathbf{w}_i^{(k)}(n-1) + \mathbf{g}_i^{(k)}(n)\,\varepsilon_i^{(k)}(n) \tag{13.249}$$

The initial conditions are given by

$$\mathbf{D}_{(i)}^{(k)}(0) = \delta\mathbf{I}, \quad \delta > 0 \tag{13.250}$$

$$\mathbf{D}_{(i)}^{(k)}(0) = \mathbf{I} \tag{13.251}$$

and the actual error $\varepsilon_i^{(k)}(n)$ is calculated by making use of formula (13.100).

13.7 Simulation results

13.7.1 Performance evaluation

We have conducted intensive simulations in order to test a family of the RLS algorithms derived in previous sections. The following typical problems are simulated: the logistic function, the circle in the square problem and the 4-b parity problem. In all simulations the learning algorithms are run one hundred times. The results will be depicted in tables with entries showing both the average number of epochs requied to meet a stopping criterion and the percentage of successful runs. In a case of unsuccessful runs a dash sign is shown.

a) The logistic function

For the first test problem networks 1-3-1, 1-5-1 and 1-7-1, with a hidden layer of sigmoid nonlinearities and a linear output layer, were trained to approximate the logistic function

$$y = 4x\,(1 - x)$$

using both classical and completely connected (CC) architectures. The training set consisted of 11 input-output pairs, where the input values are in the interval [0,1]. The network was trained until sum of squares of the errors was less than 0.01. The results for the classical and CC architectures are given in Table 13.2 and Table 13.3, respectively. It is easily seen that:

- the number of epochs required to train the network decreases when the number of hidden neurons increases
- the application of CC networks results in decrease of the number of epochs
- for the classical architecture the standard RLS algorithm (see Section 13.4.2c) appeared to be the best one (about 20 times less epochs than the BP and MBP algorithms)
- for the CC architecture the RLS, ETB RLS and ETB UD-RLS algorithms required about 40 times less epochs than the BP and MBP algorithms

TABLE 13.2. Simulation results – the logistic function

Algorithm	Structure					
	131		151		171	
BP	302.1 87	μ=0.05	235.29 97	μ=0.05	221.57 99	μ=0.05
MBP	272.58 88	μ=0.05 α=0.05	211.57 99	μ=0.05 α=0.15	202.72 100	μ= 0.05 α=0.15
RLS	20.93 100	λ=0.91 δ=1	11.17 100	λ=0.95 δ=100	8.86 100	λ=0.93 δ=100
ETB RLS	181.78	—	70.06 100	λ=0.99 δ=100	30.82 100	λ=0.95 δ=1000
QQ^T-RLS	—	—	—	—	521.57 28	λ=0.97 δ=1
ETB QQ^T -RLS	—	—	—	—	—	—
UD-RLS	697 7	λ=0.99 δ=100	837.89 9	λ=0.99 δ=10	543.8 10	λ=0.99 δ=10
ETB UD-RLS	182.28 90	λ=0.99 δ=1	81.97 100	λ=0.95 δ=100	24.23 100	λ=0.95 δ=100

TABLE 13.3. Simulation results – the logistic function

Algorithm	Structure					
	131CC		151CC		171CC	
BP	231.27 95	μ=0.05	195.86 100	μ=0.05	188.17 100	μ=0.05
MBP	184.35 98	μ=0.05 α=0.15	137.41 97	μ=0.05 α=0.35	146.11 100	μ=0.05 α=0.25
RLS	9.63 100	λ=0.92 δ=100	7.12 100	λ=0.96 δ=100	6.47 100	λ=0.91 δ=100
ETB RLS	15.2 100	λ=0.91 δ=100	2.93 100	λ=0.92 δ=1000	4.77 100	λ=0.91 δ=1000
QQ^T-RLS	51.8 98	λ=0.93 δ=1	44.38 98	λ=0.95 δ=1	43.02 100	λ=0.96 δ=0.2
ETB QQ^T -RLS	88.49 49	λ=0.99 δ=1	19.25 44	λ=0.98 δ=1	20.41 54	λ=0.99 δ=1
UD-RLS	692.17 40	λ=0.96 δ=1	454.38 55	λ=0.97 δ=1	429.04 47	λ=0.98 δ=1
ETB UD-RLS	69.08 100	λ=0.9 δ=10	9.15 100	λ=0.9 δ=100	4.85 100	λ=0.93 δ=1000

b) The circle in the square problem

In the second simulation, a neural network structure has to decide if a point of coordinate (x, y) varying from -0.5 to +0.5 is in the circle of radius equal to 0.35. The input coordinates are selected uniformly. If the distance of the training point from the origin is less than 0.35, then to the desired output the value 0.9 is assigned indicating that the point is inside the circle. A distance greater than 0.35 means that the point is outside the circle and the desired output becomes 0.1. Training patterns are presented to the network alternating between the two classes (inside and outside the circle). In one epoch we present to the network 100 input-output patterns. The results for both the classical and CC 2-2-2-1, 2-3-3-1, 2-4-4-1 and 2-5-5-1 architectures are given in Tables 13.4a, 13.4b, 13.5a and 13.5b. The network was trained until the sum of squares of the errors was less than 0.1. The conclusions from Tables 13.4a, 13.4b, 13.5a, 13.5b and 13.6 are the following:

- the number of epochs required to train the network decreases when the number of hidden neurons increases, however some experiments have failed to reach the error goal of 0.1 especially for the classical architecture
- the application of CC networks results in the decrease of the number of epochs (even 10-100 times)
- for the classical architecture the RLS, ETB RLS and ETB UD-RLS algorithms appeared to be the best ones and required about 20-50 times less epochs than the BP and MBP algorithms
- for the CC architectures also the RLS, ETB RLS and ETB UD-RLS algorithms gave the best results, however their performance was less spectacular in comparison with the BP and MBP algorithms

TABLE 13.4a. Simulation results – the circle in the square problem

Algorithm	Structure			
	2221		2331	
BP	2463.17 95	$\mu = 0.35$	942.07 100	$\mu = 1$
MBP	1710.19 90	$\mu = 0.35$ $\alpha = 0.25$	726.46 100	$\mu = 0.95$ $\alpha = 0.05$
RLS	219.4 90	$\lambda = 0.984$ $\delta = 10$	86.05 100	$\lambda = 0.998$ $\delta = 1000$
ETB RLS	346.9 91	$\lambda = 0.994$ $\delta = 100$	155.86 96	$\lambda = 0.998$ $\delta = 100$
QQ^T-RLS	544.83 6	$\lambda = 0.994$ $\delta = 1000$	—	—
ETB QQ^T-RLS	—	—	—	—
UD-RLS	—	—	—	—
ETB UD-RLS	301.69 91	$\lambda = 0.998$ $\delta = 100$	107.68 95	$\lambda = 0.986$ $\delta = 100$

TABLE 13.4b. Simulation results – the circle in the square problem

Algorithm	Structure			
	2441		2551	
BP	577.39 100	$\mu = 1$	534.94 100	$\mu = 0.85$
MBP	478.06 100	$\mu = 0.95$ $\alpha = 0.45$	419.88 100	$\mu = 0.85$ $\alpha = 0.55$
RLS	26.82 100	$\lambda = 0.998$ $\delta = 1000$	20.7 100	$\lambda = 0.99$ $\delta = 100$
ETB RLS	11.46 98	$\lambda = 0.998$ $\delta = 100$	54.62 99	$\lambda = 0.99$ $\delta = 1000$
QQ^T-RLS	—	—	17.49 99	$\lambda = 0.98$ $\delta = 10$
ETB QQ^T-RLS	122.7 23	$\lambda = 0.996$ $\delta = 1000$	475.47 17	$\lambda = 0.95$ $\delta = 10$
UD-RLS	574.63 51	$\lambda = 0.996$ $\delta = 10$	—	—
ETB UD-RLS	22.1 99	$\lambda = 0.994$ $\delta = 100$	47.98 100	$\lambda = 0.99$ $\delta = 1000$

TABLE 13.5a. Simulation results – the circle in the square problem

Algorithm	Structure			
	2221CC		2331CC	
BP	71.78 100	$\mu = 1$	58.63 100	$\mu = 1$
MBP	28.4 100	$\mu = 0.75$ $\alpha = 0.85$	26.32 100	$\mu = 0.95$ $\alpha = 0.75$
RLS	9.88 100	$\lambda = 0.98$ $\delta = 1000$	5.25 100	$\lambda = 0.98$ $\delta = 1000$
ETB RLS	39.7 100	$\lambda = 0.986$ $\delta = 10$	11.75 100	$\lambda = 0.986$ $\delta = 10$
QQ^T-RLS	5.46 100	$\lambda = 0.986$ $\delta = 100$	3.95 100	$\lambda = 0.98$ $\delta = 100$
ETB QQ^T-RLS	537.6 20	$\lambda = 0.994$ $\delta = 10$	470.06 48	$\lambda = 0.996$ $\delta = 1$
UD-RLS	—	—	—	—
ETB UD-RLS	34.01 100	$\lambda = 0.98$ $\delta = 100$	20.07 100	$\lambda = 0.992$ $\delta = 1$

TABLE 13.5b. Simulation results – the circle in the square problem

Algorithm	Structure			
	2441CC		2551CC	
BP	54.3 100	$\mu = 1$	53.73 100	$\mu = 1$
MBP	26.03 100	$\mu = 0.95$ $\alpha = 0.75$	27.15 100	$\mu = 1$ $\alpha = 0.6$
RLS	4.65 100	$\lambda = 0.98$ $\delta = 1000$	4.62 100	$\lambda = 0.99$ $\delta = 1000$
ETB RLS	3.31 100	$\lambda = 0.986$ $\delta = 1000$	47.86 100	$\lambda = 0.99$ $\delta = 1$
QQ^T-RLS	—	—	5.18 100	$\lambda = 0.98$ $\delta = 10$
ETB QQ^T-RLS	139.22 60	$\lambda = 0.998$ $\delta = 1$	32.78 99	$\lambda = 0.99$ $\delta = 1$
UD-RLS	365.34 94	$\lambda = 0.998$ $\delta = 10$	—	—
ETB UD-RLS	3.84 100	$\lambda = 0.984$ $\delta = 100$	43.83 100	$\lambda = 0.99$ $\delta = 10$

c) The 4-b parity problem

The goal of the third simulation is to determine the parity of a 4-b binary number. The neural network inputs are logic values (0.5 for the higher level; -0.5 for the lower level). In each iteration we present to the neural network 16 input combinations with their desired outputs (0.1 for the lower level and 0.9 for the higher level). The network was trained until the sum of squares of the errors was less than 0.1. The results for the classical and CC 4-2-1, 4-3-1 and 4-4-1 architectures are given in Table 13.6 and Table 13.7, respectively. From Tables 13.6 and 13.7 we conclude that:

- for the CC architecture more experiments were successful in comparison with the classical architecture
- for the classical architecture the RLS and ETB UD-RLS algorithms exhibited the best performance (about 30 times better comparing with the BP and MBP algorithms)
- for the CC architectures also the RLS, ETB RLS and ETB UD-RLS algorithms gave the best results

TABLE 13.6. Simulation results – the 4-b parity problem

Algorithm	Structure					
	421		431		441	
BP	1167.86 100	μ=0.3	6587.21 29	μ=0.05	4941.14 56	μ=0.1
MBP	1144.79 100	μ=0.45 α=0.25	3136.97 60	μ=0.65 α=0.95	4621.82 68	μ=0.05 α=0.55
RLS	379.17 76	λ=0.9 δ=100	659.5 86	λ=0.93 δ=10	499.24 92	λ=0.91 δ=100
ETB RLS	1181.38 8	λ=0.9 δ=1000	—	—	251.77 88	λ=0.99 δ=100
QQ^T-RLS	—	—	—	—	1140.53 17	λ=0.99 δ=1000
ETB QQ^T -RLS	924.15 41	λ=0.95 δ=10	841.4 20	λ=0.92 δ=1	562.18 11	λ=0.91 δ=10
UD-RLS	—	—	1008.9 20	λ=0.91 δ=100	806.36 14	λ=0.99 δ=10
ETB UD-RLS	1124.5 8	λ=0.9 δ=10	564.95 61	λ=0.99 δ=100	168.77 74	λ=0.97 δ=100

TABLE 13.7. Simulation results – the 4-b parity problem

Algorithm	Structure					
	421CC		431CC		441CC	
BP	4739.38 40	μ=0.2	4873.22 68	μ=0.1	4555.87 84	μ=0.1
MBP	5032.22 27	μ=0.25 α=0.05	2107.44 96	μ=0.45 α=0.95	1064.44 96	μ=0.65 α=0.95
RLS	144.71 100	λ=0.91 δ=100	110.87 97	λ=0.9 δ=1	89.98 99	λ=0.9 δ=10
ETB RLS	487.5 8	λ=0.95 δ=1000	149 84	λ=0.99 δ=100	104.29 97	λ=0.99 δ=100
QQ^T-RLS	925.33 6	λ=0.95 δ=1	1187.5 18	λ=0.91 δ=1000	1122.3 10	λ=0.94 δ=1000
ETB QQ^T -RLS	390.22 9	λ=0.99 δ=10	—	—	14.54 100	λ=0.91 δ=1
UD-RLS	—	—	—	—	—	—
ETB UD-RLS	567 10	λ=0.9 δ=1000	174.63 88	λ=0.99 δ=1000	72.13 75	λ=0.97 δ=100

13.8 Concluding remarks

In this chapter we have presented a family of the RLS learning algorithms. The error was calculated in the linear parts of the neurons (ETB – RLS algorithms) and "as usually" in the non-linear parts of the neurons (RLS algorithms). We have studied the classical RLS algorithms and their modifications QQ^T-RLS and UD-RLS which are less liable to numerical errors. The RLS algorithms seem to be an interesting alternative to the gradient algorithms. From simulations in Section 13.7 we draw the following conclusions:

- the CC networks required less epochs than the classical architectures
- the learning process rarely failed for the CC networks
- the algorithms from the RLS family required in most cases 10 – 100 times less epochs than the BP and MBP algorithms

- all the ETB algorithms as well as the QQ^T and UD algorithms require much less computational operations than the classical RLS algorithm
- for some structures and problems it is not possible to factorize matrices $\mathbf{P}$ and $\mathbf{R}$ (to find matrices $\mathbf{Q}$, $\mathbf{U}$ and $\mathbf{D}$), in these cases the learning process can not be successful.

In future research and simulations it would be interesting to combine the RLS learning algorithms with an evolutionary design [30], [31], [108], [136], [186], [187], [189], [254], [302] in order to determine the structure and initial weights of the neural network. The material presented in this chapter was partially published in [33] and [34].

14
Systolic Implementations of the RLS Learning Algorithms

14.1 Introduction

Systolic and wavefront arrays (see [145] – [148], [328]) belong to a new class of pipelined array architectures. A systolic array is a network of processors which rhythmically compute and pass data through the system Since it operates like the pumping action of the human heart, the name "systolic" is in a common use. This structure is well suited to neural network implementations, because of its properties of regularity, modularity and local interconnnections. The methodology for deriving systolic arrays consists of three stages (see [146]):

- Dependence Graph (DG) Design
- Signal Flow Graph (SFG) Construction
- Array Processor Design

A DG is a graph which presents the data dependencies of an algorithm. In the dependence graph nodes represent computations and arcs dependencies between computations. A recursive algorithm may be easily transformed to a DG by using proper arcs in the index space.

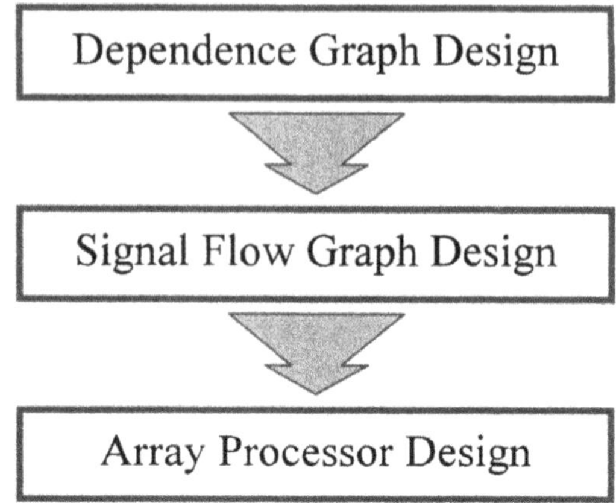

FIGURE 14.1. The methodology for deriving systolic arrays

The SFG construction consists of processing nodes, communicating edges, and delays (see [146] for details). A simple way of mapping a DG onto a SFG array is by means of projection, which assigns the operations of all nodes along a line to a single PE. For example, the three-dimensional index space of a DG may be projected onto a two-dimensional SFG array. The SFG can be viewed as a simplified graph. That means the SFG is a more concise representation than the DG. Moreover, the SFG is closer to the hardware level design. Consequently, the SFG implies the type of arrays that will be obtained. By an incorporation of pipelining into the SFG array we get a systolic structure corresponding to the original processing algorithm.

In this chapter we present systolic architectures implementing two RLS learning algorithms derived in Chapter 13: the ETB RLS learning algorithm (Section 13.4.2b) and classical RLS learning algorithm (Section 13.4.2c). In both cases we assume a traditional architecture of the multilayer feedforward neural network (not pyramid structure). In Section 14.2 we present the systolic architecture for the recall phase which is the same for both RLS learning algorithms. Systolic architectures for the learning phase of these RLS algorithms are different, and are designed in Sections 14.3 and 14.4, respectively. The performance evaluation of systolic architectures and comparison with classical sequential architectures is given in Section 14.5.

14.2 Systolic architecture for the recall phase

The systolic architecture design begins with a data dependence graph (DG) to express the recurrence and parallelism. As we mentioned in the introduction, a recursive algorithm may be easily transformed to a dependence graph by using proper arcs in the index space. Next, this description will be mapped onto a systolic array. A description of the neural network is given by

$$s_i^{(k)}(n) = \sum_{j=0}^{N_{k-1}} w_{ij}^{(k)}(n) \cdot x_j^{(k)}(n) \tag{14.1}$$

$$y_j^{(k)}(n) = f\left(s_j^{(k)}(n)\right) \tag{14.2}$$

The dependence graph for the recall phase is shown in Fig. 14.2 and the functional operation in each node of the dependence graph is presented in Fig. 14.3.

A mapping of the dependence graph to a systolic array is straightforward. The projection vector and schedule vector (see Kung [146] for details) can be taken along the vertical direction $\overrightarrow{\mathbf{d}} = [i, j] = [0, 1]$, $\overrightarrow{\mathbf{s}} = [i, j] = [0, 1]$. This leads to ring systolic architectures as shown in Fig. 14.4.

Each value $x_i^{(k)}$ at the i-th processor element, is multiplied by $w_{ij}^{(k)}$ which is stored in the memory of the i-th processor element. The product is added to the accumulator A_i which has an initial value. The value $x_i^{(k)}$ will move counterclokwise across the ring array and pass each of other processor elements in consecutive clock units. The above procedure can be executed in a pipelined fashion. Figure 14.5 shows the functional operation at PE of the systolic architecture, whereas Fig. 14.6 – Fig. 14.9 illustrate systolic architectures at different stages of data pumping.

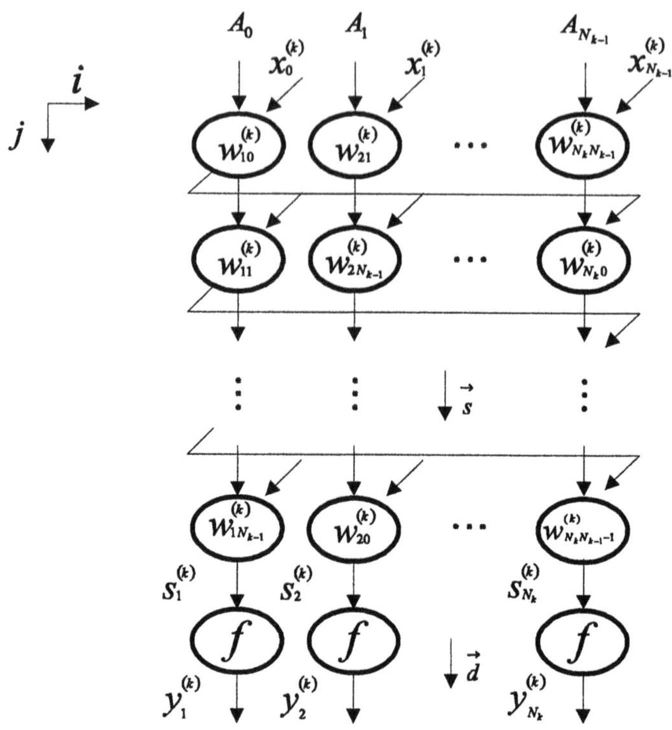

FIGURE 14.2. Dependence graph for the recall phase

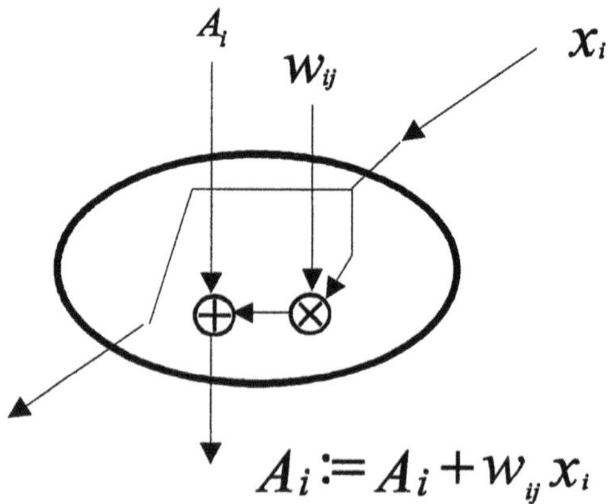

FIGURE 14.3. Functional operation at each node of the dependence graph

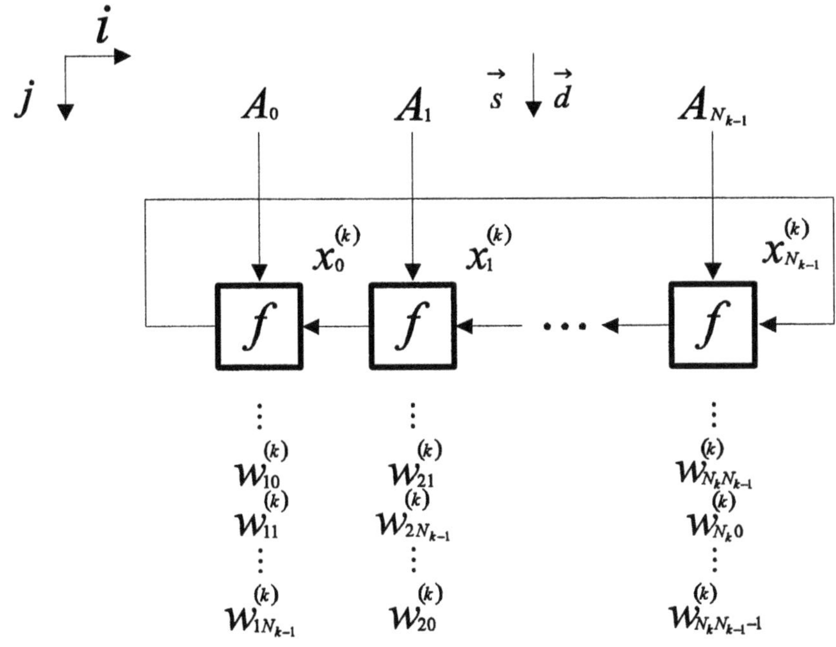

FIGURE 14.4. Systolic architecture for the recall phase

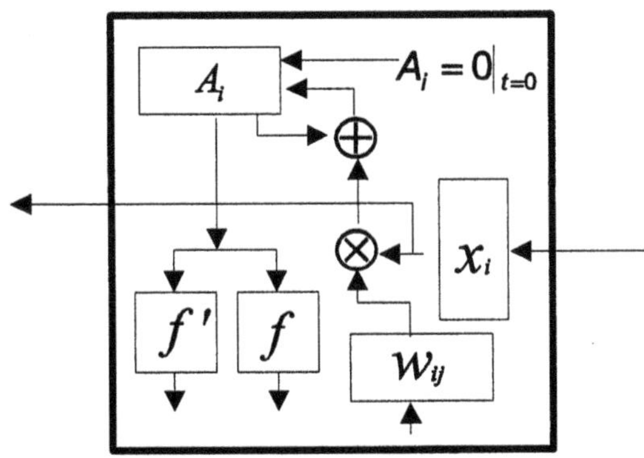

FIGURE 14.5. Functional operation at each PE of the systolic architecture for the recall phase

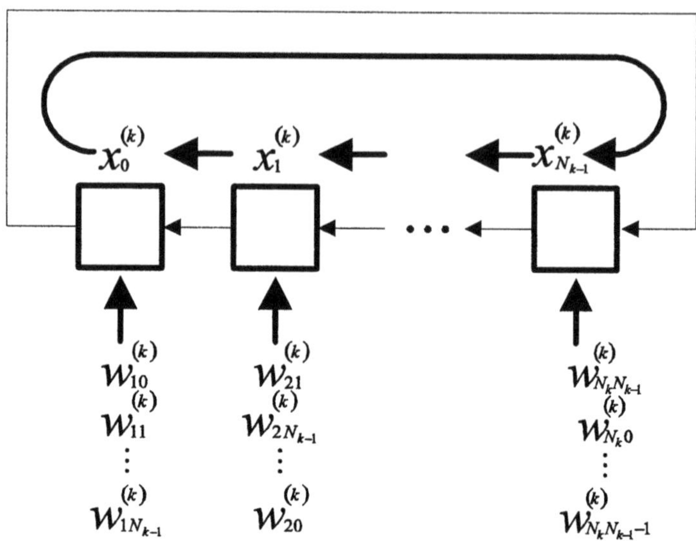

FIGURE 14.6. Systolic architecture for the recall phase – initial stage

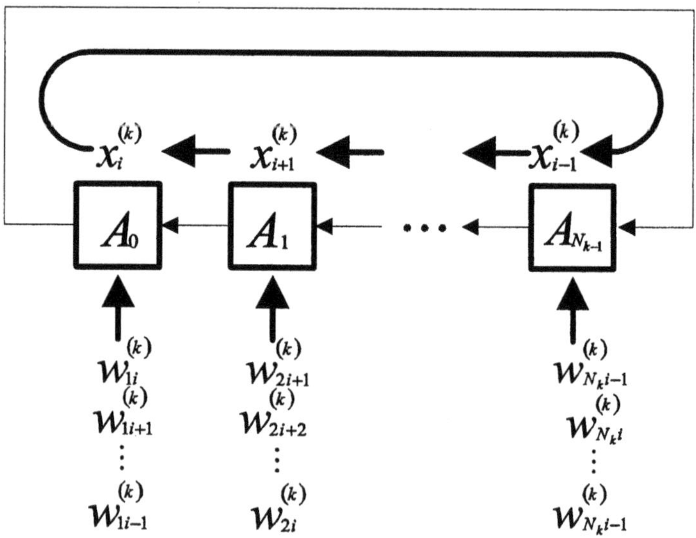

FIGURE 14.7. Systolic ring architecture for the recall phase – intermediate stage

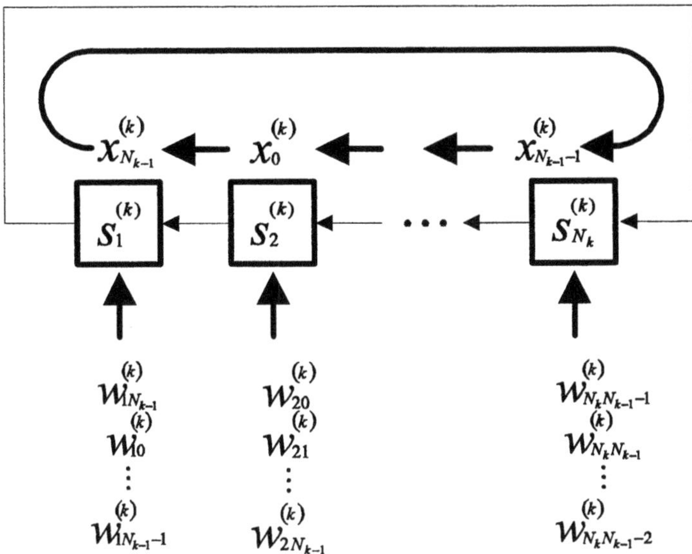

FIGURE 14.8. Systolic ring architecture for the recall phase (linear output) – final stage

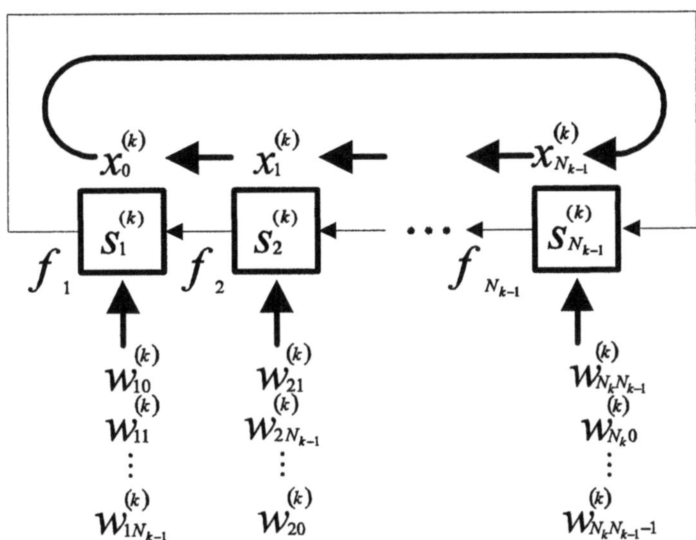

FIGURE 14.9. Systolic ring architecture for the recall phase (nonlinear output) – final stage

14.3 Systolic architectures for the ETB RLS learning algorithms

In this section we design systolic architectures for the ETB neural networks described by the following equations:

$$\begin{aligned}\gamma_i^{(k)}(n) &= b_i^{(k)}(n) - \mathbf{x}^{(k)^T}(n)\,\mathbf{w}_i^{(k)}(n-1) \\ &= b_i^{(k)}(n) - s_i^{(k)}(n)\end{aligned} \tag{14.3}$$

$$b_i^{(k)}(n) = \begin{cases} b_i^{(L)}(n) & \text{for } k = L \\ s_i^{(k)}(n) + e_i^{(k)}(n) & \text{for } k = 1, ..., L-1 \end{cases} \tag{14.4}$$

$$\begin{aligned}e_i^{(k)}(n) &= \sum_{j=1}^{N_{k+1}} f'\left(s_i^{(k)}(n)\,w_{ij}^{(k+1)}(n)\right) e_j^{(k+1)}(n) \\ &\quad \text{for } k = 1, ..., L-1\end{aligned} \tag{14.5}$$

$$\mathbf{g}^{(k)}(n) = \frac{\mathbf{P}^{(k)}(n-1)\,\mathbf{x}^{(k)}(n)}{\lambda + \mathbf{x}^{(k)^T}(n)\,\mathbf{P}^{(k)}(n-1)\,\mathbf{x}^{(k)}(n)} \tag{14.6}$$

$$\mathbf{P}^{(k)}(n) = \lambda^{-1}\left[\mathbf{I} - \mathbf{g}^{(k)}(n)\,\mathbf{x}^{(k)^T}(n)\right]\mathbf{P}^{(k)}(n-1) \tag{14.7}$$

$$\delta_i^{(k)}(n) = e_i^{(k)}(n)\,f'\left(s_i^{(k)}(n)\right) \tag{14.8}$$

$$\mathbf{w}_i^{(k)}(n) = \mathbf{w}_i^{(k)}(n-1) + \mathbf{g}^{(k)}(n)\,\gamma_i^{(k)}(n) \tag{14.9}$$

The dependence graph corresponding to the error calculation operation is shown in Fig. 14.10 and functional operations at each node of the DG are depicted in Fig. 14.11. The systolic ring architecture for the error calculating phase is shown in Fig. 14.12.

The scheme of a single PE of the systolic architecture for the error calculation phase is shown Fig. 14.13, whereas Fig. 14.14 – Fig. 14.16 show systolic architectures at different stages of data pumping.

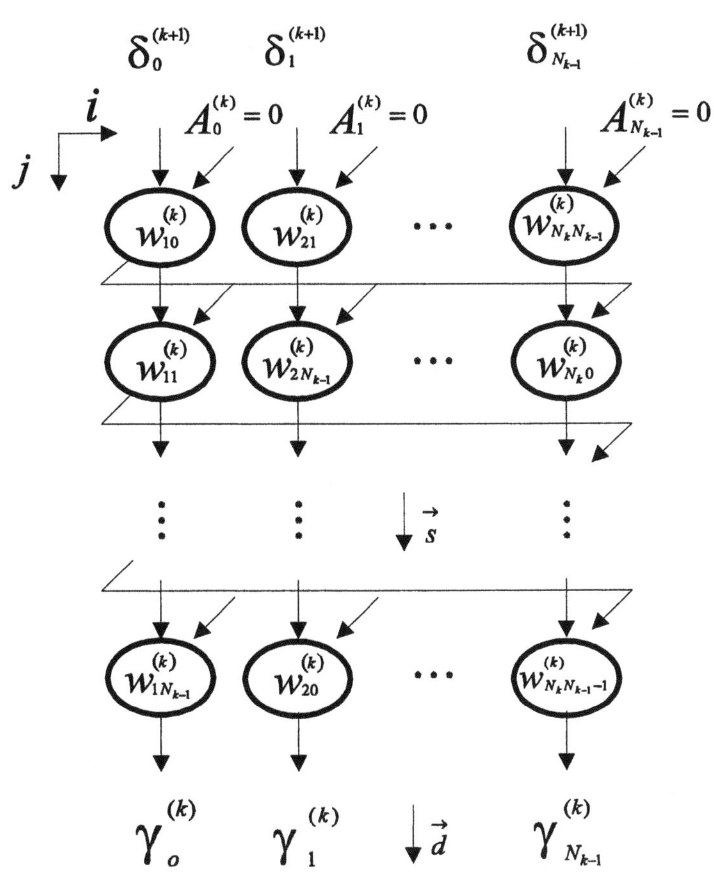

FIGURE 14.10. Dependence graph for the error calculation phase

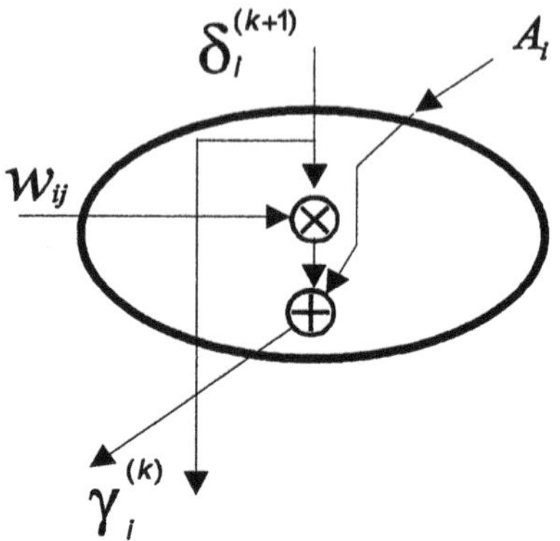

FIGURE 14.11. Functional operations at each node of the dependence graph for the error calculation phase

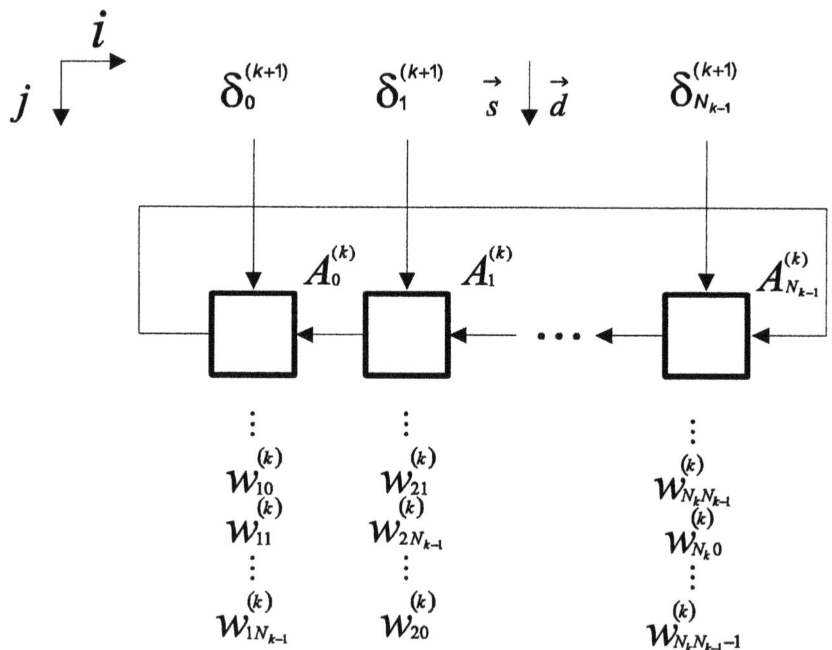

FIGURE 14.12. Systolic architecture for the error calculation phase

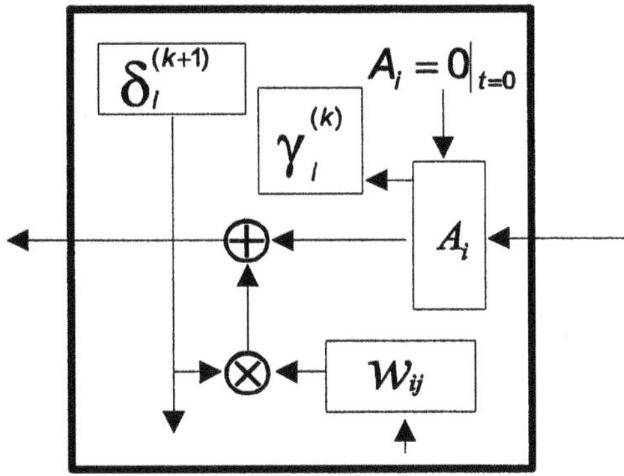

FIGURE 14.13. Functional operations at each PE of the systolic architecture for the error calculation phase

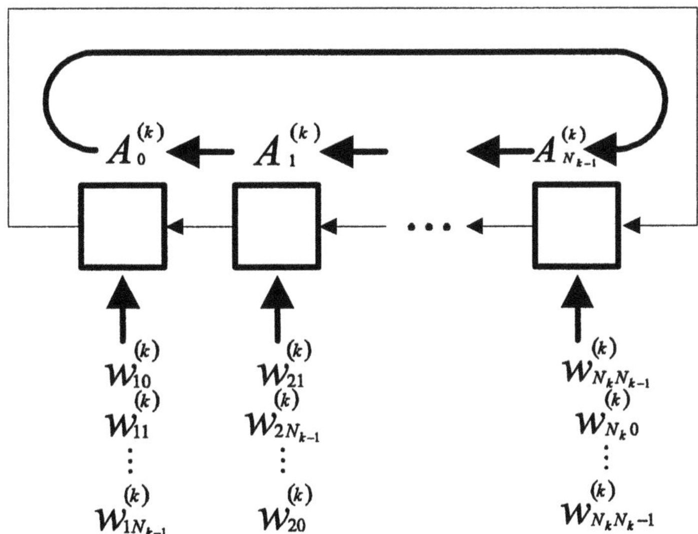

FIGURE 14.14. Systolic architecture for the error calculation phase – initial stage

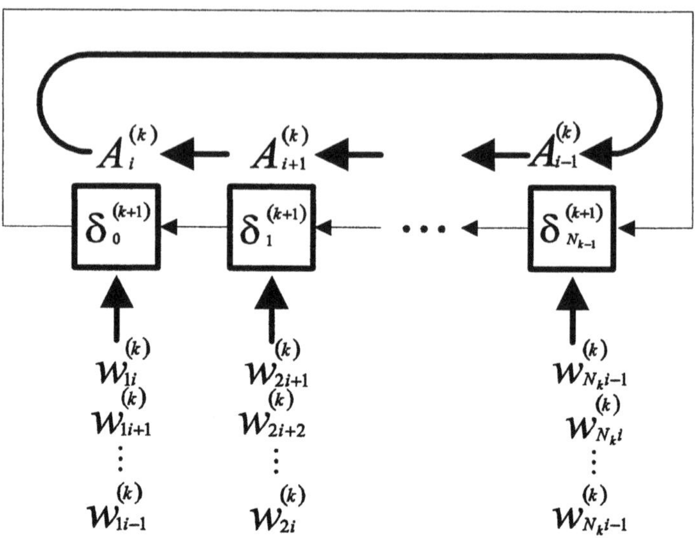

FIGURE 14.15. Systolic architecture for the error calculation phase – intermediate stage

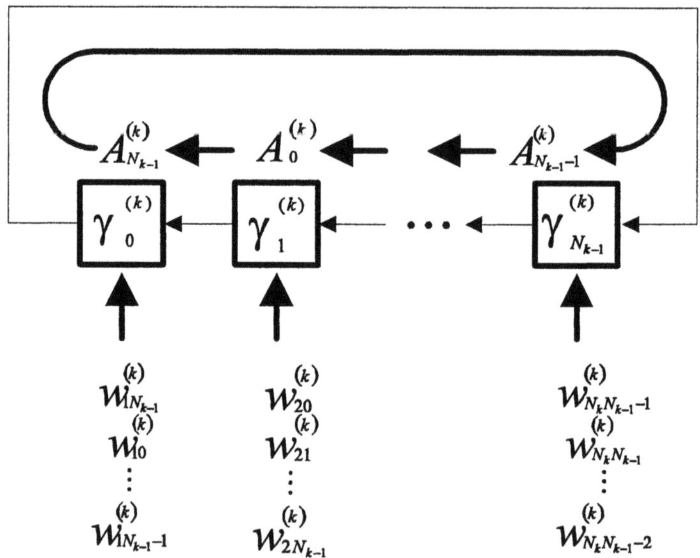

FIGURE 14.16. Systolic architecture for the error calculation phase – final stage

Systolic processing for weight updating operations is similar to that of matrix-vector multiplication. The dependence graph for this operations is shown in Fig. 14.17. The functional operation in each node is also a multiplication and accumulation opreation (see Fig. 14.18). Before the weight updating process we have to calculate $\mathbf{g}^{(k)}$ and $\mathbf{P}^{(k)}$, where $\mathbf{P}^{(k)}$ is the correlation matrix. The systolic ring architecture for the weight updating phase is shown in Fig. 14.19.

In Fig. 14.20 we present functional operations at each PE of the systolic architecture for the weight updating phase, whereas Fig. 14.21 – Fig. 14.23 illustrate the pumping of data at different stages of the design process.

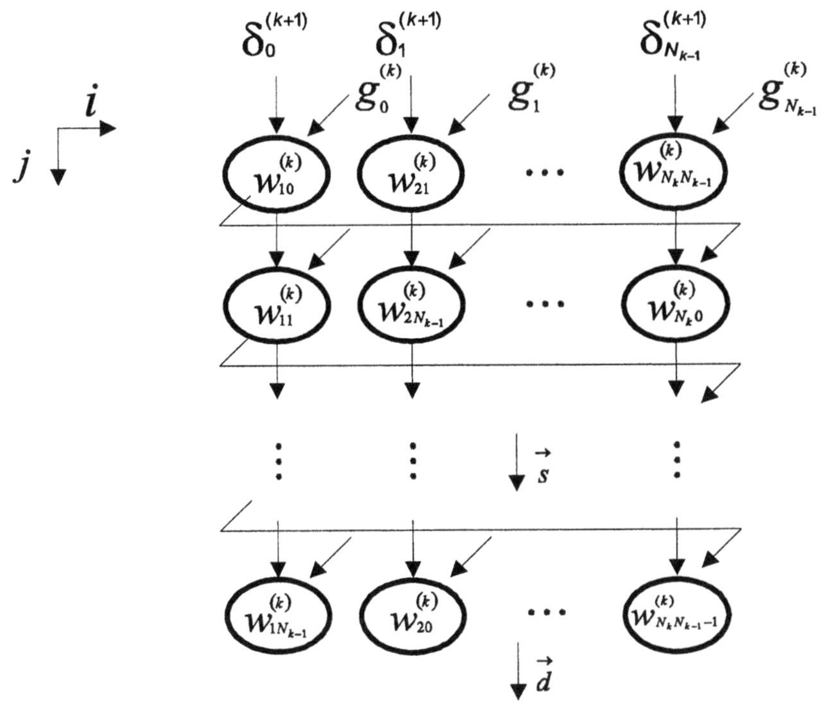

FIGURE 14.17. Dependence graph for the weight updating phase

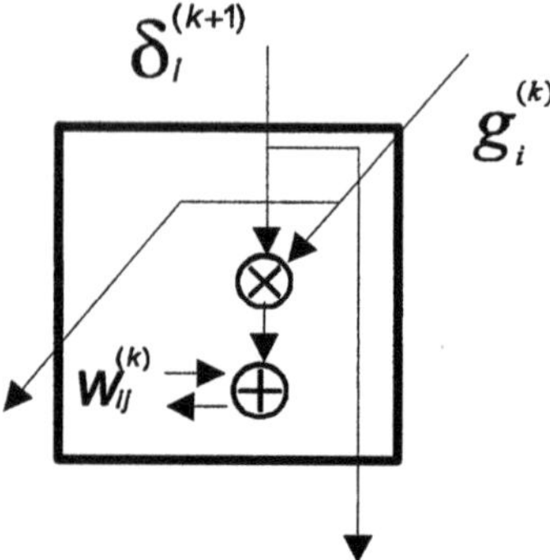

FIGURE 14.18. Functional operations at each node of the dependence graph for the weight updating phase

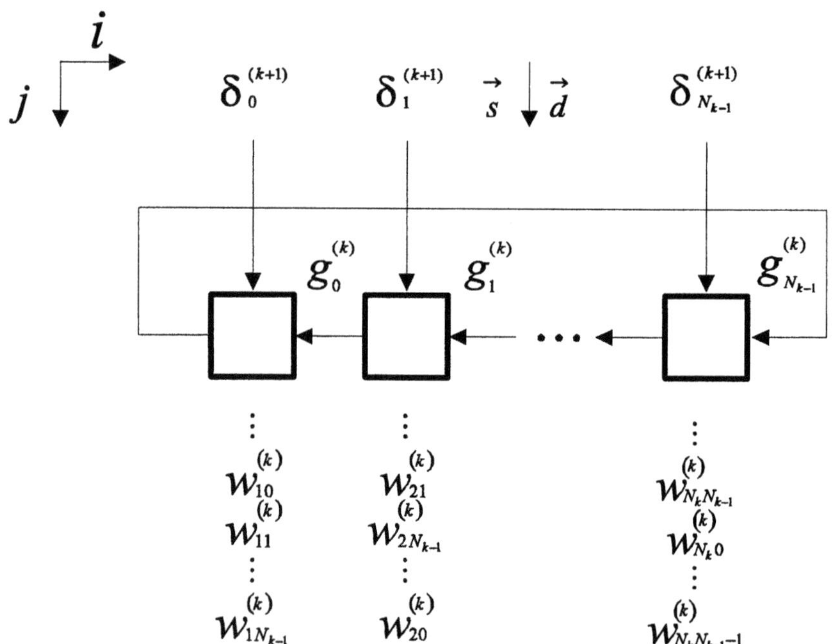

FIGURE 14.19. Systolic architecture for the weight updating phase

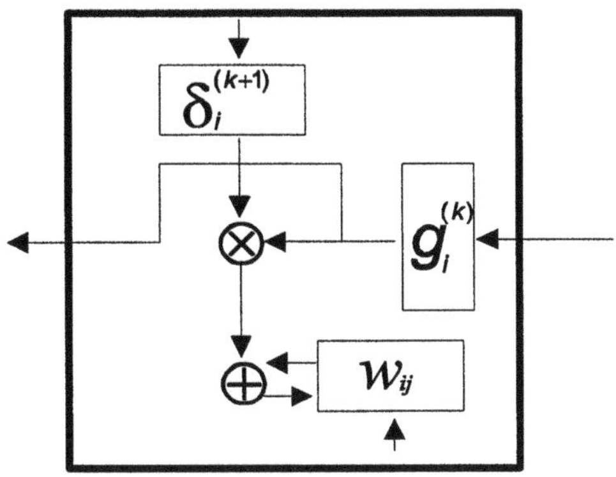

FIGURE 14.20. Functional operations at each PE of the systolic architecture for the weight updating phase

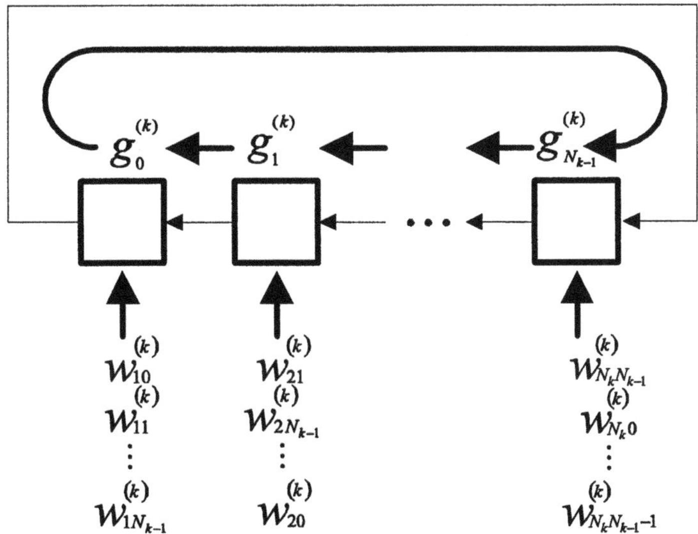

FIGURE 14.21. Systolic architecture for the weight updating phase – initial stage

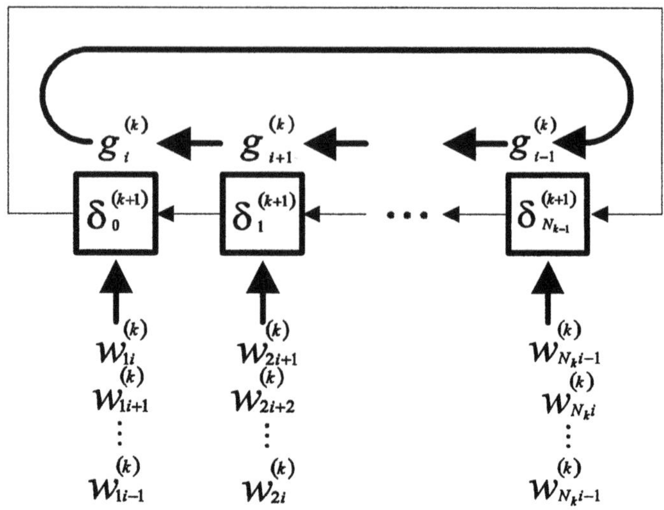

FIGURE 14.22. Systolic architecture for the weight updating phase – intermediate stage

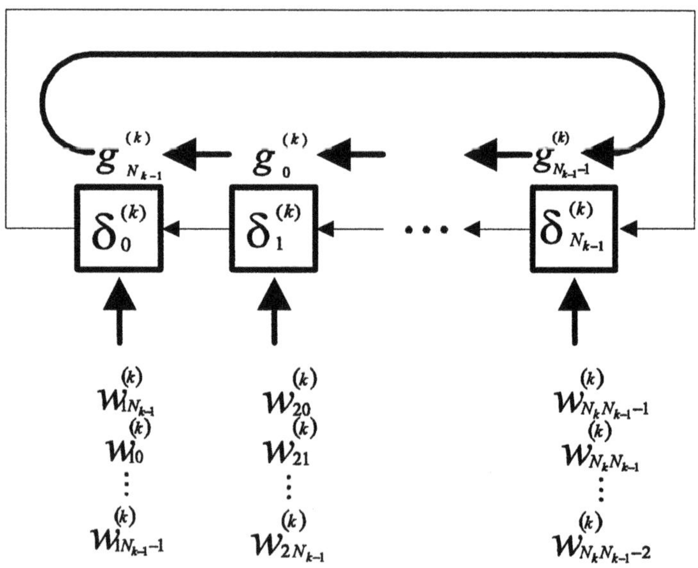

FIGURE 14.23. Systolic architecture for the weight updating phase – final stage

We will now present systolic architectures realizing formulas (14.6) and (14.7), i.e.

$$\mathbf{g}^{(k)}(n) = \frac{\mathbf{P}^{(k)}(n-1)\,\mathbf{x}^{(k)}(n)}{\lambda + \mathbf{x}^{(k)^T}(n)\,\mathbf{P}^{(k)}(n-1)\,\mathbf{x}^{(k)}(n)}$$

and

$$\mathbf{P}^{(k)}(n) = \lambda^{-1}\left[\mathbf{I} - \mathbf{g}^{(k)}(n)\,\mathbf{x}^{(k)^T}(n)\right]\mathbf{P}^{(k)}(n-1)$$

Let us define

$$\mathbf{v}^{(k)}(n) = \mathbf{P}^{(k)}(n-1)\,\mathbf{x}^{(k)}(n)$$

Then the formula for calculating $\mathbf{g}^{(k)}(n)$ takes the form

$$\mathbf{g}^{(k)}(n) = \frac{\mathbf{P}^{(k)}(n-1)\,\mathbf{x}^{(k)}(n)}{\lambda + \mathbf{x}^{(k)^T}(n)\,\mathbf{P}^{(k)}(n-1)\,\mathbf{x}^{(k)}(n)} \tag{14.10}$$

$$= \frac{\mathbf{v}^{(k)}(n)}{\lambda + \mathbf{x}^{(k)^T}(n)\,\mathbf{v}^{(k)}(n)} = \frac{\mathbf{v}^{(k)}(n)}{\lambda + \sum_{i=1}^{N_k} m_i^{(k)}(n)}$$

where $m_i^{(k)}(n)$ is the i-th component of the product $\mathbf{x}^{(k)^T}(n)\,\mathbf{v}^{(k)}(n)$.

In Fig. 14.24 we present the dependence graph for the calculating nominator of formula (14.10), whereas Fig. 14.25 shows functional operations at each node of the dependence graphs for calculating the nominator of formula (14.10).

In Fig. 14.25 we depict the functional operations at each node of the dependence graph for calculating the nominator of formula (14.10). Fig. 14.26 shows systolic architecture for the calculating nominator of formula (14.10) and Fig. 14.27 presents functional operations at each PE for calculating the nominator and denominator of formula (14.10).

In Fig. 14.28 we illustrate the ring structure for calculating $\mathbf{g}^{(k)}(n)$, Fig. 14.29 presents systolic architecture for calculating matrix $\mathbf{P}^{(k)}(n)$ and Fig. 14.30 shows the structure of PE for calculating matrix $\mathbf{P}^{(k)}(n)$.

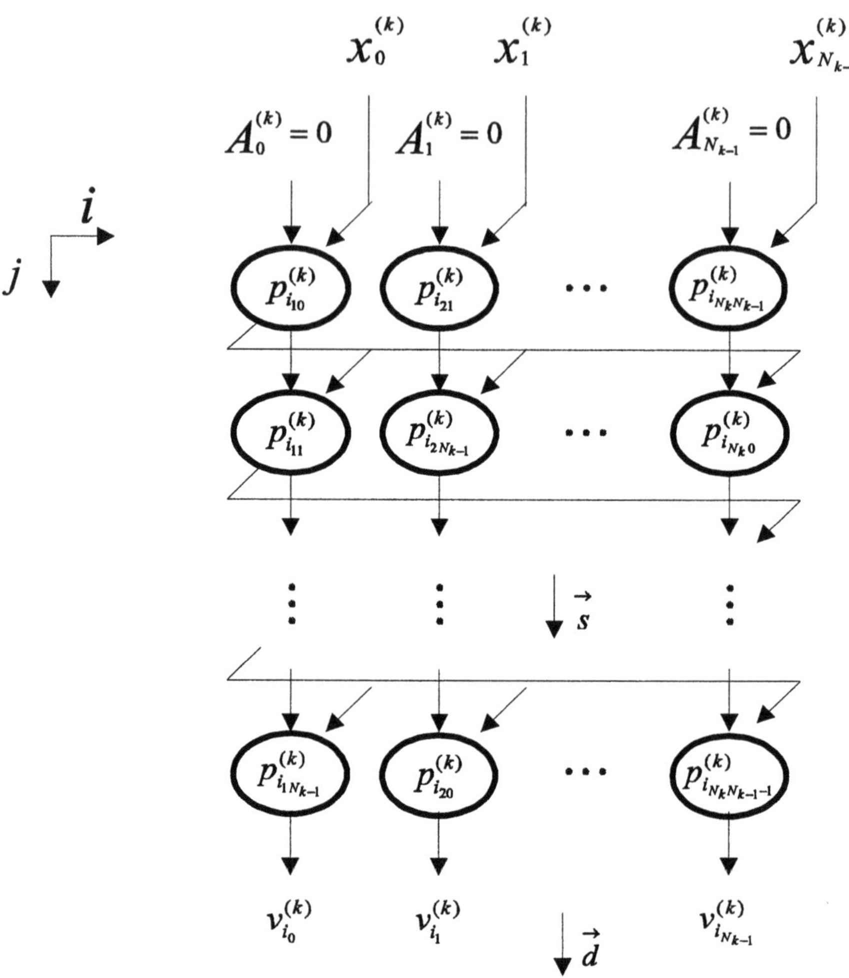

FIGURE 14.24. Dependence graph for calculating the nominator of formula (14.10)

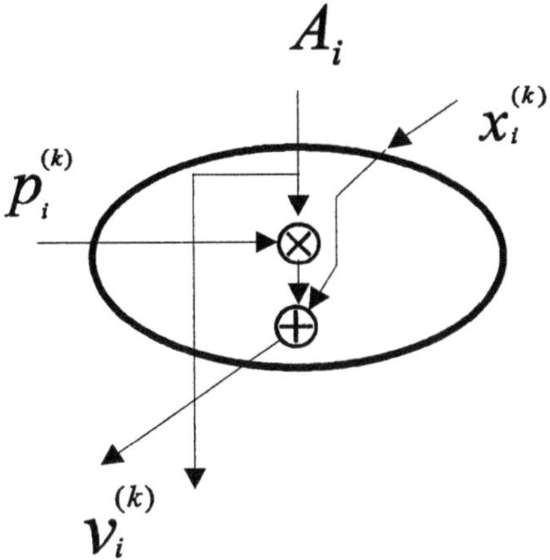

FIGURE 14.25. The functional operations at each node of the dependence graph for calculating the nominator of formula (14.10)

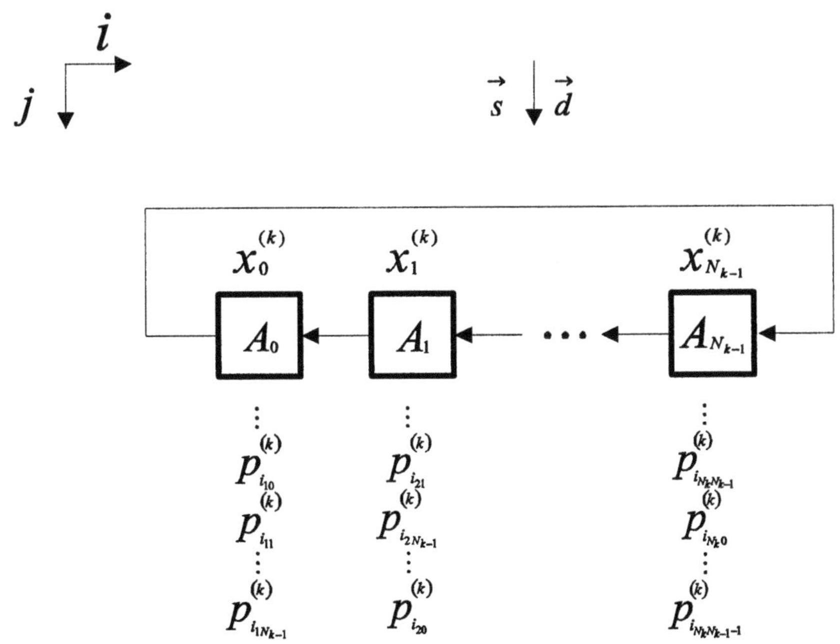

FIGURE 14.26. Systolic architecture for calculating the nominator of formula (14.10)

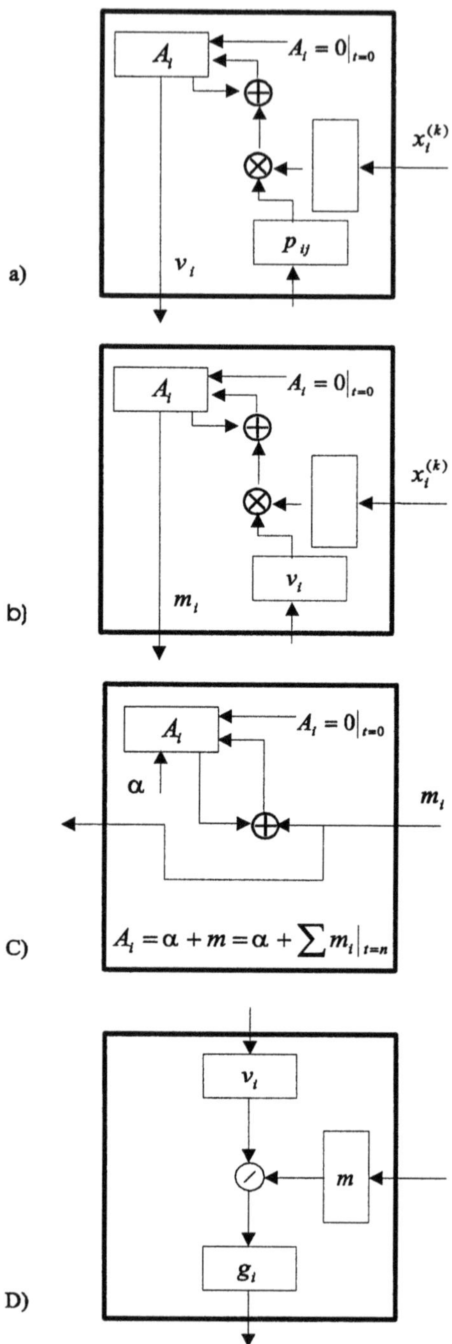

FIGURE 14.27. Functional operations at each PE for calculating the nominator and denominator of formula (14.10)

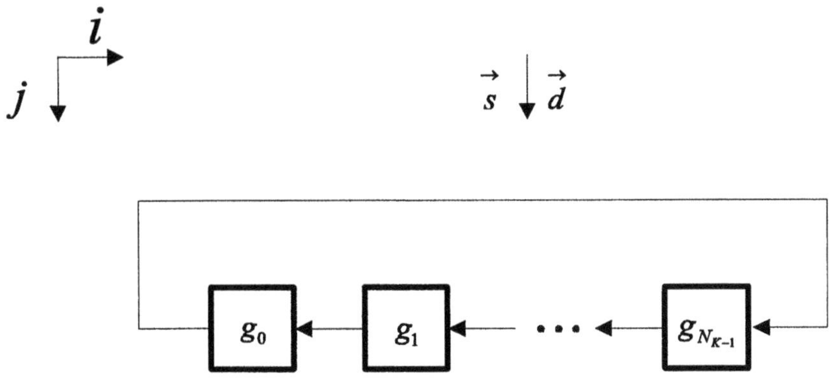

FIGURE 14.28. Ring structure for calculating $g^{(k)}(n)$

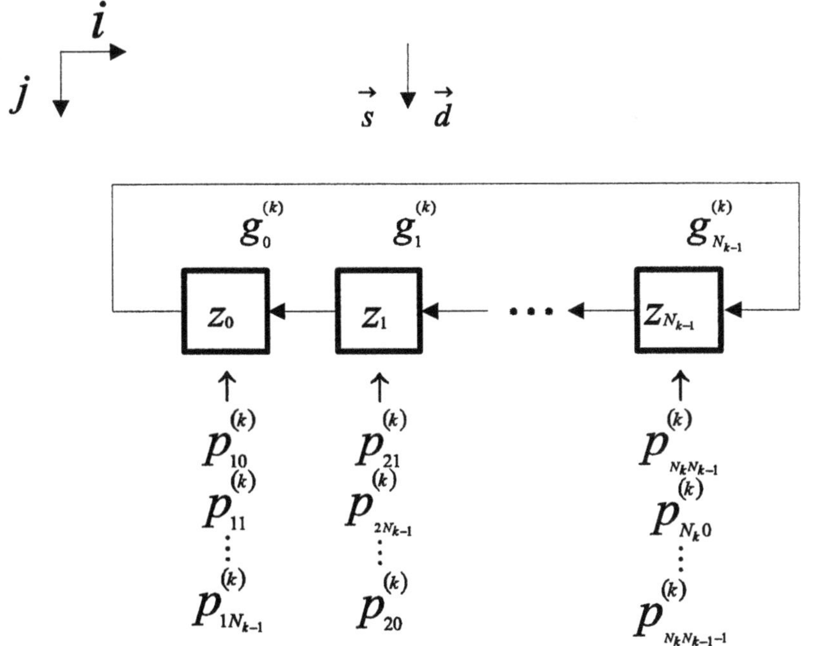

FIGURE 14.29. Systolic architecture for calculating matrix $\mathbf{P}^{(k)}n$

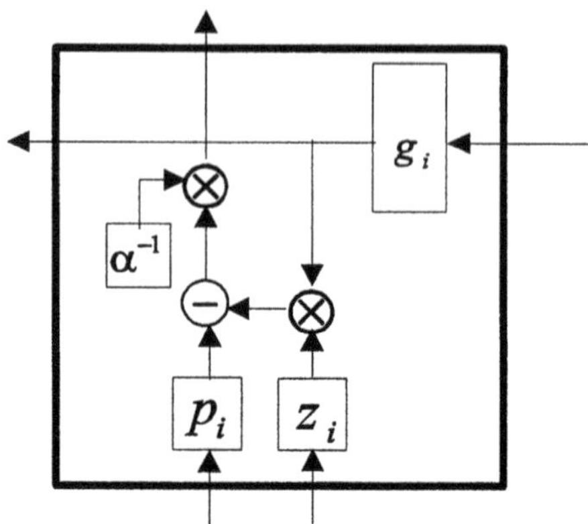

FIGURE 14.30. Structure of PE for calculating matrix $\mathbf{P}^{(k)}(n)$

14.4 Systolic architectures for the RLS learning algorithms

In this section we design the systolic architecture for the RLS learning algorithm described by the following formulas:

$$\gamma_i^{(k)}(n) = \begin{cases} d_i^{(L)}(n) - y_i^{(L)}(n) \quad \text{for} \quad k = L \\ \sum_{j=1}^{N_{k+1}} f'\left(s_j^{(k+1)}(n)\right) w_{ji}^{(k+1)}(n)\, \gamma_j^{(k+1)}(j) \\ \text{for} \quad k = 1...L-1 \end{cases} \tag{14.11}$$

$$g_i^{(k)}(n) = \frac{f'\left(s_i^{(k)}(n)\right) \mathbf{P}_i^{(k)}(n-1)\, \mathbf{x}^{(k)}(n)}{\lambda + f'^2\left(s_i^{(k)}(n)\right) \mathbf{x}^{(k)T}(n)\, \mathbf{P}_i^{(k)}(n-1)\, \mathbf{x}^{(k)}(n)} \tag{14.12}$$

$$\mathbf{P}_i^{(k)}(n) = \lambda^{-1}\left[\mathbf{I} - f'\left(s_i^{(k)}(n)\right) \cdot \mathbf{g}_i^{(k)}(n)\, x^{(k)T}(n)\right] \mathbf{P}_i^{(k)}(n-1) \tag{14.13}$$

$$\delta_i^{(k)}(n) = \gamma_i^{(k)}(n)\, f'\left(s_i^{(k)}(n)\right) \tag{14.14}$$

$$\mathbf{w}_i^{(k)}(n) = \mathbf{w}_i^{(k)}(n-1) + \mathbf{g}_i^{(k)}(n)\, \gamma_i^{(k)}(n) \tag{14.15}$$

The dependence graph corresponding to the error calculation operations is shown in Fig. 14.31 and functional operations at each node of the DG in Fig. 14.32. In Fig. 14.33 we present the ring systolic architecture for the error calculation phase and in Fig. 14.34 the functional operations at each PE of this architecture.

In Fig. 14.35, 14.36 and 14.37 we depict ring systolic architectures for the error calculation phase in initial, intermediate and final stages of data pumping, respectively.

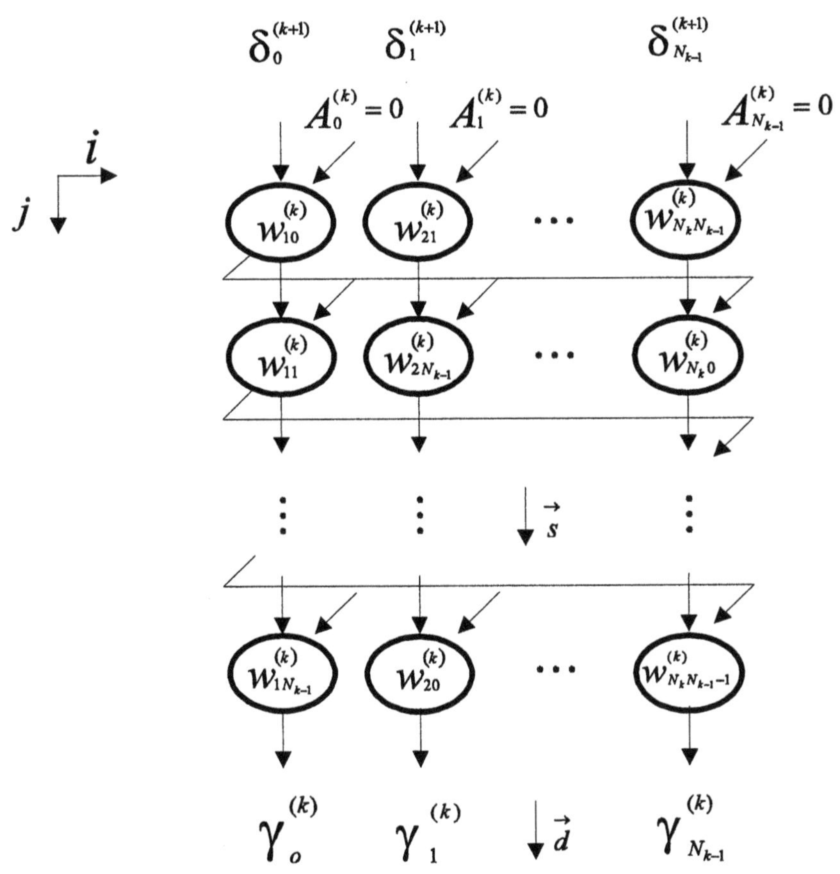

FIGURE 14.31. Dependence graph for the error calculation phase

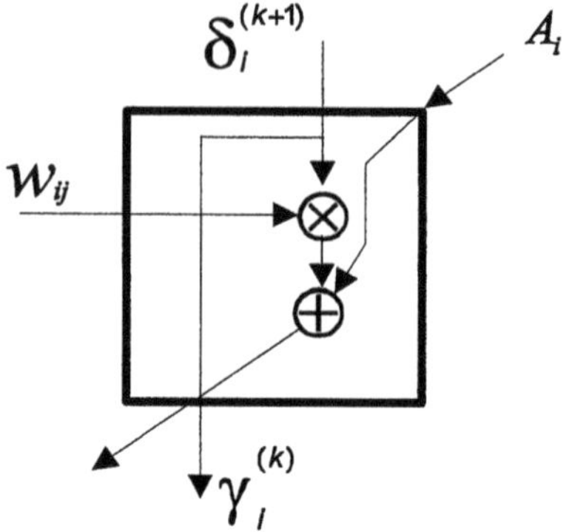

FIGURE 14.32. The functional operations at each node of the dependence graph for the error calculation phase

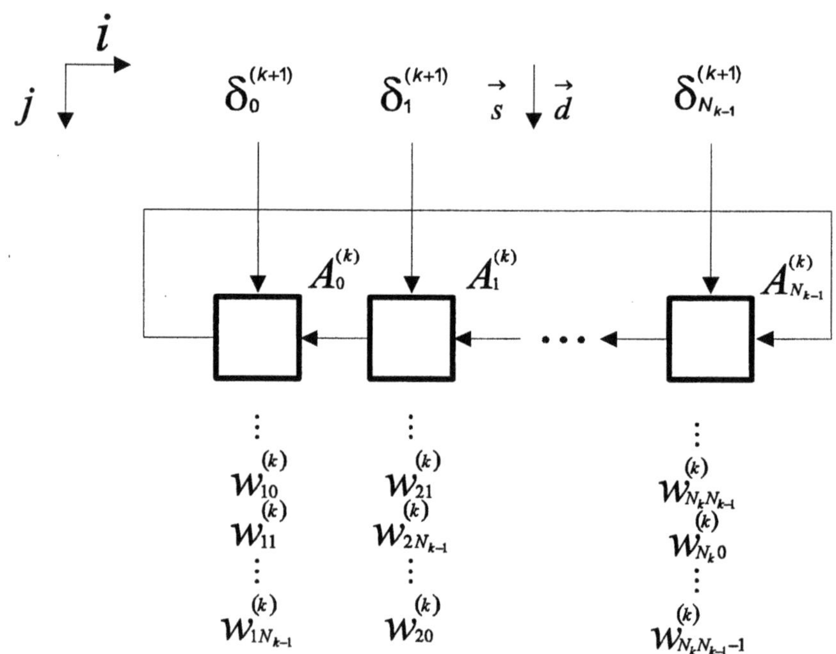

FIGURE 14.33. Ring systolic architecture for the error calculation phase

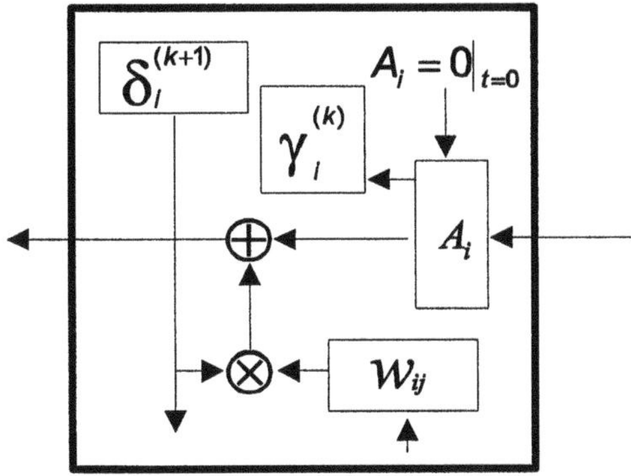

FIGURE 14.34. The functional operations at each PE of the systolic architecture for the error calculation phase

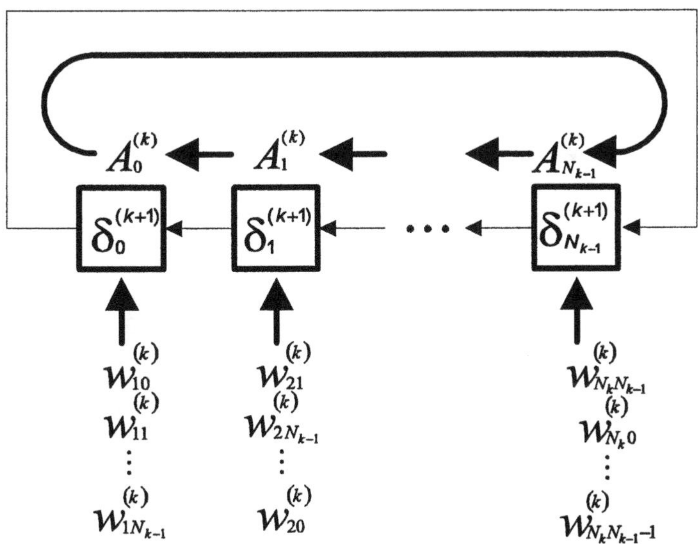

FIGURE 14.35. Ring systolic architecture for the error calculation phase – initial stage

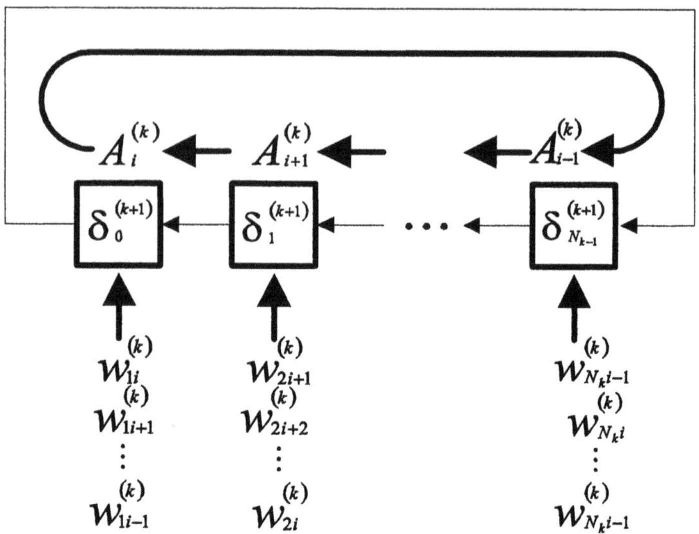

FIGURE 14.36. Ring systolic architecture for the error calculation phase – intermediate stage

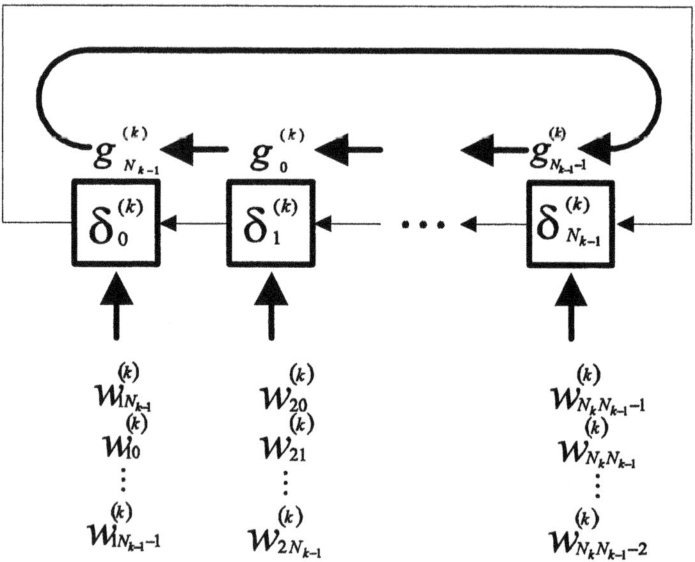

FIGURE 14.37. Ring systolic architecture for the error calculation phase – final stage

Systolic architecture for the weight updating operations again is similar to that of matrix-vector multiplication. The dependence graph for these operations is presented in Fig. 14.38.

The functional operation at each node is also multiplication and accumulation as it is shown in Fig. 14.39. In Fig. 14.40 we illustrate the systolic ring architecture for the weight updating phase, whereas Fig. 14.41 shows the functional operations at each PE of this architecture. In Fig. 14.42, Fig. 14.43 and Fig. 14.44 we present ring systolic architectures for the weight updating in initial, intermediate and final stages of the signal pumping.

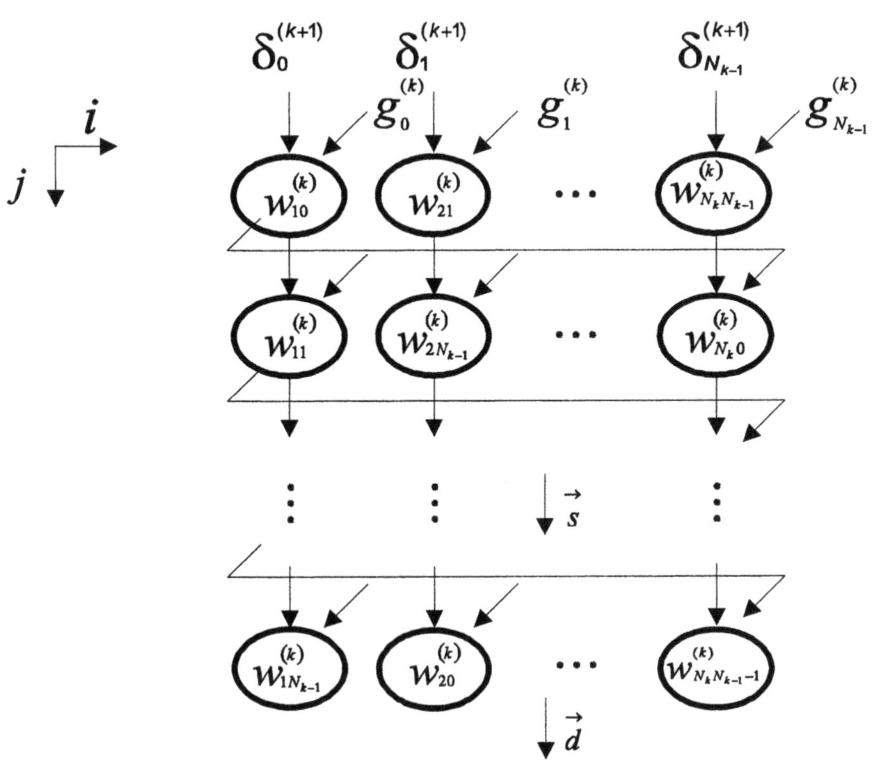

FIGURE 14.38. Dependence graph for the weight updating phase

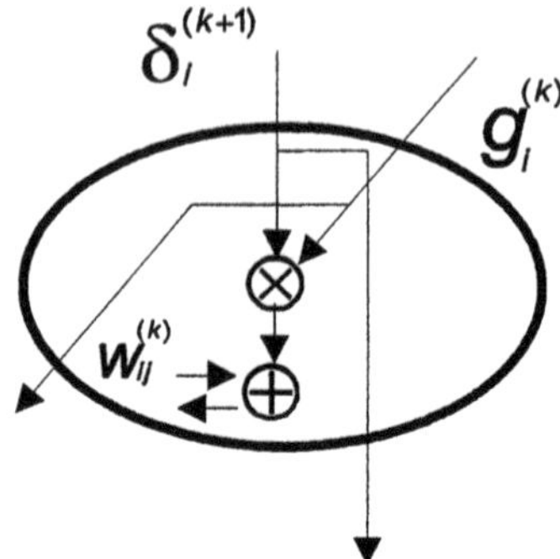

FIGURE 14.39. The functional operations at each node of the dependence graph for the weight updating phase

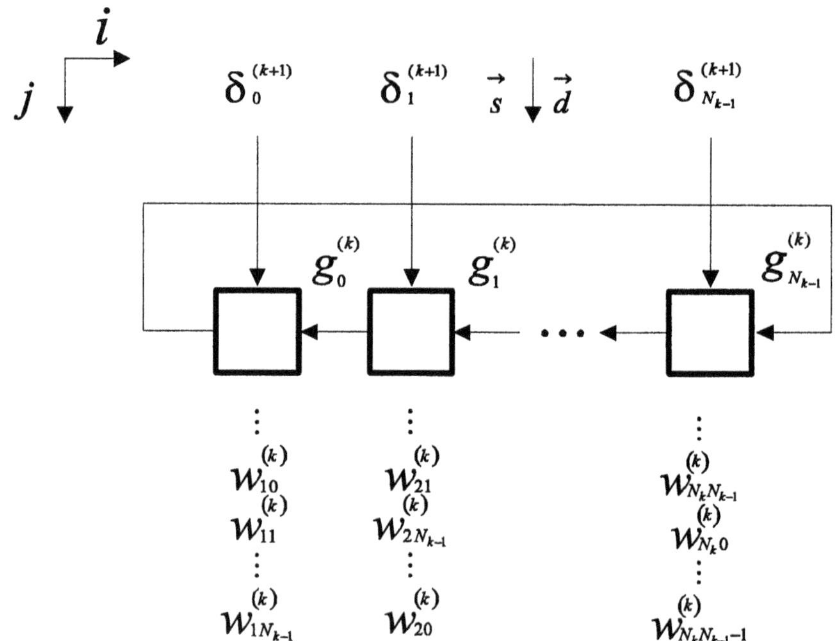

FIGURE 14.40. Systolic ring architecture for the weight updating phase

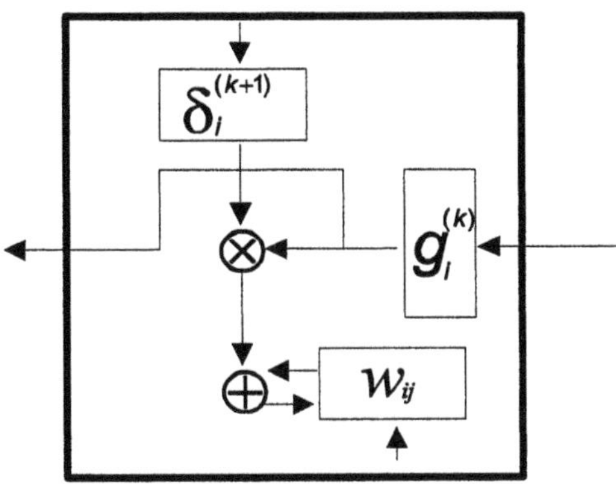

FIGURE 14.41. The functional operations at each PE of the systolic architecture for the weight updating phase

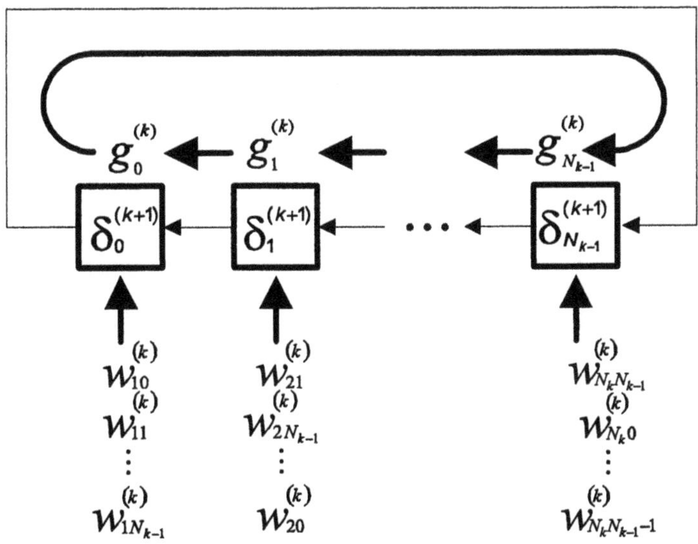

FIGURE 14.42. Ring systolic architectures for the weight updating – initial stage

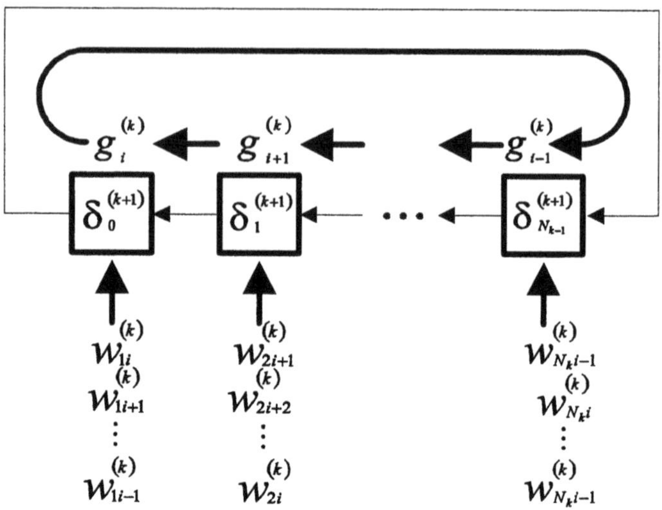

FIGURE 14.43. Ring systolic architectures for the weight updating – intermediate stage

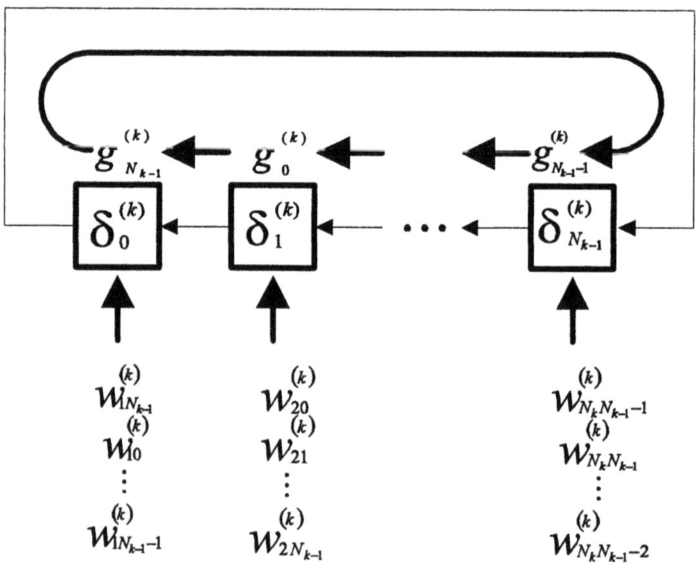

FIGURE 14.44. Ring systolic architectures for the weight updating – final stage

We will now present systolic architectures for calculating matrix $\mathbf{P}_i^{(k)}(n)$ and vector $\mathbf{g}_i^{(k)}(n)$. Contrary to the ETB RLS algorithm studied in Section 14.3, matrix $\mathbf{P}_i^{(k)}(n)$ should be calculated separately for each neuron in a layer. Let us define

$$\mathbf{v}_i^{(k)}(n) = \mathbf{P}_i^{(k)}(n-1)\,\mathbf{x}^{(k)}(n) \tag{14.16}$$

We rewrite formula (14.12) as follows:

$$\begin{aligned}
\mathbf{g}_i^{(k)}(n) &= \frac{f'\left(s_i^{(k)}(n)\right)\mathbf{P}_i^{(k)}(n-1)\,\mathbf{x}^{(k)}(n)}{\lambda + f'^2\left(s_i^{(k)}(n)\right)\mathbf{x}^{(k)^T}(n)\,\mathbf{P}_i^{(k)}(n-1)\,\mathbf{x}^{(k)}(n)} \\
&= \frac{f'\left(s_i^{(k)}(n)\right)\mathbf{v}_i^{(k)}(n)}{\lambda + f'^2\left(s_i^{(k)}(n)\right)\mathbf{x}^{(k)^T}(n)\,\mathbf{v}_i^{(k)}(n)} \qquad (14.17) \\
&= \frac{f'\left(s_i^{(k)}(n)\right)\mathbf{v}_i^{(k)}(n)}{\lambda + f'^2\left(s_i^{(k)}(n)\right)\sum_{i=1}^{N_k} m_i^{(k)}(n)}
\end{aligned}$$

where $m_i^{(k)}(n)$ is the i-th component of product $\mathbf{x}^{(k)^T}(n)\,\mathbf{v}^{(k)}(n)$.

The dependence graph for the calculation of the nominator of formula (14.17) is given in Fig. 14.45 and the functional operations at each node of this graph are presented in Fig. 14.46.

The systolic ring architecture for the calculation of the nominator of formula (14.17) is depicted in Fig. 14.47, whereas the functional operation at each PE of this architecture is shown in Fig. 14.48.

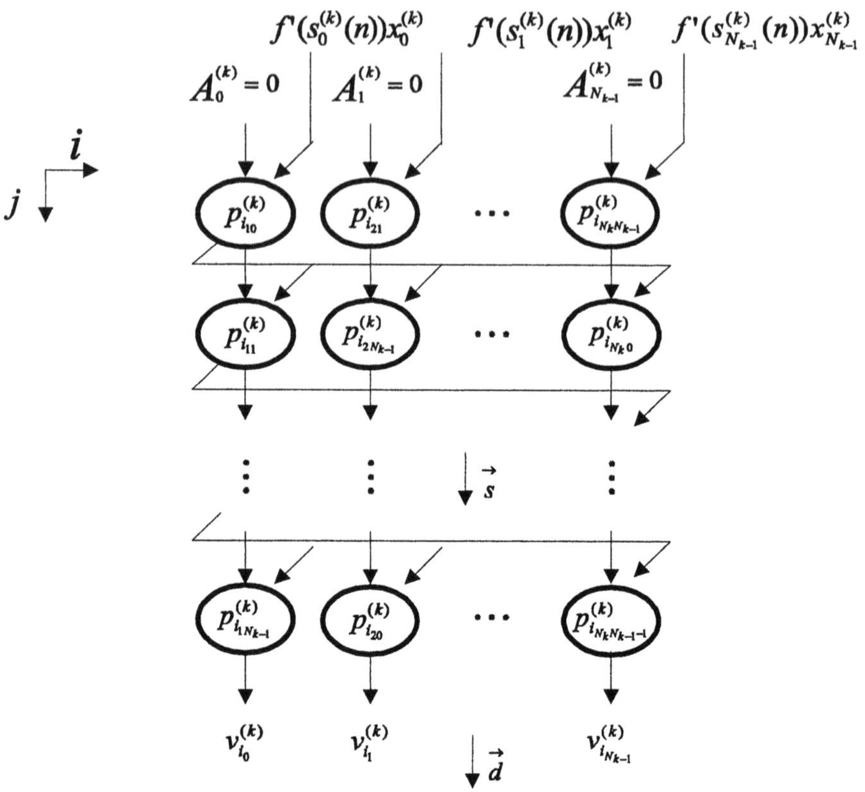

FIGURE 14.45. Dependence graph for calculation of nominator of formula (14.17)

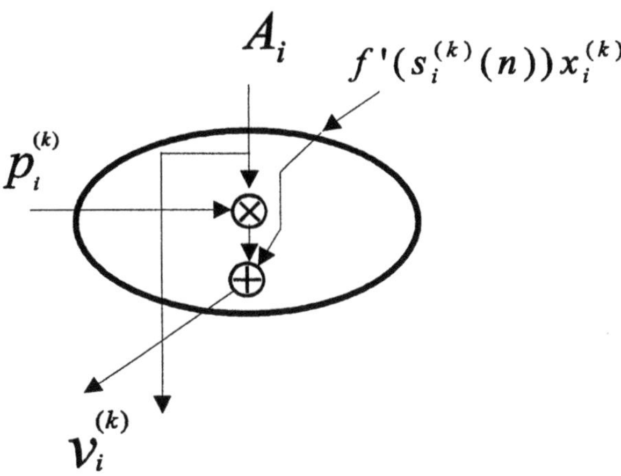

FIGURE 14.46. The functional operations at each node of the dependence graph for calculation of nominator of formula (14.17)

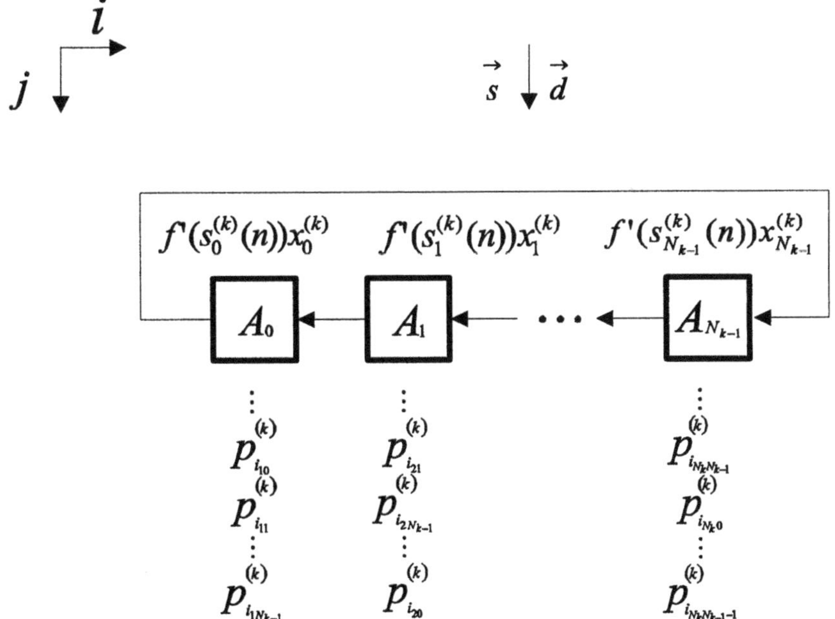

FIGURE 14.47. Systolic ring architecture for calculation of nominator of formula (14.17)

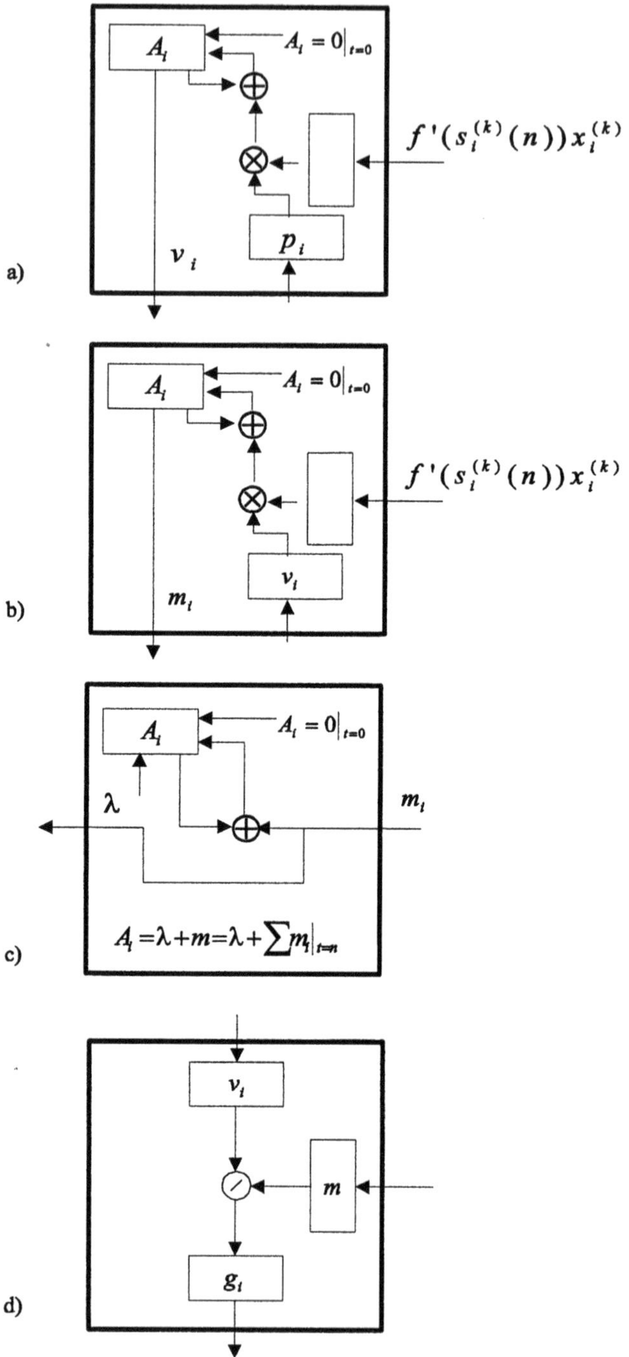

FIGURE 14.48. The functional operations at each PE for calculation of nominator and denominator of formula (14.17)

Finally we design the systolic ring architecture for calculating matrix $\mathbf{P}_i^{(k)}$. It can be done by a multiple use of the structure shown in Fig. 14.29.

Alternatively we propose a simultaneous application of the structures shown in Fig. 14.49 (each structure corresponds to one neuron). We denote $\mathbf{z}_i = f'\left(s_i^{(k)}(n)\right)\mathbf{x}^{(k)^T}$ for a fixed k. In Fig. 14.50 we depict the PE for calculating matrix $\mathbf{P}_i^{(k)}$.

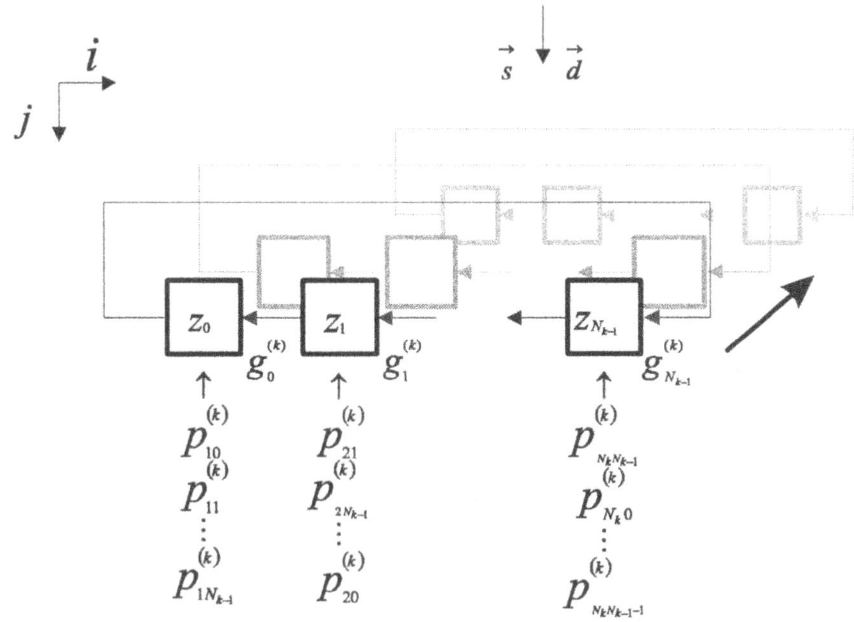

FIGURE 14.49. Systolic ring architecture for calculating matrix $\mathbf{P}_i^{(k)}$

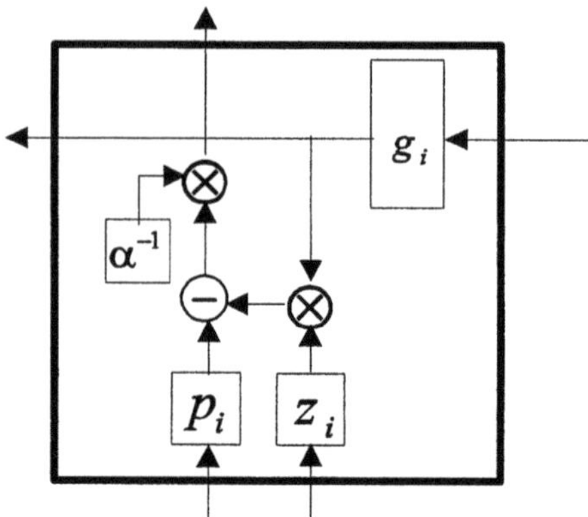

FIGURE 14.50. Structure of PE for calculating matrix $\mathbf{P}_i^{(k)}$

14.5 Performance evaluation of systolic architectures

In this section we analyse the computational performance of ring systolic architectures developed in previous sections and compare them with classical sequential architectures. For simplicity we consider single-layered neural networks having a inputs and c outputs. Moreover, it is assumed that fundamental operations, i.e. multiplications, additions and function (or derivative) value calculations, take the same unit of time.

In Fig. 14.51 – 14.60 we plot numbers of cycles (time units), required to compute the RLS algorithm, versus number of inputs and outputs.

14.5.1 The recall phase

The recall phase requires $2ac$ cycles in the case of the classical architectures (Fig. 14.51) or $a + 1$ cycles in the case of the ring systolic architecture (Fig. 14.52).

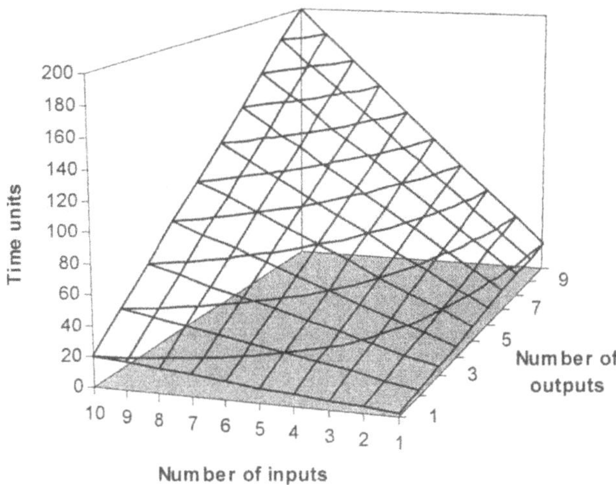

FIGURE 14.51. Performance evaluation of the classical architecture – recall phase

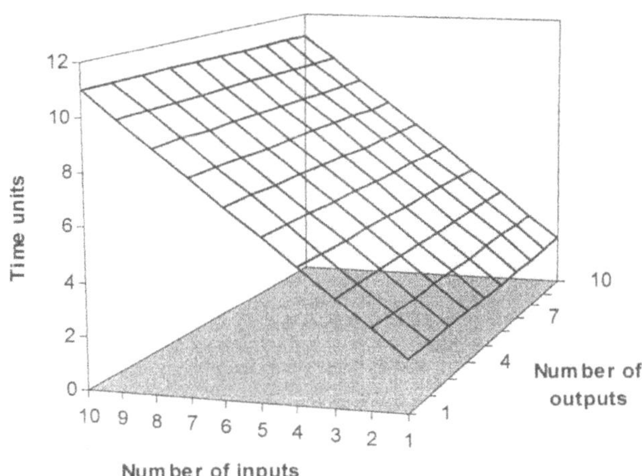

FIGURE 14.52. Performance evaluation of the ring systolic architecture – recall phase

14.5.2 *The learning phase: the ETB RLS algorithm*

We will present performance evaluation of the ETB RLS algorithm (see Section 14.4.2.b).

a) Error calculation

The error calculation requires $2c(a-1)$ cycles in the case of the classical architecture (Fig. 14.53) or $c+1$ cycles in the case of the ring systolic architectures (Fig. 14.54).

b) Weight updating

The weight updating requires $7a^2+2+2ac$ cycles in the case of the classical architecture (Fig. 14.55) or $3(a+1)$ cycles in the case of the ring systolic architecture (Fig. 14.56).

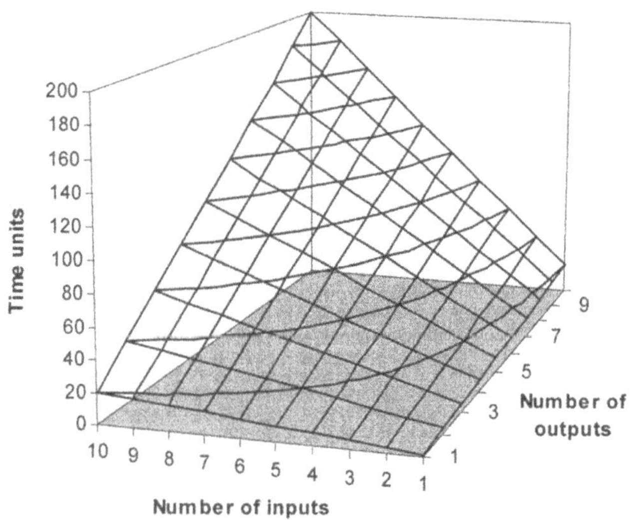

FIGURE 14.53. Performance evaluation of the classical architecture – error calculation

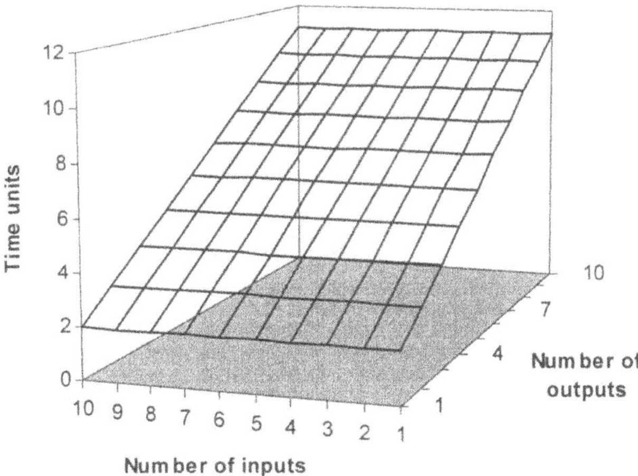

FIGURE 14.54. Performance evaluation of the ring systolic architecture – error calculation

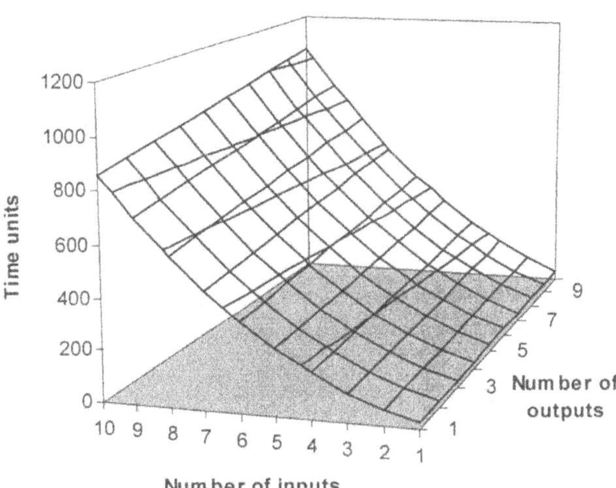

FIGURE 14.55. Performance evaluation of the classical architecture – weight updating

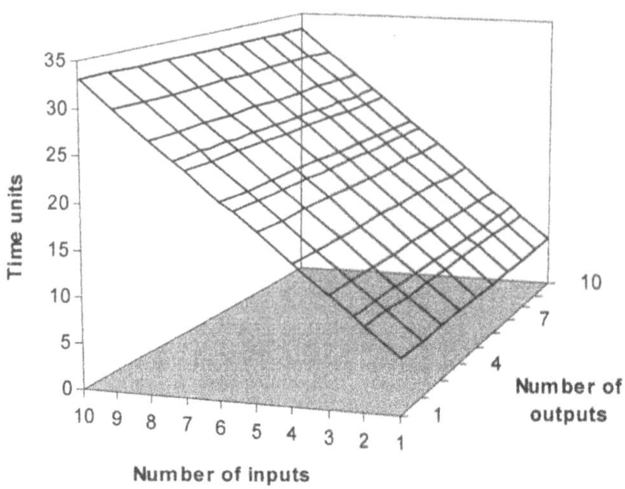

FIGURE 14.56. Performance evaluation of the ring systolic architecture – weight updating

14.5.3 The learning phase: the RLS algorithm

We will present the performance evaluation of the RLS algorithm (see Section 14.4.2.c).

a) Error calculation

The error calculation requires $(c + ca) + (c - 1)\,a + c$ cycles in the case of the classical architecture (Fig. 14.57) or $c + 1$ cycles in the case of the ring systolic architectures (Fig. 14.58).

b) Weight updating

The weight updating requires $\left(7a^2 + 2a\right) c + 2ac$ cycles in the case of the classical architecture (Fig. 14.59) or $3c\,(a + 1)$ cycles in the case of the ring systolic architecture (Fig. 14.60).

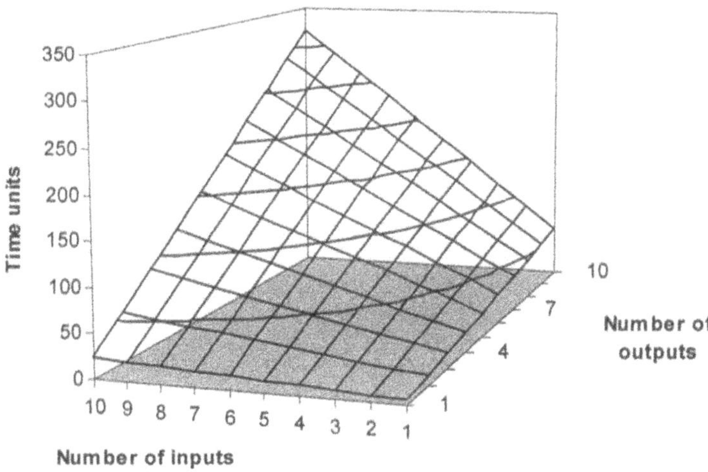

FIGURE 14.57. Performance evaluation of the classical architecture – error calculation

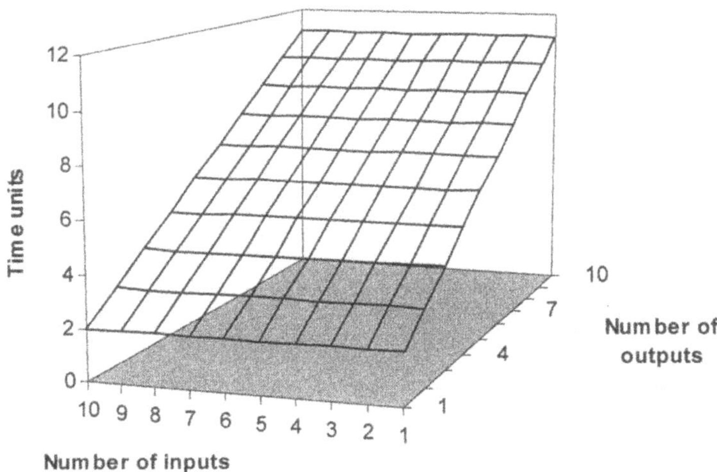

FIGURE 14.58. Performance evaluation of the ring systolic architecture – error calculation

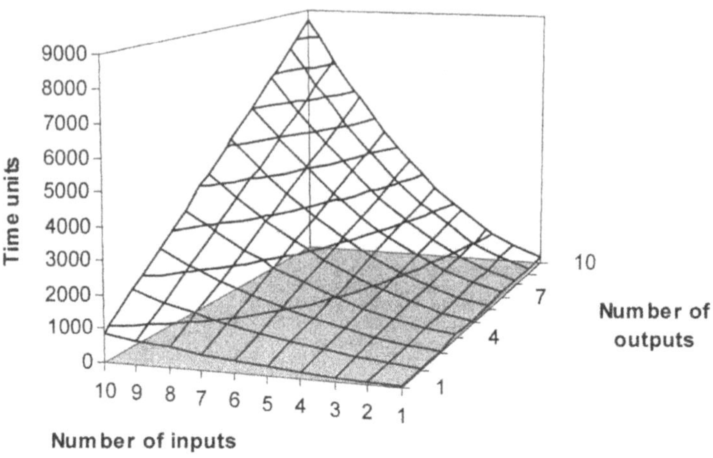

FIGURE 14.59. Performance evaluation of the classical architecture – weight updating

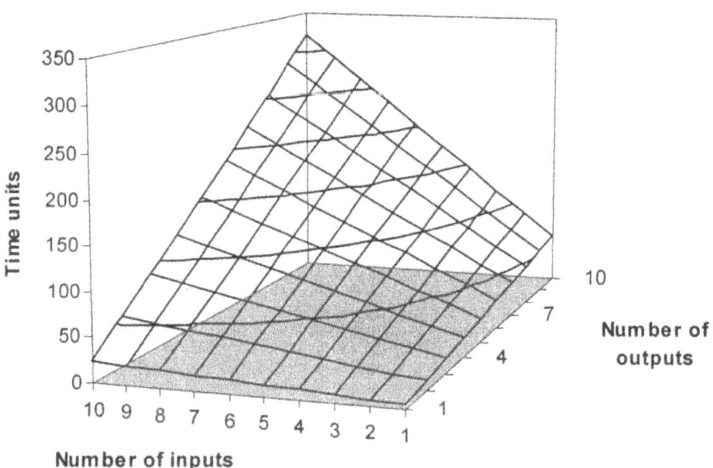

FIGURE 14.60. Performance evaluation of the ring systolic architecture – weight updating

In Fig. 14.61 – Fig. 14.65 we illustrate the following ratios

$$k/s = \frac{\text{time units for classical architectures}}{\text{time units for systolic architectures}}$$

for: recall phase (Fig. 14.61), error calculation – ETB RLS algorithm (Fig. 14.62), error calculation – RLS algorithm (Fig. 14.63), weight updating – ETB RLS algorithm (Fig. 14.64) and weight updating – RLS algorithm (Fig. 14.65).

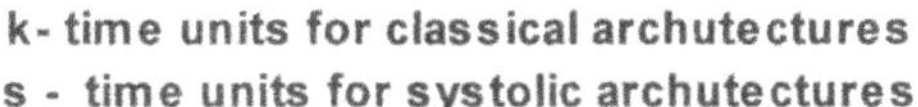

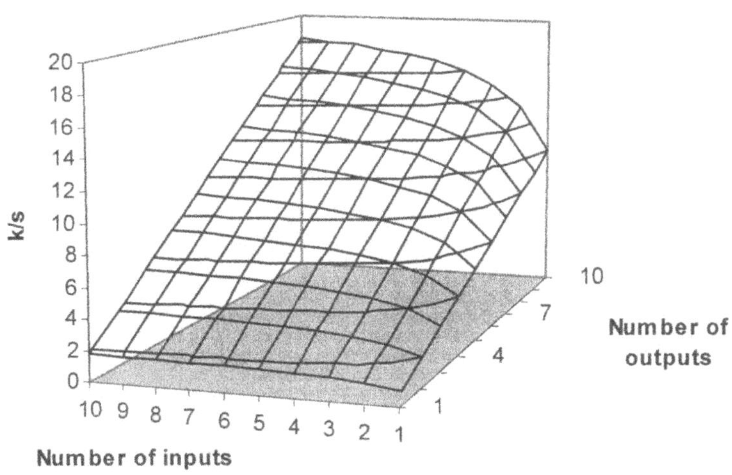

FIGURE 14.61. Ratio k/s for the recall phase

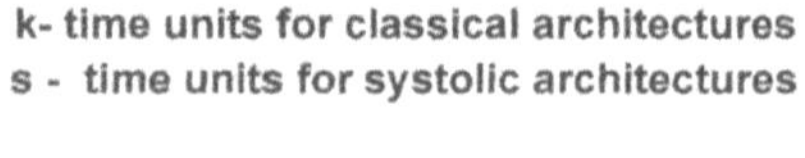

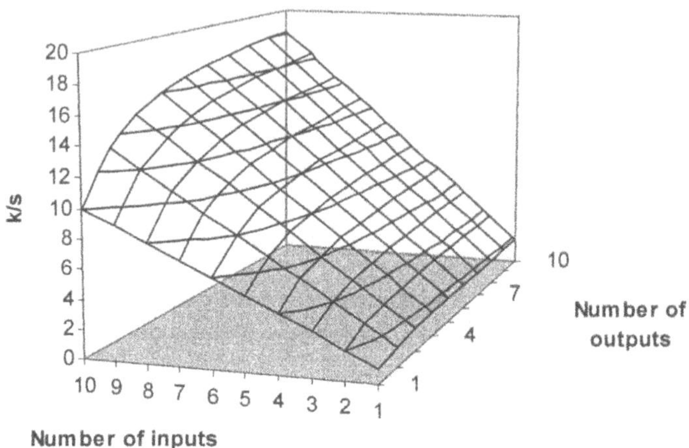

FIGURE 14.62. Ratio k/s for the error calculation – the ETB RLS algorithm

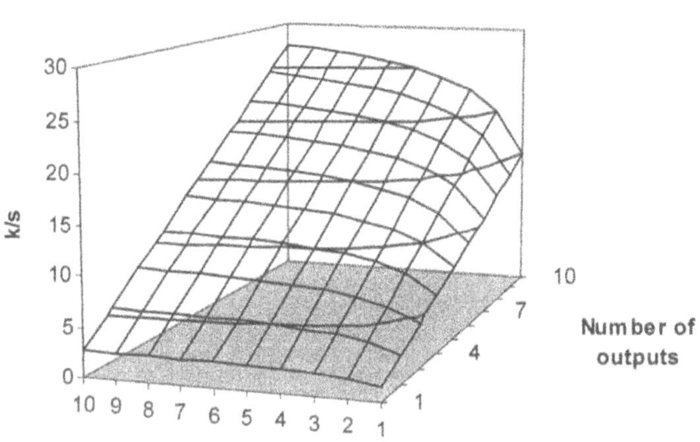

FIGURE 14.63. Ratio k/s for the error calculation – the RLS algorithm

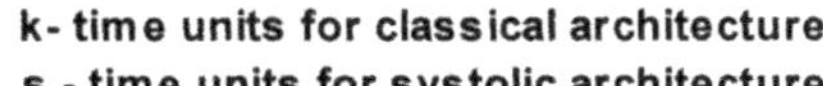

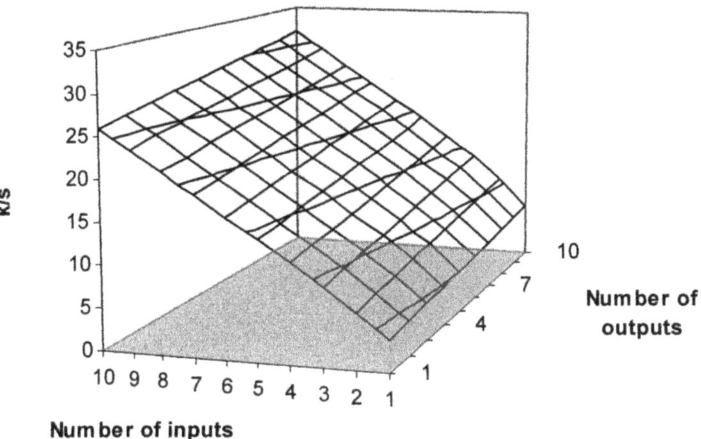

FIGURE 14.64. Ratio k/s for the weight updating – the ETB RLS algorithm

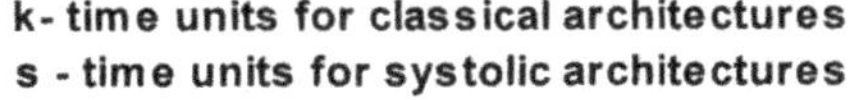

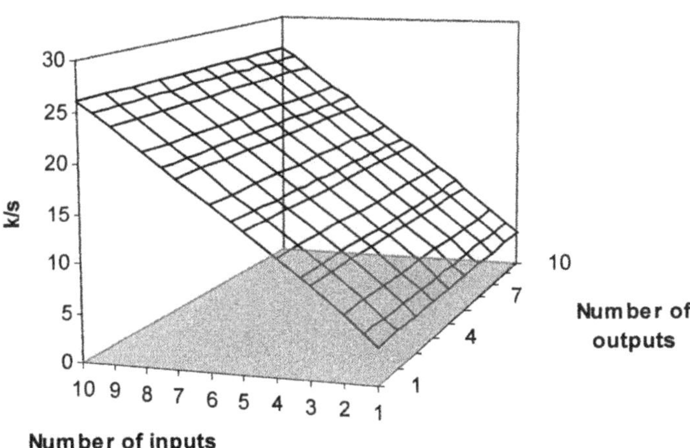

FIGURE 14.65. Ratio k/s for the weight updating – the RLS algorithm

14.6 Concluding remarks

In this chapter we have presented systolic architectures for the implementation of the ETB RLS algorithm and the RLS algorithm. The results are potentially very useful in designing VLSI microelectronics technology for a family of the RLS algorithms studied in Chapter 13. We have also shown that systolic architectures outperform classical computer architectures even 20 – 30 times (see Fig. 14.61 – 14.65).

Appendix

Proof of Theorems 4.1 and 4.2: Observe that

$$\left(\widehat{R}_n(x) - R_n(x)\right)^2 \leq 2\left(\widehat{R}_n(x) - r_n(x)\right)^2 + 2\left(r_n(x) - R_n(x)\right)^2.$$

By making use of (4.3) and (4.12) we get

$$E\left[\left(\widehat{R}_{n+1}(x) - r_{n+1}(x)\right)^2 | X_1, Y_1, ..., X_n, Y_n\right]$$

$$= (1 - a_{n+1})^2 \left(\widehat{R}_n(x) - r_n(x)\right)^2$$

$$+ a_{n+1}^2 E\left[Y_{n+1} K_{n+1}(x, X_{n+1}) - r_{n+1}(x)\right]^2$$

$$+ (1 - a_{n+1})^2 \left(r_{n+1}(x) - r_n(x)\right)^2$$

$$+ 2(1 - a_{n+1})^2 \left(r_n(x) - r_{n+1}(x)\right)\left(\widehat{R}_n(x) - r_n(x)\right).$$

Of course,

$$E\left[Y_{n+1} K_{n+1}(x, X_{n+1}) - r_{n+1}(x)\right]^2 = \operatorname{var}\left[Y_{n+1} K_{n+1}(x, X_{n+1})\right]$$

Using the inequality $2ab \leq a^2 k + b^2 k^{-1}$, true for any $k > 0$ and setting $k = (a_{n+1} c_1)^{-1}$, $0 < c_1 < 1$, we obtain

$$2\left(r_{n+1}(x) - r_n(x)\right)\left(\widehat{R}_n(x) - r_n(x)\right)$$

$$\leq c_1 a_{n+1} \left(\widehat{R}_n(x) - r_n(x)\right)^2 + c_1^{-1} a_{n+1}^{-1} (r_{n+1}(x) - r_n(x))^2$$

The following inequality is true

$$(r_{n+1}(x) - r_n(x))^2 \leq 3(r_{n+1}(x) - R_{n+1}(x))^2$$
$$+3(R_{n+1}(x) - R_n(x))^2 + 3(R_n(x) - r_n(x))^2$$

Consequently,

$$E\left[\left(\widehat{R}_{n+1}(x) - r_{n+1}(x)\right)\right]^2 [X_1, Y_1, ..., X_n, Y_n]$$
$$\leq (1 - a_{n+1}(1 - c_1))\left(\widehat{R}_n(x) - r_n(x)\right)^2$$
$$+a_{n+1}^2 \mathrm{var}\left[Y_{n+1} K_{n+1}(x, X_{n+1})\right] + c_2 a_{n+1}^{-1} \quad (1)$$
$$\cdot (r_{n+1}(x) - R_{n+1}(x))^2 + c_3 a_{n+1}^{-1} (R_{n+1}(x) - R_n(x))^2$$
$$+c_4 a_{n+1}^{-1} (R_n(x) - r_n(x))^2$$

We will now use the following lemma:
(Braverman, Rozonoer [37]). Let W_n be a certain sequence of random variables. Let us introduce a sequence of functions $U_n = U_n(W_1, ..., W_n)$. Let a_n, s_n and t_n be sequences of numbers. Let us assume that
(i)

$$U_n \geq 0, \quad n = 1, 2, ..., \text{ with pr. } 1$$

(ii)

$$EU_1 < \infty$$

(iii)

$$a_n \geq 0, \quad a_n \xrightarrow{n} 0, \quad \sum_{n=1}^{\infty} a_n = \infty$$

a) If

$$E\left[U_{n+1} \left| W_1, ..., W_n\right.\right] \leq (1 - a_n) U_n + a_n s_n,$$

where

$$s_n \xrightarrow{n} 0,$$

then

$$EU_n \xrightarrow{n} 0.$$

b) If

$$E\left[U_{n+1} \left| W_1, ..., W_n\right.\right] \leq (1 - a_n) U_n + t_n,$$

where

$$\sum_{n=1}^{\infty} t_n < \infty,$$

then

$$U_n \xrightarrow{n} 0 \text{ with pr. } 1$$

Applying the above lemma to inequality (1), we obtain the conclusion of Theorems 4.1 and 4.2.■

Proof of Theorems 4.3 and 4.4: Of course,

$$\int \left(\widehat{R}_n(x) - R_n(x)\right)^2 dx \le 2 \int \left(\widehat{R}_n(x) - r_n(x)\right)^2 dx$$
$$+2 \int_A (r_n(x) - R_n(x))^2 dx.$$

Using an argumentation similar to the previous proof, we obtain

$$E\left[\int_A \left(\widehat{R}_{n+1}(x) - r_{n+1}(x)\right)^2 dx \,|X_1, Y_1, ..., X_n, Y_n\right]$$
$$\le (1 - a_{n+1}(1 - c_5)) \int_A \left(\widehat{R}_n(x) - r_n(x)\right)^2 dx$$
$$+a_{n+1}^2 \int \text{var } [Y_{n+1} K_{n+1}(x, X_{n+1})] \, dx$$
$$+c_6 a_{n+1}^{-1} \int (r_{n+1}(x) - R_{n+1}(x))^2 dx \tag{2}$$
$$+c_7 a_{n+1}^{-1} \int (R_{n+1}(x) - R_n(x))^2 dx$$
$$+c_8 a_{n+1}^{-1} \int (r_n(x) - R_n(x))^2 dx.$$

The application of the lemma that was quoted above concludes the proof. ■

Proof of Theorems 4.5 and 4.6: These theorems are a consequence of the application of Chung's [56] lemma (for $0 < a < 1$) or

Watanabe's [291] (for $0 < a \leq 1$) to expressions (1) and (2). ∎

Proof of Theorem 4.7: On the basis of a well-known theorem about arithmetic means [275]

$$r_n(x) \xrightarrow{n} R(x) \Longrightarrow n^{-1} \sum_{i=1}^{n} r_i(x) \longrightarrow R(x).$$

The first part of the theorem follows from the following inequality:

$$E\left[\widehat{R}_n(x) - R_n(x)\right]^2 \leq 3n^{-2} \sum_{i=1}^{n} \operatorname{var}\left[Y_i K_i(x, X_i)\right]$$

$$+3\left(n^{-1} \sum_{i=1}^{n} r_i(x) - R(x)\right)^2 + 3\left(R(x) - R_n(x)\right)^2$$

Moreover, note that

$$\left|\widehat{R}_n(x) - R_n(x)\right| \leq \left|n^{-1} \sum_{i=1}^{n} \left(Y_i K_i(x, X_i) - E Y_i K_i(x, X_i)\right)\right|$$

$$+\left|n^{-1} \sum_{i=1}^{n} r_i(x) - R(x)\right| + \left|R(x) - R_n(x)\right|.$$

Now the second part of the theorem is a consequence of the application of Kołmogorow's theorem (see e.g. [275]). ∎

Proof of Corollary 4.1: From the obvious inequalities

$$I_{n,k}(x) \leq I_n(x) + \left|R_{n+k}(x) - R_n(x)\right| \tag{3}$$

and

$$I_{n,k} \leq 2I_n + 2 \int \left(R_{n+k}(x) - R_n(x)\right)^2 dx \tag{4}$$

it follows that in order to ensure the convergence of procedure (4.3) for the prediction problem, the conditions of theorems presented in Sections 4.4 and 4.5 should be supplemented with

$$\left|R_{n+k}(x) - R_n(x)\right| \xrightarrow{n} 0 \tag{5}$$

and

$$\int \left(R_{n+k}(x) - R_n(x)\right)^2 dx \xrightarrow{n} 0 \tag{6}$$

It is easily seen that for $k=1$, conditions (5) and (6) are implied by assumptions (4.17) and (4.27).
For $k \geq 2$, the following holds

$$|R_{n+k}(x) - R_n(x)| \leq |R_{n+k}(x) - R_{n+k-1}(x)| \tag{7}$$

$$+ |R_{n+k-1}(x) - R_{n+k-2}(x)| + \ldots + |R_{n+1}(x) - R_n(x)|.$$

Moreover, applying inequality $(a+b)^2 \leq 2a^2 + 2b^2$ many times, we obtain

$$\int (R_{n+k}(x) - R_n(x))^2 \, dx$$

$$\leq c_1 \int (R_{n+k}(x) - R_{n+k-1}(x))^2 \, dx$$

$$+ c_2 \int (R_{n+k-1}(x) - R_{n+k-2}(x))^2 \, dx \tag{8}$$

$$+ \ldots + c_k \int (R_{n+1}(x) - R_n(x))^2 \, dx.$$

It means that for $k \geq 2$, conditions (5) and (6) are implied by assumptions (4.17) and (4.27) which concludes the proof. ■

Proof of Corollary 4.2: The first part of the corollary results immediately from inequalities (3) and (7). In a similar way, the second part is a direct consequence of inequalities (4) and (8). ■

Proof of Theorem 5.1: The conclusion of this theorem results immediately from the following inequality:

$$\left|\widehat{\phi}_n(x) - \phi_n^*(x)\right| \leq \frac{1}{\widehat{f}_n(x)} \left(\left|\widehat{R}_n(x) - R_n(x)\right|\right. \tag{9}$$

$$\left. + \left|\phi_n^*(x)\left(\widehat{f}_n(x) - f(x)\right)\right|\right) ■$$

Proof of Corollaries 5.1 and 5.2: Observe that

$$\operatorname{var}\left[Y_n K_n(x, X_n)\right] \leq E Y_n^2 K_n^2(x, X_n)$$

$$= h_n^{-2p} \int E\left[Y_n^2 \,|X_n = u\right] f_n(u) K^2\left(\frac{x-u}{h_n}\right) du \tag{10}$$

$$\leq 2 \sup_u |K(u)| h^{-p} m_n',$$

Assessment (10) can be alternatively carried out in the following manner

$$\mathrm{var}\left[Y_n K_n (x, X_n)\right] \leq E Y_n^2 K_n^2 (x, X_n) \leq 2(\sup K(x))^2 h_n^{-2p} m_n''$$

Of course,

$$r_n(x) = h_n^{-p} E Y_n K\left(\frac{x - X_n}{h_n}\right)$$

Denote

$$P_n(x) = |r_n(x) - R_n(x)|, \tag{11}$$

The expression (11) can be written as

$$P_n(x) = \left| \int K(u) \left[R_n(x - h_n u) - R_n(x)\right] du \right|$$

Expanding functions $R_n(x - h_n u)$ in the multidimensional Taylor's series, we obtain

$$\begin{aligned} P_n(x) &= h_n^r \int \dots \int \prod_{i=1}^{p} H\left(u^{(i)}\right) \left[u^{(1)} \frac{\partial}{\partial x^{(1)}} + \dots + u^{(p)} \frac{\partial}{\partial x^{(p)}}\right]^r \\ &\quad \times R_n(x - h_n u \theta_n) \, du^{(1)} \dots du^{(p)}, \end{aligned}$$

where $0 < \theta_n < 1$.As a result

$$|P_n(x)| \leq \text{const. } h_n^r \sum_{i_1=1}^{p} \dots \sum_{i_r=1}^{p} \sup_x \left| \frac{\partial^r}{\partial x^{(i_1)} \dots \partial x^{(i_r)}} R_n(x) \right|$$

assuming that functions R_n have continuous partial derivatives up to the r-th order.

Now, Corollaries 5.1 and 5.2 are consequences of Theorems 5.1 and 5.2. ■

Proof of Corollaries 5.3 and 5.4: Observe that

$$\mathrm{var}\left[Y_n K_n (x, X_n)\right] \leq E Y_n^2 K_n^2 (x, X_n)$$

$$= \int E\left[Y_n^2 | X_n = u\right] f(u) \left(\sum_{|\underline{j}| \leq q} \Psi_{\underline{j}}(x) \Psi_{\underline{j}}(u) \right)^2 du \tag{12}$$

$$\leq 2\left(\sum_{j=0}^{q(n)} G_j^2\right)^p m_n'$$

Assessment (12) can be alternatively expressed as

$$\operatorname{var}\left[Y_n K_n\left(x, X_n\right)\right] \leq E Y_n^2 K_n^2\left(x, X_n\right) \leq \left(\sum_{j=0}^{q(n)} G_j^2\right)^{2p} 2m_n''$$

Observe that

$$r_n(x) = \sum_{|\underline{j}| \leq q} b_{\underline{j}n} \Psi_{\underline{j}}(x),$$

where

$$b_{\underline{j}n} = E Y_n \Psi_{\underline{j}}\left(X_n\right).$$

Now both corollaries follow directly from Theorems 4.1 and 4.2. ■

Proof of conditions (5.45) and (5.46): Let B_{jn} be the coefficient of the expansion of function t_n^l, $l \geq 1$, $n = 1, 2, \ldots$, into multidimensional Hermite series. If $t_n^l \in L_2(R^p)$, then

$$b_{j_p, \ldots, j_p, n} \leq \frac{\left|B_{j_1+1,n}, \ldots, B_{j_p+1,n}\right|}{j_1^{1/2} \ldots j_p^{1/2}}. \tag{13}$$

For $p = 1$ the above inequality was presented in work [289] and its generalization to the multidimensional case is straightforward. Assuming that (5.44) is true, under (13) and (2.10), we obtain

$$\begin{aligned} S_n(x) &\leq \left|\sum_{|\underline{j}| \leq q} b_{jn} \Psi_{\underline{j}}(x)\right| \leq \left\|t_n^l\right\|_{L_2} \left(\sum_{i=q+1}^{\infty} i^{-l+1/6}\right)^{\frac{p}{2}} \\ &\leq \left\|t_n^l\right\|_{L_2} [q(n)]^{\frac{-p(l-5/6)}{2}} \end{aligned}$$

Carrying out similar considerations for the multidimensional Fourier series, using [100] we obtain (5.45) and (5.46). ■

Proof of Theorem 5.2: We will use ideas presented in work [99]. Let us consider the following occurrences:

$$A_1 : \left|\widehat{f}_n(x) - f(x)\right| < f(x) \frac{\varepsilon}{\varepsilon + 2},$$

$$A_2 : \left|\widehat{R}_n(x) - R_n(x)\right| < f(x)\frac{\varepsilon}{\varepsilon+2},$$

$$A_3 : \left|\phi_n^*(x)\left(\widehat{f}_n(x) - f(x)\right)\right| < f(x)\frac{\varepsilon}{\varepsilon+2}.$$

Under inequality (9), occurrences A_1, A_2 and A_3 imply the occurrence

$$B : \left|\widehat{\phi}_n(x) - \phi_n^*(x)\right| < \varepsilon.$$

As a result,

$$P\left(\overline{B}\right) \leq P\left(\overline{A_1 \cap A_2 \cap A_3}\right) \leq P\left(\overline{A}_1\right) + P\left(\overline{A}_2\right) + P\left(\overline{A}_3\right),$$

which concludes the proof of Theorem 5.2. ■

Proof of Corollary 5.5: With reference to symbols of Theorem 4.7, let us denote

$$R(x) = F(x) f(x)$$

Obviously,

$$|r_n(x) - R(x)| \leq \left|r_n(x) - \overline{R}_n(x)\right| + \left|\overline{R}_n(x) - R(x)\right|,$$

where

$$\overline{R}_n(x) = \int K(u) R(x - h_n u)\, du$$

Observe that

$$\left|r_n(x) - \overline{R}_n(x)\right| = \left|h_n^{-p} \int K\left(\frac{x-z}{h_n}\right)(R_n(z) - R(z))\, dz\right|$$

If function K satisfies conditions (5.27) then the following inequalities are true

$$\left|r_n(x) - \overline{R}_n(x)\right| \leq \sup |R_n(x) - R(x)|$$

$$\left|r_n(x) - \overline{R}_n(x)\right| \leq h_n^{-p} \sup |K(x)| \int |R_n(x) - R(x)|\, dx$$

$$\left|r_n(x) - \overline{R}_n(x)\right| \leq (\sup |K(x)|)^{1/2} h_n^{-p/2} \|R_n(x) - R(x)\|_{L_2}$$

Since $R \in L_1$, then

$$\left|\overline{R}_n(x) - R(x)\right| \xrightarrow{n} 0$$

for almost all x (see [260], [299]), which concludes the proof of this corollary. ■

Proof of Corollary 5.6: Note that

$$|r_n(x) - R(x)| \le |r_n(x) - S_n(x)| + |S_n(x) - R(x)|$$

and

$$|r_n(x) - S_n(x)| = \left| \sum_{|\underline{j}| \le q} \Psi_{\underline{j}}(x) \left(d_{\underline{j}n} - D_{\underline{j}} \right) \right|$$

The following inequalities are true:

$$|r_n(x) - S_n(x)| \le \sup |\phi_n^*(x) - F(x)| \left(\sum_{j=0}^{q} G_j^2 \right)^p$$

$$|r_n(x) - S_n(x)| \le \|R_n(x) - R(x)\|_{L_2} \left(\sum_{j=0}^{q} G_j \right)^p$$

$$|r_n(x) - S_n(x)| \le \int |R_n(x) - R(x)| \, dx \left(\sum_{j=0}^{q} G_j^2 \right)^p$$

Now, this corollary follows directly from Theorem 4.7. ■

Proof of Theorem 6.1: Slightly modifying the proof of the theorem in work [300], we obtain

$$0 \le L_{n+k}(\widehat{\varphi}_n) - L_{n+k}(\varphi_{n+k}^*)$$

$$\le \sum_{m=1}^{M} \int_{\chi_n} \left| \widehat{d}_{mn}(x) - d_{m,n+k}(x) \right| dx \tag{14}$$

$$+ \sum_{m=1}^{M} \int_{\chi_n} d_{m,n+k}(x) \, dx$$

Under Schwartz's inequality

$$\begin{aligned} & E\sum_{m=1}^{M}\int_{\chi_n}\left|\widehat{d}_{mn}(x)-d_{m,n+k}(x)\right|dx \\ \leq\ & \mu^{\frac{1}{2}}(\chi_n)\sum_{m=1}^{M}\left[\int E\left(\widehat{d}_{mn}(x)-d_{m,n+k}(x)\right)^2\right]^{\frac{1}{2}} \end{aligned}$$

Thus, from the inequality

$$\sum_{m=1}^{M}\int_{\chi_n}d_{m,n+k}(x)\,dx\leq\frac{\varepsilon}{2}$$

follows the first part of the theorem.

The second part can be proved in a similar way. ■

Proof of Theorem 6.2: From inequality (14) it follows that

$$\begin{aligned} 0\ \leq\ & EL_{n+k}(\widehat{\varphi}_n)-L_{n+k}\left(\varphi^*_{n+k}\right) \\ \leq\ & \mu^{\frac{1}{2}}(\chi_n)\sum_{m=1}^{M}\left[\int E\left(\widehat{d}_{mn}(x)-d_{m,n+k}(x)\right)^2\right]^{\frac{1}{2}} \\ & +\sum_{m=1}^{M}\int_{\chi_n}f_{m,n+k}(x)\,dx. \end{aligned}$$

Let χ_n be a p-dimensional cube with the middle in point zero and let the length of its side be e_n. Then, for $s>0$,

$$\int_{\chi_n}f_{m,n+k}(x)\,dx\leq e_n^{-sp}t_{m,n+k}$$

Consequently,

$$0\leq EL_{n+k}(\widehat{\varphi}_n)-L_{n+k}\left(\varphi^*_{n+k}\right)\leq\text{const.}\ e_n^{p/2}u_n^{1/2}+e_n^{-sp}t_{n+k}$$

Choosing $e_n=0\left(u_n^{-1/(2s+1)p}\right)$, we obtain the conclusion of Theorem 6.2. Let us point out that if $f_{mn}=f_m(x-c_{mn})$, then as χ_n we may choose a p-dimensional cube with the middle in c_{mn}. ■

Proof of Corollaries 6.1 and 6.2: Let us point out that

$$ET_{mn}K_n(x, X_n) = p_{mn} \int K_n(x, u) f_{mn}(u)\, du$$

Hence

$$\int (ET_{mn}K_n(x, X_n) - d_{mn}(x))^2\, dx = p_{mn}^2 \tag{15}$$

$$\cdot \int \left(\int K(u) (f_{mn}(x - uh_n) - f_{mn}(x))\, du \right)^2 dx$$

Assuming that function K is of type (5.13) and conditions (5.14) – (5.17) are satisfied, we obtain

$$\int (ET_{mn}K_n(x, X_n) - d_{mn}(x))^2\, dx \leq \text{const. } p_{mn}^2 h_n^{2r} \delta_{mn}^{\underline{i}}(x)\,.$$

Moreover,

$$\int \operatorname{var}[T_{mn}K_n(x, X_n)]\, dx$$

$$\leq h_n^{-p} \iint K^2(z) f_n(x - zh_n)\, dx\, dz$$

$$\leq h_n^{-p} \int K^2(z)\, dz$$

and

$$\int (d_{m,n+1}(x) - d_{mn}(x))^2\, dx \leq 2 (p_{m,n+1} - p_{mn})^2 \int f_{m,n}^2(x)\, dx$$

$$+2p_{m,n+1}^2 \int (f_{m.n+1}(x) - f_{mn}(x))^2\, dx$$

Now, Corollaries 6.1 and 6.2 are a direct consequence of Theorems 4.3 and 4.4. ■

Proof of Corollaries 6.3. and 6.4: Taking into consideration the following facts

$$\operatorname{var}[T_{mn}K_n(x, X_n)] \leq \left(\sum_{j=0}^{q(n)} G_j^2 \right)^p,$$

$$\int \left(ET_{mn}K_n\left(x, X_n\right) - d_{mn}\left(x\right)\right)^2 dx = p_{mn}^2 S_{mn}.$$

we proceed in a similar manner like in the previous proof. ■

Proof of Corollary 6.5: Let us denote

$$d_m\left(x\right) = p_m f_m\left(x\right).$$

Observe that condition (4.55) now takes the form

$$ET_{mn}K_n\left(x, X_n\right) \xrightarrow{n} d_m\left(x\right), \quad m = 1, ..., M \tag{16}$$

where

$$ET_{mn}K_n\left(x, X_n\right) = p_{mn} \int f_{mn}K_n\left(x, u\right) du.$$

In order to prove convergence (16), we will use inequalities

$$\left|ET_{mn}K_n\left(x, X_n\right) - d_m\left(x\right)\right|$$

$$\leq \left|\left(p_{mn} - p_m\right) \int f_{mn}\left(u\right) K_n\left(x, u\right) du\right|$$

$$+p_m \left|\int K_n\left(x, u\right) f_{mn}\left(u\right) du - f_m\left(x\right)\right|,$$

$$\left|\int K_n\left(x, u\right) f_{mn}\left(u\right) du - f_m\left(x\right)\right|$$

$$\leq \left|\int K_n\left(x, u\right) f_{mn}\left(u\right) du - \overline{f}_m\left(x\right)\right|$$

$$+\left|\overline{f}_m\left(x\right) - f_m\left(x\right)\right|,$$

where

$$\overline{f}_m\left(x\right) = \int K_n\left(x, u\right) f_m\left(u\right) du.$$

Obviously,

$$\int f_{mn}\left(u\right) K_n\left(x, u\right) du \leq h_n^{-p} \sup K\left(x\right).$$

Moreover, the following inequalities are true

$$\left|K_n\left(x, u\right)\left(f_{mn}\left(u\right) - \overline{f}_m\left(u\right)\right) du\right|$$

$$\leq h_n^{-p} \int |f_{mn}(u) - f_m(u)|\, du \ \sup K(x)$$

$$\left|K_n(x,u)\left(f_{mn}(u) - \overline{f}_m(u)\right) du\right|$$

$$\leq \sup \left|f_{mn}(u) - \overline{f}_m(u)\right|,$$

$$\left|K_n(x,u)\left(f_{mn}(u) - \overline{f}_m(u)\right) du\right|$$

$$\leq h_n^{-p/2} \left\|f_{mn}(u) - \overline{f}_m(u)\right\|_{L_2} \sup{}^{1/2} K(x).$$

If kernel K satisfies condition (5.32) then

$$\overline{f}_m(x) \xrightarrow{n} f_m(x)$$

for almost all x ([260], [299]). That concludes the proof. ■

Proof of Corollary 6.6: We proceed in a similar manner like in the previous proof. Let us point out that

$$\int f_{mn}(u) K_n(x,u)\, du \leq \left(\sum_{j=0}^{q(n)} G_j^2\right)^p.$$

The following inequalities are true

$$\left|\int K_n(x,u)\left(f_{mn}(u) - \overline{f}_m(u)\right) du\right|$$

$$\leq \left(\sum_{j=0}^{q(n)} G_j^2\right)^p \int |f_{mn}(x) - f_m(x)|\, dx,$$

$$\left|\int K_n(x,u)\left(f_{mn}(u) - \overline{f}_m(u)\right) du\right|$$

$$\leq \left(\sum_{j=0}^{q(n)} G_j\right)^p \|f_{mn}(x) - f_m(x)\|_{L_2}.$$

Moreover,

$$\overline{f}_n(x) = \sum_{|\underline{j}| \leq q} B_{\underline{j}}^m \Psi_{\underline{j}}(x)$$

where

$$B_{\underline{j}}^m = \int f_m(x) \Psi_{\underline{j}}(x)\, dx.$$

This completes the proof. ■

References

[1] Abid S., Fnaiech F. and Najim M. (2001): *A fast feedforward training algorithm using a modified form of the standard backpropagation algorithm,* IEEE Transactions on Neural Networks, Vol. 12, pp. 424 – 434.

[2] Ahalt S.C., Krishnamurthy A.K., Chen P. and Melton D.E. (1990): *Competitive learning algorithms for vector quantization*, Neural Networks, Vol. 3, pp. 277 – 290.

[3] Ahmad I.A. and Lin P.E (1976): *Nonparametric sequential estimation of a multiple regression function,* Bulletin of Mathematics, Vol. 17, pp. 63 – 75.

[4] Ahmad I.A. and Lin P.E (1984): *Fitting a multiple regression*, Journal of Statistical Planning and Inference, Vol. 2, pp.163 – 176.

[5] Ahmad R. and Salam F. (1992): *Error back propagation learning using the polynomial energy function*, Proceedings of the International Joint Conference on Neural Networks, Iizuka, Japan.

[6] Aizerman M., Braverman E. and Rozonoer L. (1964): *Theoretical foundations of the potential function method in pattern*

recognition learning, Automation and Remote Control, Vol. 25, pp. 821 – 837.

[7] Albert A.E. and Gardner L.A. (1967): *Stochastic Approximation and Nonlinear Regression*, The MIT Press.

[8] Alexits G. (1961): *Convergence Problems of Orthogonal Series*, Akademiai Kiado, Hungary, Budapest.

[9] Algoet P. and Györfi L. (1999): *Strong universal pointwise consistency of some regression function estimation*, Journal of Multivariate Analysis, Vol. 71, pp. 125 – 144.

[10] Aliev R.A. and Aliev R.R. (2001): *Soft Computing and its Applications*, World Scientific Publishing, Singapore – New Jersey – London – Hong Kong.

[11] Amari S.– I. (1967): *Theory of adaptive pattern classifiers*, IEEE Trans. on Electronic Computers, EC-16, pp. 299 – 307.

[12] Amari S.– I. (1971): *Characteristics of randomly connected threshold-element networks and network systems*, IEEE Proceedings, Vol. 59, No. 1, pp. 35 – 47.

[13] Amari S.– I. (1972): *Learning patterns and pattern sequences by self-organizing nets of threshold elements*, IEEE Trans. on Computers, C-21, pp. 1197 – 1206.

[14] Amari S.– I. (1972): *Characteristics of random nets of analog neuron-like elements*, IEEE Trans. on Systems, Man and Cybernetics, SMC-2, No 5, pp. 643 – 657.

[15] Amari S.– I. (1974): *A method of statistical neurodynamics*, Biological Cybernetics, Vol. 14, pp. 201 – 215.

[16] Amari S.– I. (1977): *Neural theory of association and concept-formation*, Biological Cybernetics, Vol. 26, pp. 175 – 185.

[17] Amari S.– I. (1977): *Dynamics of pattern formation in lateral-inhibition type neural fields*, Biological Cybernetics, Vol. 27, pp. 77 – 87.

[18] Amari S.– I. (1980): *Topographic organization of nerve fields*, Bulletin of Mathematics and Biology, 42, 339 – 364.

[19] Amari S.– I. (1983): *Field theory of self-organizing neural nets*, IEEE Trans. on Systems, Man, Cybernetics, SMC-13, pp. 741 – 748.

[20] Amari S.– I. (1989): *Characteristics of sparsely encoded associative memory*, Neural Networks, Vol. 2, No. 6, pp. 451 – 457.

[21] Amari S.– I. (1990): *Mathematical foundations of neurocomputing*, Proceedings of the IEEE, Vol. 78, No. 9, pp. 1443 – 1463.

[22] Amari S.– I., Fujita N. and Shinomoto S. (1992): *Four types of learning curves*, Neural Computation, Vol. 4, No. 2, pp. 605 – 618.

[23] Amari S.– I. (1993): *A universal theorem on learning curves*, Neural Networks, Vol. 6, No. 2, pp. 161 – 166.

[24] Ampazis N. and Perantonis J. (2002): *Two higly efficient second-order algorithms for training feedforward networks*, IEEE Transactions on Neural Networks, Vol. 13, No. 5, pp. 1064 – 1074.

[25] Antos A., Györfi L. and Kohler M. (2000): *Lower bounds on the rate of convergence of nonparametric regression estimates*, J. Statistical Planning and Inference, Vol. 83, pp. 91 – 100.

[26] Azimi-Sadjadi M.R. and Liou R.J. (1992): *Fast learning process of multi-layer neural network using recursive least squares method*, IEEE Transactions on Signal Processing, Vol. 40, No. 2.

[27] Baermann F. and Biegler-Koening F. (1992): *On a class of efficient learning algorithm for multi-layered neural networks*, Neural Networks, Vol. 5, No. 1.

[28] Baldi P. (1995): *Gradient learning algorithm overview: A general dynamical systems perspective*, IEEE Trans. on Neural Networks, Vol. 6, pp. 182 – 195.

[29] Bartlett P.L. and Anthony M. (1999): *Neural Network Learning: Theoretical Foundations*, Cambridge University Press, Cambridge.

[30] Baxter J. (1992): *The evolution of learning algorithms for artificial neural networks*, Complex Systems (D. Green and T. Bossomaier, eds.), pp. 313 – 326, IOS Press, Amsterdam.

[31] Belew R., McInerney J. and Schraudolph N. (1991): *Evolving networks: Using the genetic algorithm with connectionism learning*, Proceedings Second Artificial Life Conference, pp. 511 – 547, Addison Wesley.

[32] Bendat J.S. and Piersol A.G. (1971): *Random Data Analysis and Measurement Procedures*, John Wiley & Sons – Interscience, New York.

[33] Bilski J. and Rutkowski L. (1998): *A fast training algorithm for neural networks*, IEEE Trans. on Circuits and Systems II, June 1998, pp.749 – 753.

[34] Bilski J. and Rutkowski L.(2003): *A family of the RLS neural network learning algorithms*, Techn. Report, Dept. Comp. Eng., Techn. University of Czestochowa, submitted to IEEE Transactions on Neural Networks.

[35] Bishop C.M. (1995): *Neural Networks for Pattern Recognition*, Clarendon Press, Oxford.

[36] Bojarczak O.S.P. and Stodolski M. (1996): *Fast second-order learning algorithm for feedforward multilayer neural networks and its application*, Neural Networks Vol. 9, pp. 1583 – 1596.

[37] Braverman E.M. and Rozonoer L.I. (1969): *Convergence of random processes in machine learning theory*, Autom. Remote Control, Vol. 30, pp. 44 – 64.

[38] Bubnicki Z. (1980): *Identification of Control Plants*, Elsevier, Oxford – Amsterdam – New York.

[39] Bubnicki Z. (1999): *Uncertain variables and learning algorithms in knowledge-based control systems*, Artificial Life and Robotics, Vol. 3, pp. 155 – 159.

[40] Bubnicki Z. (2001): *Uncertain variables and their applications for a class of uncertain systems*, International Journal of Systems Science, Vol. 32, pp. 651 – 659.

[41] Bubnicki Z. (2001): *Uncertain variables and their application to decision making*, IEEE Trans. on SMC, Part A: Systems and Humans, Vol. 31.

[42] Bubnicki Z. (2002): *Uncertain Logics, Variables and Systems*, Springer-Verlag, Berlin – London – New York.

[43] Bubnicki Z. (2002): *A unified approach to descriptive and prescriptive concepts in uncertain decision systems*, Systems Analysis Modelling Simulation Vol. 42, pp. 331 – 342.

[44] Burrascano P. (1991): *Learning vector quantization for the probabilistic neural network*, IEEE Trans. on Neural Networks, Vol. 2, pp. 458 – 461.

[45] Cacoullos T. (1965): *Estimation of a multivariate density*, Ann. Inst. Statist. Math. Vol. 18, pp. 179 – 189.

[46] Carleson L. (1966): *On convergence and growth of partial sums of Fourier series*, Acta Math., Vol. 116, pp. 135 – 137.

[47] Chang P.C. and Gray R.M. (1986): *Gradient algorithms for designing predictive vector quantizers*, IEEE Trans. on Acoust. Speech, Signal Processing, Vol. 34, pp. 679 – 690.

[48] Chang H.– M. and Woods J.W. (1985): *Predictive vector quantization of images*, IEEE Trans. on Communications, COM-33, pp. 1208 – 1219.

[49] Chen X.– H.Y.G.– A. (1992): *Efficient backpropagation learning using optimal learning rate and momentum*, Neural Networks, Vol. 10, No. 3, pp. 517 – 527.

[50] Chen S., Cowan C.F.N. and Grant P.M. (1991): *Orthogonal least squares learning algorithm for radial basis networks*, IEEE Trans. on Neural Networks, Vol. 2, pp. 302 – 309.

[51] Chen J.– H. and Gersho A. (1986): *Covariance and autocorrelation methods for vector linear prediction*, in Proceedings of the International Conference on Acoustics, Speech, and Signal Precessing, Dallas, Texas, pp. 1545 – 1548.

[52] Chen J.– H. and Gersho A. (1987): *Gain-adaptive vector quantization with application to speech coding*, IEEE Trans. on Communications, Vol. 35, pp.918 – 930.

[53] Chien Y.T. and Fu K.S. (1969): *Stochastic learning of time-varying parameters in random environment*, IEEE Trans. Syst. Sc. Cybern., Vol. 5, pp. 237 – 246.

[54] Cherkassky V. and Shepherd R. (1998): *Regularization effect of weight initialisation in back propagation networks*, IEEE International Joint Conference of Neural Networks (IJCNN'98), (Anchorage, Alaska), pp. 2258 – 2261.

[55] Chui C. (1992): *Wavelets: a Tutorial in Theory and Applications*, Academic Press, Boston.

[56] Chung K.L. (1954): *On a stochastic approximation method*, Annals Mathematics of Statistics., Vol. 25, pp. 463 – 483.

[57] Cover T.M. and Hart P.E. (1967): *Neareset neighbor pattern classification*, IEEE Trans. on Information Theory, Vol. 13, pp. 21 – 27.

[58] Čencov N.N. (1962): *Evaluation of an unknown distribution density from observations*, Soviet Mathematics, Vol. 3, pp. 1559 – 1562.

[59] Davies H.I. (1973): *Strong consistency of a sequential estimator of a probability density function*, Bull. Math. Statist., Vol. 15, pp. 49 – 53.

[60] Denoeux T. and Lengelle R. (1993): *Initialising back propagation networks using prototypes*, Neural Networks, Vol. 6, No. 3, pp. 351 – 363.

[61] Devroye L.P. and Wagner T.J. (1976): *Nonparametric discrimination and density estimation*, Technical Report 183, Electronics Research Center, University of Texas.

[62] Devroye L.P. (1978): *Universal consistency in nonparametric regression and nonparametric discrimination*, Tech. Report. School of Computer Science, Mc Gill Univ.

[63] Devroye L.P. (1979): *On the pointwise and the integral convergence of recursive kernel estimates of probability densities*, Utilitias Math., Vol. 15, pp. 113 – 128.

[64] Devroye L.P. and Wagner T.J. (1980): *On the L_1 convergence of kernel estimators of regression functions with applications in discrimination*, Zeitschrift für Wahrscheinlichkeitstheorie und verwandte Gebiete, Vol. 51. 15 – 21.

[65] Devroye L.P. (1981): *On the almost everywhere convergence of nonparametric regression function estimates*, Annals of Statistics, Vol. 9, pp. 1310 – 1309.

[66] Devroye L.P. (1982): *Necessary and sufficient conditions for the almost everywhere convergence of nearest neighbor regression function estimates*, Zeitschrift für Wahrscheinlichkeitstheorie und verwandte Gebiete, Vol. 61, pp. 467 – 481.

[67] Devroye L.P. and Györfi L. (1985): *Nonparametric Density Estimation: The L_1 View*, John Wiley, New York.

[68] Devroye L.P. and Krzyżak A. (1989): *An equivalence theorem for L_1 convergence of the kernel regression estimate*, Journal of Statistical Planning and Inference, Vol. 23, pp. 71 – 82.

[69] Devroye L.P., Györfi L., Krzyżak A. and Lugosi G. (1994): *On the strong universal consistency of nearest neighbor regression function estimates*, Annals of Statistics, Vol. 22, pp. 1371 – 1385.

[70] Devroye L.P., Györfi L. and Lugosi G. (1996): *Probabilistic Theory of Pattern Recognition*, Springer-Verlag, New York.

[71] Devroye L. and Lugosi G. (2001): *Combinatorial Methods in Density Estimation*, New York: Springer-Verlag.

[72] Dianat S.A., Nasrabadi N.M. and Venkataraman S. (1991): *A nonlinear predictor for differential pulse-code encoder (DPCM) using neural networks*, Proc IEEE Int. Conf. on Acoustics. Speech and Signal Processing, pp. 2793 – 2796.

[73] Dony R.D. and Haykin S. (1995): *Neural Network Approaches to Image Compression*, Precessing IEEE, Vol. 83, pp. 288 – 303.

[74] Duch W. and Jankowski N. (1999): *Survey of neural transfer functions*, Neural Computing Surveys, Vol. 2, pp. 163 – 213.

[75] Duda R.O., Hart P.E. and Stork D.G. (2001): *Pattern Classification*, John Wiley & Sons, London.

[76] Dupač V. (1965): *A dynamic stochastic approximation method,* Annals Mathematics of Statistics., Vol. 36, pp. 1695 – 1702.

[77] Dupač V. (1966): *Stochastic approximations in the presence of trend,* Czechosl. Mat. Żur., Vol. 16, pp. 454 – 461.

[78] Eaton H. and Oliver T. (1992): *Improving the convergence of the back propagation algorithm*, Neural Networks, Vol. 5, pp. 283 – 288.

[79] Efromovich S. (1999): *Nonparametric Curve Estimation. Methods, Theory and Applications*, Springer-Verlag, New York.

[80] Eubank R.L. (1988): *Spline Smoothing and Nonparametric Regression*, Marcel Dekker, INC, New York.

[81] Eubank R.L. (1999): *Nonparametric Regression and Spline Smoothing*, Marcel Dekker, New York.

[82] Fahlman S. (1988): *An empirical study of learning speed in back propagation networks*, Tech. Rcp. CMU-CS-88-162, Carnegie Mellon University, Pittsburgh, PA, 1988.

[83] Fang W.C., Sheu B.J., Chen O.T.C. and Choi J. (1992): *A VLSI neural processor for image data compression using self-organization networks*, IEEE Trans. on Neural Networks, Vol. 3, pp. 507 – 515.

[84] de Figueiredo R.J.P. (1968): *Convergent algorithms for pattern recognition in nonlinearly evolving nonstationary environment*, Proc. IEEE, Vol. 56, pp. 188 – 189.

[85] Fogel D.B. (1995): *Evolutionary Computation: Towards a New Philosophy of Machine Intelligence*, IEEE Press, New York.

[86] Fowler J.E., Carbonara M.R. and Ahalt S.C. (1993): *Image coding using differential vector quantization*, IEEE Trans. on Circuits and Syst. for Video Tech., Vol. 3, pp. 350 – 367.

[87] Földes A. and Révész P. (1974): *A general method for density estimation*, Studia Sci. Math. Hungar., Vol. 9, pp. 81 – 92.

[88] Frumkin M.A. (1992): *Systolic Computations*, Kluwer Academic Publishers.

[89] Fu K.S. (1968): *Sequential Methods in Pattern Recognition and Machine Learning*, Academic Press: New York.

[90] Fukunaga K. (1990): *Introduction to Statistical Pattern Recognition*, Academic Press, New York, second edition.

[91] Gałkowski T. and Rutkowski L. (1983): *New algorithms for shape analysis - the multidimensional case*, in: Proceedings of the Third Scandinavian Conference on Image Analysis, Copenhagen, July 12 – 14, pp. 49 – 53.

[92] Gałkowski T. and Rutkowski L. (1985): *Nonparametric recovery of multivariate functions with applications to system identification*, Proceedings of the IEEE, Vol. 73, pp. 942 – 943, New York.

[93] Gałkowski T. and Rutkowski L. (1986): *Nonparametric fitting of multivariable functions*, IEEE Trans. on Automatic Control, Vol. AC-31, pp. 785 – 787.

[94] Georgiev A.A. (1988): *Consistent nonparametric multiple regression: the fixed design case*, Journal of Multivariate Analysis, Vol. 25, pp. 100 – 110.

[95] Gersho A. and Gray R.M. (1992): *Vector Quantization and Signal Compression*, Kluwer Academic Publishers, Boston.

[96] Goldberg D. (1989): *Genetic Algorithm in Search, Optimisation and Machine Learning*, Addison-Wesley.

[97] Gori M. and Tesi A. (1992): *On the problem of local minima in backpropagation*, IEEE Trans. on PAMI, Vol. 14, No. 1, pp. 76 – 86.

[98] Greblicki W. (1978): *Asymptotically optimal pattern recognition procedures with density estimates*, IEEE Trans. on Information Theory, Vol. 24, pp. 250 – 251.

[99] Greblicki W. and Krzyżak A. (1980): *Asymptotic properties of kernel estimates of a regression function*, J. Statist. Planning Inference, Vol. 4, pp. 81 – 90.

[100] Greblicki W. and Pawlak M. (1981): *Classification using the Fourier series estimate of multivariate density function*, IEEE Trans. Syst. Mann. Cybernetics., Vol. 11, pp. 726 – 730.

[101] Greblicki W. and Rutkowski L. (1981): *Density-free Bayes risk consistency of nonparametric pattern recognition procedures*, Proceedings of the IEEE, Vol. 69, No. 4, pp. 482 – 483.

[102] Greblicki W., Krzyżak A. and Pawlak M. (1984): *Distribution-free pointwise consistency of kernel regression estimate*, Annals of Statistics, Vol. 12, 1570 – 1575.

[103] Gray R.M. (1984): *Vector quantization*, IEEE ASSP, Mag. 1, pp. 4 – 29.

[104] Györfi L., Härdle W., Sarda P. and Vieu P. (1989): *Nonparametric Curve Estimation from Time Series*, Springer Verlag, New York.

[105] Györfi L., Kohler M., Krzyżak A. and Walk H. (2002): *A Distribution-Free Theory of Nonparametric Regression*, Springer Verlag, New York.

[106] Hagan M. and Menhaj M.B. (1994): *Training feed forward networks with the Marquardt algorithm*, IEEE Trans. on Neural Networks, Vol. 5, pp. 989 – 993.

[107] Hagan M., Demuth H. and Beale M. (1996): *Neural Network Design*, PWS Publishing Company, Boston.

[108] Harp S., Samad T. and Guha A. (1990): *Designing application-specific neural networks using the genetic algorithm*, Advances in Neural Information Processing Systems II (D. Touretzky, eds.), pp. 447 – 454, San Mateo, California, Morgan Kaufmann.

[109] Hassoun M.H. (1995): *Fundamentals of Artificial Neural Networks*, The MIT Press, Cambridge, MA.

[110] Haykin S. (1994): *Neural Networks: A Comprehensive Foundation*, IEEE Society Press, Macmillan College Publishing.

[111] Haykin S. and Bhattacharya T.K. (1997): *Modular learning strategy for signal detection in a nonstationary environment*, IEEE Trans. on Signal Proceesing, Vol. 45, Vol. 6. pp. 1619 – 1637.

[112] Härdle W. and Marron J.S. (1985): *Optimal bandwidth selection in nonparametric regression function estimation*, Annals of Statistics, Vol. 13, pp. 1465 – 1481.

[113] Härdle W. (1990): *Applied Nonparametric Regression*, Cambridge Univ. Press, Cambridge.

[114] Härdle W., Kerkyacharian G., Picard D. and Tsybakov A.B. (1998): *Wavelets, Approximation, and Statistical Applications*, Springer Verlag, New York.

[115] Hirose Y., Yamashit K. and Hijiya S. (1991): *Back propagation algorithm which varies the number of hidden units*, Neural Networks, Vol. 4, No. 1, pp. 61 – 66.

[116] Holland J.H. (1992): *Adaptation in Natural and Artificial Systems*, Cambridge, Masachusetts, MIT Press, 2nd ed.

[117] Holt M. and Semnani S. (1990): *Convergence of back-propagation in neural networks using a log-likelihood cost function*, Electronics Letters, Vol. 26, No. 23, pp. 1964 – 1965.

[118] Hornik K. (1991): *Approximation capabilities of multilayer feedforward networks*, Neural Networks, Vol. 4, pp. 251 – 257.

[119] Hornik K. (1993): *Some new results on neural network approximation*, Neural Networks, Vol. 6, pp.1069 – 1072.

[120] Hoshino M. and Chao J.(2001): *On representation and generalization capability of pyramid neural networks,* Proceedings of the IEEE Conference, Melbourne.

[121] Hwang N. and Kung S.Y. (1989): *Parallel algorithms architectures for neural networks*, Journal of VLSI Signal Processing, Vol. 1, pp. 221 – 251.

[122] Ibragimov I.A. and Khasminskii R.Z. (1981): *Statistical Estimation: Asymptotic Theory*, Springer-Verlag, New York.

[123] Iiguni Y., Sakai H. and Tokumaru H. (1992): *A real-time learning algorithm for a multilayered neural network based on the extended Kalman filter*, IEEE Trans. Signal Processing, Vol. 40, pp. 959 – 966.

[124] Jacobs R. (1989): *Increased rates of convergence through learning rate adaptation*, Neural Networks, Vol. 1, pp. 295 – 307.

[125] Jain A.K. (1989): *Fundamentals of Digital Image Processing*, Englewood Cliffs, NJ: Prentice-Hall, England.

[126] Jayant N., Johnston J. and Safranek R. (1993): *Signal compression based on models of human perception*, Proc. IEEE 81, pp. 1385 – 1422.

[127] Jones M.C., Marron J.S. and Sheather S.J. (1996): *A brief survey of bandwidh selection for density estimation*, Journal of the American Statistical Association, Vol. 91, pp. 401 – 407.

[128] Joost M. and Schiffmann W. (1998): *Speeding up backpropagation algorithms by using cross-entropy combined with pattern normalization*, International Journal of Uncertainty, Fuzziness and Knowledge-Based Systems, Vol. 6, No. 2, pp. 117 – 126.

[129] Kacprzyk J. (1997): *Multistage Fuzzy Control*, John Wiley & Sons, Chichester.

[130] Kaczorek T. (1995): *Adaptation algorithms for 2-D feedforward neural networks*, IEEE Trans. on Neural Networks, Vol. 6, pp. 519 – 521.

[131] Karras D. and Perantonis S. (1995): *An efficient constrained training algorithm for feedforward networks*, IEEE Trans. on Neural Networks, Vol. 6, pp. 1420 – 1434.

[132] Kay S.A. (1988): *Modern Spectral Estimation*, Englewood Cliffs: Prentice Hall.

[133] Kecman V. (2001): *Learning and Soft Computing*, MIT, Cambridge.

[134] Kenyon S.C. (1991): *Hyperspace organization for classification of non-stationary patterns*, Proceedings of the IEEE International Conference on Systems, Man and Cybernetics, Vol. 1, pp. 185 – 190.

[135] Kiefer J. and Wolfowitz J. (1952): *Stochastic estimation of the maximum of a regression function*, Annals Mathematic of Statistics, Vol. 23, No. 3, pp. 462 – 466.

[136] Kitano H. (1994): *Neurogenetic learning: an integrated method of designing and training neural networks using genetic algorithms*, Physica D., Vol. 75, pp. 225 – 238.

[137] Kollias S. and Anastassiou D. (1989): *An adaptive least squares algorithm for the efficient training of artificial neural networks*, IEEE Trans. on Circuits Syst., Vol. 36, pp. 1092 – 1101.

[138] Kozietulski M. and Rutkowski L. (1989): *A nonparametric procedure for identification of the step response function and its microprocessor implementation*, Advances in Modelling and Simulation, Vol. 17, No 1, pp. 25 – 36.

[139] Kramer C., Mckay B. and Belina J. (1995): *Probabilistic neural network array architecture for ECG classification*, in Proc. Ann. Int. Conf. IEEE Eng. Medicine Biol., Vol. 17, pp. 807 – 808.

[140] Krishnamurthy A.K., Ahalt S.C., Melton D.E. and Chen P. (1990): *Neural networks for vector quantization of speech and images*, IEEE J. Selected Areas of Commun., SAC-8, pp. 1449 – 1457.

[141] Kronmal R. and Tarter M. (1968): *The estimation of probability densities and cumulatives by Fourier series methods*, J. Amer. Statist. Assoc. Vol. 63, pp. 925 – 952.

[142] Kruschke J. and Movellan J. (1991): *Benefits of gain: Speeded learning and minimal hidden layers in back propagation networks*, IEEE Trans. on Systems, Man and Cybernetics, Vol. 21, No. 1, pp. 273 – 280.

[143] Kruschke J. and Movellan J. (1992): *Fast learning algorithms for neural networks*, IEEE Trans. on Circuits and Systems-II: Analog and Digital Signal Processing, Vol. 39, No. 7, pp. 453 – 473.

[144] Kuncheva L.I. (1996): *On the equivalence between fuzzy and statistical classifiers*, International Journal of Uncertainty,

Fuzziness and Knowledge-Based Systems, Vol. 4, No. 3, pp. 245 – 253.

[145] Kung S.Y. and Leiserson C.E. (1979): *Systolic arrays (for VLSI)*, Sparse Matrix Proc., Soc. Industrial and Appl. Math., pp. 256 – 286.

[146] Kung S.Y. (1988): *VLSI Array Processors*, Englewood Cliffs, NJ: Prentice-Hall, England.

[147] Kung S.Y. and Hwang J.N. (1989): *A unified systolic architecture for artificial neural networks*, Journal of Parallel and Distributed Computing, Vol. 6, pp. 358 – 387.

[148] Kung S.Y. and Hwang J.N. (1989): *Parallel algorithms / architectures for neural networks*, Journal of VLSI Signal Processing, Vol. 1, pp. 221 – 251.

[149] Kwon T.M. and Cheng H. (1996): *Contrast enhancement for backpropagation*, IEEE Trans. on Neural Networks, Vol. 7, No. 1, pp. 515 – 524.

[150] Lau C. (1992): *Neural Networks: Theoretical Foundations and Analysis*, IEEE Press New York.

[151] Lera G. and Pinzolas M. (2002): *Neighborhood based Levenberg-Marquardt algorithm for neural network training*, IEEE Trans. on Neural Networks, Vol. 13, No. 5.

[152] Leung Ch.S., Tsoi Ah.Ch. and Chan L. W. (2001): *Two regularizers for recursive least squared algorithms in feedforward multilayered neural networks*, IEEE Trans. on Neural Networks, Vol. 12, pp. 1314 – 1332.

[153] Li J. and Manikopoulos C.N. (1990): *Nonlinear prediction in image coding with DPCM*, Electronics Lett, Vol. 26, pp. 1357 – 1359.

[154] Linde Y., Buzo A. and Gray R.M. (1980): *An algorithm for vector quantizer design*, IEEE Trans. on Communications, Vol. 28, pp. 84 – 95.

[155] Ljung L. (1987): *System Identification: Theory for the User*, Englewood Cliffs, NJ: Prentice-Hall.

[156] Loftsgaarden D.O. and Quesenberry C.P. (1965): *A nonparametric estimate of a multivariate density function*, Annals Mathematic of Statistics, Vol. 36, pp. 1049 – 1051.

[157] Lu C.– Ch. and Shin Y.H. (1992): *Neural networks for classified vector quantization of images*, Eng. Applic. Artif. Intell. Vol. 5, pp. 451 – 456.

[158] Mack Y.P. and Silverman B.W. (1982): *Weak and strong uniform consistency of kernel regression estimates*, Zeitschrift für Wahrscheinlichkeitstheorie und verwandte Gebiete, Vol. 61, pp. 405 – 415.

[159] Manikopoulos C.N.: *Neural network approach to DPCM system design for image coding*, IEEE Proc. Part I, Vol. 139, pp. 501 – 507.

[160] Mao K.Z., Tan K.– C. and Ser W. (2000): *Probabilistic neural-network structure determination for pattern classification*, IEEE Trans. on Neural Networks, Vol. 11, No. 4, pp. 1009 – 1016.

[161] Marple S.L. (1987): *Digital Spectral Analysis with Application*, Englewood Cliffs: Prentice Hall.

[162] Meyer Y. (1993): *Wavelets: Algorithms and Applications*, SIAM, Philadelphia.

[163] Mithell R., Bishop J. and Low W. (1993): *Using a genetic algorithm to find the rules of a neural network*, Artificial Neural Nets and Genetic Algorithms (R. Albrecht, C. Reeves and Steele N. eds.).

[164] Mohsenian N., Rizvi S.A. and Nasrabadi N.M. (1993): *Predictive vector quantization using a neural network approach*, Optical Eng., Vol. 32, pp. 1503 – 1513.

[165] Moller M. (1993): *A scaled conjugate gradient algorithm for fast supervised learning*, Neural Networks, Vol. 6, pp. 525 – 533.

[166] Mougeot M., Azencott R. and Angeniol B. (1991): *Image compression with back propagation: Improvement of the visual*

restoration using different cost functions, Neural Networks, Vol. 4, pp. 467 – 476.

[167] Musavi M.T., Chan K.H., Hummels D.M. and Kalantri K. (1994): *On the generalization ability of neural-network classifier*, IEEE Trans. Pattern Anal. Machine Intell., Vol. 16, pp. 659 – 663.

[168] Nadaraya E.A. (1964): *On estimating regression*, Theory of Probability and its Applications, Vol. 9, pp. 141 – 142.

[169] Namphol A., Arozullah M. and Chin S. (1991): *Higher order data compression with neural networks*, Proc. Int. Joint Conf. on Neural Networks, Baltimore I, pp. 55 – 59.

[170] Nasrabadi N.M. and Feng Y. (1988): *Vector quantization of images based upon the Kohonen self-organizing feature maps*, Proc. IEEE Int. Conf. Neural Networks, pp. 1101 – 1108.

[171] Nasrabadi N.M. and King R.A. (1988): *Image coding using vector quantization: A review*, IEEE Trans. on Communications, COM-36, pp. 957 – 971.

[172] Niemann H. and Wu J.K. (1993): *Neural network adaptive image coding*, IEEE Trans. on Neural Networks, Vol. 4, pp. 615 – 627.

[173] Nikolsky S.M. (1977): *A Course of Mathematical Analysis*, Mir Publishers, Moscow.

[174] Ooyen A. and Nienhuis B. (1992): *Improving the convergence of the back propagation algorithm*, Neural Networks, Vol. 5, pp. 465 – 471.

[175] Pagan A. and Ullah A. (1999): *Nonparametric Econometrics*, Cambridge Univ. Press., London.

[176] Parlos A., Fermandez B., Atiya A., Muthusami J. and Tsai W. (1994): *An accelerated learning algorithm for multilayer perceptron networks*, IEEE Trans. on Neural Networks, Vol. 5, No. 3, pp. 493 – 497.

[177] Parodi G. and Passaggio F. (1994): *Size-adaptive neural network for image compression*, Proceedings IEEE International Conference on Neural Networks, pp. 945 – 947.

[178] Parzen E. (1962): *On estimation of probability density function and mode*, Annals Mathematic of Statistics, Vol. 33, pp. 1065 – 1076.

[179] Passaggio F., Anguita D. and Zunino R. (1994): *Human visual system for image compression by BP*, Proceedings IEEE International Conference on Neural Networks, pp. 1221 – 1224.

[180] Patterson D.W. (1995): *Artificial Neural Networks, Theory and Applications*, Prentice Hall, New York.

[181] Pawlak Z. (1982): *Rough sets*, International Journal of Information and Computer Science, Vol. 11, No. 341.

[182] Pawlak Z. (1991): *Rough set – Theoretical Aspect of Reasoning About Data*, Kluwer Academic Publishers, London.

[183] Pawlak Z. (1991): *Rough sets, decision algorithms and Bayes theorem*, European Journal of Operational Research, Vol. 136, pp. 181 – 189.

[184] Pedrycz W. and Gomide F. (1998): *An Introduction to Fuzzy Sets: Analysis and Design*, The MIT Press, Cambridge.

[185] Perantonis S. and Karras D. (1995): *An efficient constrained learning algorithm with momentum acceleration*, Neural Networks, Vol. 8, pp. 237 – 249.

[186] Pujol J. and Poli R. (1998): *Evolving neural networks using a dual representation with a combined crossover operator*, IEEE International Conference Evolutionary Computation (ICEC'98), (Anchorage, Alaska), pp. 416 – 421.

[187] Pujol J. and Poli R. (1998): *Evolving the topology and the weights of neural networks using a dual representation*, Special Issue Evolutionary Learning of the Applied Intelligence Journal, Vol. 8, No. 1, pp. 73 – 84.

[188] Puskorius G.V. and Feldkamp L.A. (1994): *Neurocontrol of nonlinear dynamical systems with Kalman filter trained recurrent networks*, IEEE Trans. on Neural Networks, Vol. 5, pp. 279 – 297.

[189] Radi A. and Poli R. (1998): *Discovery of optimal backpropagation learning rules using genetic programming*, IEEE International Conference on Evolutionary Computation (Anchorage, Alaska), IEEE Press, pp. 371 – 375.

[190] Rafajłowicz E. (1987): *Nonparametric orthogonal series estimators of regression: a class attaining the optimal convergence rate in L_2*, Statistics and Probability Letters, Vol. 5, pp. 213 – 224.

[191] Rafajłowicz E. (1997): *Consistency of orthogonal series density estimators based on grouped observations*, IEEE Trans. on Information Theory, Vol. 43, No. 1, pp. 283 – 285.

[192] Rafajłowicz E. and Pawlak M. (1997): *On function recovery by neural networks based on orthogonal expansions*, Nonlinear Analysis – Theory and Methods, Vol. 30, No. 3, pp. 1343 – 1354.

[193] Raghu P.P. and Yegnanarayana B. (1998): *Supervised texture classification using a probabilistic neural network and constraint satisfaction model*, IEEE Trans. on Neural Networks, Vol. 9, pp. 516 – 522.

[194] Ramamurthi B. and Gersho A. (1986): *Classified vector quantization of images.*, IEEE Trans. on Communications, COM-34, pp. 1105 – 1115.

[195] Rao B.L.S.P. (1983): *Nonparametric Functional Estimation*, Academic Press, New York.

[196] Rao P.V. and Thornby J.I. (1969): *A robust point estimate in a generalized regression model*, Annals Mathematic of Statistics, Vol. 40, pp. 1784 – 1790.

[197] Révész P. (1977): *How to apply the method of stochastic approximation in the nonparametric estimation of a regression function*, Mathematische Operationsforschung und Statistik Series Statistics, Vol. 8. pp. 119 – 126.

[198] Riedmiller M. and Braun H. (1993): *A direct method for faster backpropagation learning: The RPROP Algorithm*, IEEE International Conference on Neural Networks (ICNN93), San Francisco, pp. 586 – 591.

[199] Ripley B.D. (1996): *Pattern Recognition and Neural Networks*, Cambridge University Press, Cambridge.

[200] Rizvi S.A. and Nasrabadi N.M. (1994): *Predictive vector quantizer using constrained optimization*, IEEE Signal Proc. Letters Vol. 1, pp. 15 – 18.

[201] Robbins H. and Monro S.A. (1951): *A stochastic approximation method*, Annals Mathematic of Statistics., Vol. 22, No. 1.

[202] Roberts G., Zoubir A.M. and Boashash B. (1996): *Time-Frequency and Time-Scale Analysis.*, Proceedings of the IEEE-SP International Symposium, pp. 245 – 248.

[203] Romero R.D., Touretzky D.S. and Thibadeau G.H. (1997): *Optical Chinese character recognition using probabilistic neural netwoks*, Pattern Recognit., Vol. 3, pp. 1279 – 1292.

[204] Rosenblatt M. (1956): *Remarks on some estimates of a density function*, Annals Mathematic of Statistics, Vol. 27, pp. 832 – 837.

[205] Rubanov N.S. (2000): *The layer-wise method and the backpropagation hybrid approach to learning a feedforward neural network*, IEEE Trans. on Neural Networks, Vol. 11, pp. 295 – 305.

[206] Rumelhart D., Hinton G and Williams R. (1986): *Parallel Distributed Processing*, MIT Press, Cambridge, MA.

[207] Rutkowska D. (2002): *Neuro-Fuzzy Architectures and Hybrid Learning*, Physica-Verlag, Springer-Verlag Company, Heidelberg, New York, 2002.

[208] Rutkowski L. (1980): *Sequential estimates of probability densities by orthogonal series and their application in pattern classification*, IEEE Trans. on Systems, Man and Cybernetics, Vol. SMC-10, No. 12, pp. 918 – 920.

[209] Rutkowski L. (1981): *Sequential estimates of a regression function by orthogonal series with applications in discrimination*, in: Lectures Notes in Statistics, Springer, New YorkVol. 8, pp. 236 – 244.

[210] Rutkowski L. (1982): *On system identification by nonparametric function fitting*, IEEE Trans. on Automatic Control, Vol. AC-27, pp. 225 – 227.

[211] Rutkowski L. (1982): *On-line identification of time-varying systems by nonparametric techniques*, IEEE Trans. on Automatic Control, Vol. AC-27, pp. 228 – 230.

[212] Rutkowski L. (1982): *Orthogonal series estimates of a regression function with applications in system identification*, in: Probability and Statistical Inference, D. Reidel Publishing Company, London, pp. 343 – 347.

[213] Rutkowski L. (1982): *Nonparametric identification of weighting function by orthogonal series method*, in: Proceedings of 1982 American Control Conference, Arlington, Virginia, June 14 – 16, pp. 170 – 171.

[214] Rutkowski L. (1982): *Nonparametric identification of dynamic systems*, in: Proceedings of 27-th International Scientific Colloquium, The Ilmenau Institute of Technology, pp. 181 – 183.

[215] Rutkowski L. (1982): *Nonparametric classification of time-varying patterns by multivariate Hermite series method*, in: Proceedings of the 6th International Conference on Pattern Recognition, Munich, Oct. 19 – 22, IEEE Computer Society Press, pp. 879 – 881.

[216] Rutkowski L. (1982): *On stochastic processes with learning properties in a time-varying environment and some applications*, in: Proceedings of the Third Pannonian Symposium on Mathematical Statistics, Visegrad, Sept. 13 – 18, pp. 297 – 304.

[217] Rutkowski L. (1982): *On Bayes risk consistent pattern recognition procedures in a quasi-stationary environment*, IEEE Trans. on Pattern Analysis and Machine Intelligence, PAMI-4. No 1, pp. 84 – 87.

[218] Rutkowski L. (1984): *On nonparametric identification with prediction of time-varying systems*, IEEE Trans. on Automatic Control, Vol. AC-29, pp. 58 – 60.

[219] Rutkowski L. (1985): *Nonparametric identification of quasi-stationary systems*, Systems and Control Letters, Vol. 6, pp. 33 – 35.

[220] Rutkowski L. (1985): *The real-time identification of time-varying systems by nonparametric algorithms based on the Parzen kernels*, International Journal of Systems Science, Vol. 16, pp. 1123 – 1130.

[221] Rutkowski L. (1985): *Nonparametric identification of the CO conversion process*, in: Proceedings of the IFAC Workshop, Adaptive Control of Chemical Processes, pp. 64 – 66, Pergamon Press, Oxford.

[222] Rutkowski L. (1988): *Sequential pattern recognition procedures derived from multiple Fourier series*, Pattern Recognition Letters, Vol. 8, pp. 213 – 216.

[223] Rutkowski L. (1988): *Nonparametric procedures for identification and control of linear dynamic systems*, Proceedings of 1988 American Control Conference, June 15 – 17, pp. 1325 – 1326.

[224] Rutkowski L. (1989): *Nonparametric learning algorithms in the time-varying environments*, Signal Processing, Vol. 18, pp. 129 – 137.

[225] Rutkowski L. (1989): *An application of multiple Fourier series to identification of multivariable nonstationary systems*, International Journal of Systems Science, Vol. 20, No.10, pp. 1993 – 2002.

[226] Rutkowski L. and Rafajłowicz E. (1989): *On global rate of convergence of some nonparametric identification procedures*, IEEE Trans. on Automatic Control, Vol. AC-34, No.10, pp. 1089 – 1091.

[227] Rutkowski L. (1991): *Identification of MISO nonlinear regressions in the presence of a wide class of disturbances*, IEEE Trans. on Information Theory, IT-37, pp. 214 – 216.

[228] Rutkowski L. (1993): *Multiple Fourier series procedures for extraction of nonlinear regressions from noisy data*, IEEE Trans. on Signal Processing, Vol 41, pp. 3062 – 3065.

[229] Rutkowski L. and Rutkowska D. (1993): *On nonparametric identification of linear circuits by Walsh orthogonal series*, Proceedings of the 7-th International Conference, System-Modelling-Control, Zakopane, Maj 17 – 21, pp. 146 – 148.

[230] Rutkowski L. and Gałkowski T. (1994): *On pattern classification and system identification by probabilistic neural networks*, Applied Mathematics and Computer Science, Vol. 4, No.3, pp. 413 – 422.

[231] Rutkowski L. and Cierniak R. (1996): *Image compression by competitive learning neural networks and predictive vector quantization*, Appl. Math. and Comp. Science, Vol. 6, No. 3, pp. 431 – 445.

[232] Rutkowski L. and Cpałka K. (2002): *A neuro-fuzzy controller with a compromise fuzzy reasoning*, Control and Cybernetics, Vol. 31, No. 2, pp. 297 – 308.

[233] Rutkowski L. and Cpałka K. (2003): *Flexible neuro-fuzzy systems*, IEEE Trans. on Neural Networks, Vol. 14, No. 3, pp. 554 – 574.

[234] Rutkowski L. (2004): *Adaptive probabilistic neural networks for pattern classification in time-varying environment*, IEEE Transactions on Neural Network, Vol. 15, No. 2.

[235] Rutkowski L. (2003): *Generalized Regression Neural Networks in Time-Varying Environment*, submitted to IEEE Transactions on Neural Networks.

[236] Ryzin J.V. (1966): *Bayes risk consistency of classification procedures using density estimation*, Sankhya Series A, Vol. 28, pp. 161 – 170.

[237] Sansone G. (1959): *Orthogonal Functions*, Interscience Publishers Inc., New York.

[238] Sarkar D. (1995): *Methods to speed up error back propagation learning algorithm*, ACM Computing Surveys, Vol. 27, No. 4, pp. 519 – 542.

[239] Scalero R.S. and Tepedelenlioglu N. (1992): *A fast new algorithm for training feedforward neural networks*, IEEE Trans. on Signal Processing, Vol. 40, No. 1, pp. 202 – 210.

[240] Schittenkopf C., Deco G. and Brauer W. (1997): *Two strategies to avoid overfitting in feedforward neural networks*, Neural Networks, Vol. 10, No. 3, pp. 505 – 516.

[241] Schwartz S.C. (1967): *Estimation of probability density by an orthogonal series*, Annals Mathematic of Statistics,Vol. 38, pp. 1261 – 1265.

[242] Shah S., Palmieri F. and Datum M. (1991): *Optimal filtering algorithms for fast learning in feedforward neural networks*, Neural Networks, Vol. 5, pp. 779 – 787.

[243] Sicuranzi G.L., Ramponi G. and Marsi S. (1990): *Artificial neural network for image compression*, Electronics Lett, Vol. 26, pp. 477 – 479.

[244] Silva F. and Almeida L. (1990): *Acceleration techniques for the back-propagation algorithm parallel networks*, Lecture Notes in Computer Science, Vol. 412, pp. 110 – 119.

[245] Silva F. and Almeida L. (1990): *Speeding up back-propagation*, Advanced Neural Computers (R. Eckmiller, ed.), (Amsterdam), pp. 151 – 158, Elsevier Science Publishers.

[246] Sjölin P. (1971): *Convergence almost everywhere of certain singular integrals and multiple Fourier series*, Ark. Math., Vol. 9, pp. 65 – 90.

[247] Slowinski R. (eds.) (1992): *Intelligent Decisin Support, Handbook of Applications and Advances of the Rough Sets Theory*, Kluwer Academic Publishers, Dordrecht.

[248] Smoląg J., Rutkowski L. and Bilski J. (1997): *Systolic architectures for neural networks – Part I*, Proceedings of the Third Conference Neural Networks and Their Applications, Kule, 14 – 18 X, pp. 614 – 621.

[249] Smoląg J., Bilski J. and Rutkowski L. (1997): *Systolic architectures for neural networks – Part II*, Proceedings of the Third Conference Neural Networks and Their Applications, Kule, 14 – 18 X, pp. 622 – 625.

[250] Smoląg J., Rutkowski L. and Bilski J. (1999): *Systolic array for neural networks*, Proceedings of the Fourth Conference Neural Networks and Their Applications, pp. 487 – 497, Zakopane.

[251] Solla S., Levin E. and Fleisher M. (1988): *Accelerated learning in layered neural networks*, Complex System, Vol. 2, pp. 625 – 640.

[252] Sorour E. (1978): *On the convergence of the dynamic stochastic approximation method for stochastic non-linear multidimensional dynamic systems*, Cybernetica, Vol.14. pp. 28 – 37.

[253] Söderström T. and Stoica P. (1989): *System Identification*, Englewood Cliffs, NJ: Prentice-Hall, England.

[254] Spears W., Jong K.D., Bäck T., Fogel D. and de Garis H. (1993): *An overview of evolutionary computation*, Proceeedings of the European Conference on Machine Learning (ECML-93) (P.B. Brazdil, ed.), Vol. 667 of LNAI, (Vienna, Austria), pp. 442 – 259, Springer Verlag

[255] Specht D.F. (1990): *Probabilistic neural networks and the polynomial adaline as complementary techniques for classification*, IEEE Trans. on Neural Networks, Vol. 1, pp. 111 – 121.

[256] Specht D.F. (1990): *Probabilistic neural networks*, Neural Networks, Vol. 3, pp. 109 – 118.

[257] Specht D.F. (1991): *A general regression neural network*, IEEE Trans. on Neural Networks, Vol. 2, pp. 568 – 576.

[258] Specht D.F. (1992): *Enhancements to the probabilistic neural networks*, in Proc. IEEE Int. Joint Conf., Neural Networks: Baltimore, MD, pp. 761 – 768.

[259] Stan O. and Kamen E. (2000): *A local linearized least squares algorithm for training feedforward neural networks*, IEEE Trans. on Neural Networks, Vol. 11, pp. 487 – 495.

[260] Stein E.M. (1970): *Singular Integrals and Differentiability Properties of Functions*, New Jersey: Princeton Univ. Press Princeton, New Jersey.

[261] Stone C.J. (1977): *Consistent nonparametric regression,* Annals of Statistics, Vol. 5, pp. 595 – 645.

[262] Stone C.J. (1980): *Optimal rates of convergence for nonparametric estimators*, Annals of Statistics, Vol. 8, pp. 1348 – 1360.

[263] Stone C.J. (1982): *Optimal global rates of convergence for nonparametric regression*, Annals of Statistics, Vol. 10, pp. 1040 – 1053.

[264] Streit R. L. and Luginbuhl T.E. (1994): *Maximum likelihood training of probabilistic neural network*, IEEE Trans. on Neural Networks, Vol. 5, pp. 764 – 783.

[265] Strobach P. (1990): *Linear Prediction Theory – A Mathematical Basis for Adaptive Systems*, Springer-Verlag, New York.

[266] Sum J., Chan L.W., Leung C.S. and Young G. (1998): *Extended Kalman filter-based pruning method for recurrent neural networks*, Neural Computing, Vol. 10, pp. 1481 – 1505.

[267] Sum J., Leung C., Young G. H. and Kan W. (1999): *On the Kalman filtering method in neural-network training and pruning*, IEEE Trans. on Neural Networks, Vol. 10, pp. 161 – 166.

[268] Szegö G. (1959): *Orthogonal Polynomials*, Amer. Math. Soc. Coll. Publ., Vol. 23.

[269] Tadeusiewicz R. (1993): *Neural Networks*, RM Academic Publishing House, Warsaw (in Polish).

[270] Tadeusiewicz R. (1998): *Elementary Introduction to Neural Networks with Computer Programs*, Academic Publishing House, Warsaw (in Polish).

[271] Tadeusiewicz R. and Ogiela M.R. (2002): *Automatic understanding of medical images – new achievements in syntactic analysis of selected medical images*, Biocybernetics and Biomedical Engineering, Vol. 22, No. 4, pp. 17 – 29.

[272] Thompson J.R. and Tapia R.A. (1990): *Nonparametric Function Estimation and Simulation*, Philadelphia: SIAM.

[273] Tollenaere T. (1990): *Super SUB: Fast adaptive back propagation with good scaling properties*, Neural Networks, Vol. 3, No. 5, pp. 561 – 573.

[274] Traven H.G.C. (1991): *A neural-network approach to statistical pattern classification by semiparametric estimation of a probability density functions*, IEEE Trans. on Neural Networks, Vol. 2, pp. 366 – 377.

[275] Tucker H.G. (1967): *A Graduate Course in Probability*, Academic Press.

[276] Tzypkin J.Z. (1970): *Introduction to the Self-Learning Systems Theory*, Nauka Publishers, Moscow.

[277] Tzypkin J.Z. (1972): *Learning algorithms of pattern recognition in non-stationary conditions*, in: Frontiers of Pattern Recognitions, Watanabe S. (eds.), Academic Press: New York, pp. 527 – 542.

[278] Uosaki K. (1973): *Application of stochastic approximation to the tracking of a stochastic non-linear dynamic systems*, Int. J. Control, Vol. 18, pp. 1233 – 1247.

[279] Uosaki K. (1974): *Some generalizations of dynamic stochastic approximation process*, Annals of Statistics, Vol. 2, pp. 1042 – 1048.

[280] Vajda I., Györfi L. and Györfi Z. (1977): *A strong law of large numbers and some applications*, Studia Scient. Math. Hung., Vol. 12, pp. 233 – 244.

[281] Vitthal R., Sunthar P., Rao and Durgaprasada C. (1995): *The generalized proportional-integral-derivative (PID) gradient descent back propagation algorithm*, Neural Networks, Vol. 8, No. 4, pp. 563 – 569.

[282] Vogl T., Mangis J., Rigler A., Zink W. and Alkon D. (1988): *Accelerating the convergence of the backpropagation method*, Biological Cybernetics, Vol. 59, pp. 257 – 263.

[283] Wahba G. (1973): *Interpolating spline methods for density estimation, variable knots*, Technical Report 337, Dept. of Statistics, University of Wisconsin, Madison.

[284] Wahba G. (1975): *Optimal convergence properties of variable knot, kernel, and orthogonal series methods for density estimation*, Annals of Statistics, Vol. 3, pp. 15 – 29.

[285] Wahba G. (1975): *Smoothing noisy data with spline functions*, Numer. Math., Vol. 24, pp. 383 – 293.

[286] Wahba G. (1975): *Interpolating spline methods for density estimation, equi-spaced knots*, Annals of Statistics, Vol. 3, pp. 30 – 44.

[287] Wahba G. (1976): *A survey of some smoothing problems and the method of generalized cross-validation for solving them*, TR-457, Department of Statistics, University of Wisconsin.

[288] Wahba G. (1990): *Spline Models for Observational Data*, SIAM, Philadelphia.

[289] Walter G.G. (1977): *Properties of Hermite series estimation of probability density*, Annals of Statistics, Vol. 5, pp. 1258 – 1264.

[290] Watanabe M. (1974): *On Robbins-Monro stochastic approximation method with time varying observations,* Bull. Math. Statist., Vol. 16, pp. 73 – 91.

[291] Watanabe M. (1974): *On convergence of asymptotically optimal discriminant functions for pattern classification problems*, Bull. Math. Statist., Vol. 16, pp. 23 – 34.

[292] Watson G.S. (1964): *Smooth regression analysis*, Sankhya Series A, Vol. 26, pp. 359 – 372.

[293] Webb A.R. (2002): *Statistical Pattern Recognition*, John Wiley & Sons, England.

[294] Weir M. (1991): *A method for self determination of adaptive learning rate in back propagation*, Neural Networks, Vol. 4, pp. 371 – 379.

[295] Wellstead P.E. and Zarrop M.B. (1991): *Self-tuning Systems Control and Signal Processing*, John Wiley & Sons, England.

[296] Wertz W. (1978): *Statistical Density Estimation: a Survey*, Vandenhoeck & Ruprecht, Germany.

[297] Wertz W. and Schneider B. (1979): *Statistical density estimation: a bibliography*, Internat. Statist. Rev., Vol. 47, pp. 155 – 175.

[298] Wessels L. and Barnard E. (1992): *Avoiding false local minima by proper initialisation of connections*, IEEE Trans. on Neural Networks, Vol. 3, No. 6, pp. 899 – 905.

[299] Wheeden R.L. and Zygmunnd A. (1977): *Measure and Integral*, New York and Basel: Marcel Dekker. INC.

[300] Wolverton C.T. and Wagner T.J. (1969): *Asymptotically optimal discriminant functions for pattern classification*, IEEE Trans. Inform. Theory, Vol. 15, pp. 258 – 265.

[301] Yamato H. (1971): *Sequential estimation of a continuous probability density function and the mode*, Bull. Math. Statist., Vol. 14, pp. 1 – 12.

[302] Yao X. (1999): *Evolving artificial neural networks*, Proceeding of the IEEE, Vol. 87, pp. 1423 – 1447.

[303] Yee P.V. and Haykin S. (2001): *Regularized Radial Basis Function Networks, Theory and Applications*, John Wiley & Sons, New York.

[304] Young T.Y. and Westerberg R.A. (1972): *Stochastic approximation with a non-stationary regression function*, IEEE Trans. Inform. Theory, Vol. 18, pp. 518 – 519.

[305] Zadeh L.A. (1965): *Fuzzy sets*, Information and Control, Vol. 8, No. 3, pp. 338 – 353.

[306] Zadeh L.A. (1971): *Similarity relations and fuzzy orderings*, Information Science, Vol. 3, pp. 177 – 200.

[307] Zadeh L.A. (1971): *Towards a theory of fuzzy systems*, In: Kalman R.E. and DeClaris N. (eds.), Aspects of Network and System Theory, Holt, Rinehart and Winston, New York.

[308] Zadeh L.A.(1972): *A fuzzy-set theoretic interpretation of linguistic hedges*, Journal of Cybernetics, Vol. 2, pp. 4 – 34.

[309] Zadeh L.A. (1973): *Outline of a new approach to the analysis of complex systems and decision processes*, IEEE Trans. on Systems, Man, and Cybernetics, Vol. SMC-3, No. 1, pp. 28 – 44.

[310] Zadeh L.A. (1974): *On the analysis of large scale systems*, In: Gottinger H. (eds.), Systems Approaches and Environment Problems, Vandenhoeck and Ruprecht, pp. 23 – 37.

[311] Zadeh L.A (1974): *Fuzzy logic and its application to approximate reasoning*, Information Processing, Vol. 74, pp. 591 – 594.

[312] Zadeh L.A. (1975): *Calculus of fuzzy restrictions*, In: Zadeh L.A., Fu K.– S., Tanaka K. and Shimura M. (eds.), Fuzzy Sets and Their Applications to Cognitive and Decision Processes, Academic Press, New York, pp. 1 – 39.

[313] Zadeh L.A. (1975): *The concept of a linguistic variable and its application to approximate reasoning*, Information Science, Part I, Vol. 8, pp. 199 – 249, Part II, Vol. 8, pp. 301 – 357, Part III, Vol. 9, pp. 43 – 80.

[314] Zadeh L.A. (1976): *A fuzzy-algorithmic approach to the definition of complex or imprecise concepts*, International Journal of Man-Machine Studies, Vol. 8, pp. 246 – 291.

[315] Zadeh L.A. (1979): *Fuzzy sets and information granularity*, In: Gupta M., Ragade R., and Yager R. (eds.), Advances in Fuzzy Set Theory and Applications, North Holland, Amsterdam, pp. 3 – 18.

[316] Zadeh L.A. (1981): *Test-score semantics for natural languages and meaning representation via PRUF*, In: Rieger B. (eds.), Empirical Semantics, Germany, pp. 281 – 349.

[317] Zadeh L.A. (1983): *The role of fuzzy logic in the management of uncertainty in expert systems*, Fuzzy Sets and Systems, Vol. 11, pp. 199 – 227.

[318] Zadeh L.A. (1986): *Outline of a computational approach to meaning and knowledge representation based on a concept of a generalized assignment statement*, In: Thoma M. and Wyner A. (eds.), Proceedings of the International Seminar on Artificial Inteligence and Man-Machine Systems, Springer-Verlag, Heidelberg, pp. 198 – 211.

[319] Zadeh L.A. (1994): *Fuzzy logic, neural networks and soft computing*, Communications of the ACM, Vol. 37, No. 3, pp. 77 – 84.

[320] Zadeh L.A. (1994): *Soft Computing and Fuzzy Logic*, IEEE Software, Vol. 11, No. 6, pp. 48 – 58.

[321] Zadeh L.A. (1996): *Fuzzy logic = computing with words*, IEEE Trans. on Fuzzy Systems, Vol. 4, pp. 103 – 111.

[322] Zadeh L.A. (1997): *Toward a theory of fuzzy information granulation and its centrality in human reasoning and fuzzy logic*, Fuzzy Sets and Systems, Vol. 90, pp. 111 – 127.

[323] Zadeh L.A. (1999): *New Frontiers in Fuzzy Logic and Soft Computing*, Proc. of Fourth Conference, Neural Networks and Their Application, Zakopane, pp. 1 – 4.

[324] Zadeh L.A. (1999): *From computing with numbers to computing with words – from manipulation of measurements to manipulation of perceptions*, IEEE Trans. on Circuits and Systems – I: Fundamental Theory and Applications, Vol. 45, No. 1, pp. 105 – 119.

[325] Zadeh L.A. (2000): *Outline of a computational theory of perceptions based on computing with words*, In: Sinha N.K. and Gupta M.M. (eds.), Soft Computing and Intelligent Systems: Theory and Applications, Academic Press, San Diego, New York, Tokio, pp. 3 – 22.

[326] Zadeh L.A. (2001): *A new direction in AI. Toward a computational theory of perceptions*, AI Magazine, Vol. 22, No. 1, pp. 73 – 84.

[327] Zaknich A. (1997): *A vector quantization reduction method for the probabilistic neural network*, in Proc. IEEE Int. Conf. Neural Networks: Piscataway, NJ, pp. 1117 – 1120.

[328] Zhang D. and Pal S.K., (eds.) (2003): *Neural Networks and Systolic Array Design*, World Scientific Publishing.

[329] Zhang Y.Q., Li W. and Liou M.L.(eds.) (1995): *Special issue on Advances in Image and Video Compression*, Proc. IEEE, Vol. 83.

[330] Zhang Y. and Li R. (1999): *A fast U-D factorization-based learning algorithm with applications to nonlinear system modeling and identification*, IEEE Trans. on Neural Networks, Vol. 10, No. 4, pp. 930 – 938.

[331] Zurada J.M. (1992): *Introduction to Artificial Neural Systems*, West Publishing Company.

[332] Zurada J.M. (1993): *Lambda learning rule for feedforward neural networks*, Proc. IEEE Int. Conf. on Neural Networks, San Francisco (USA), pp. 1808 – 1811.

[333] Zurada J.M. and Malinowski A. (1994): *Multilayer perceptron networks: selected aspects of training optimization*, Appl. Math. and Comp. Scien., Vol. 4, No. 3, pp. 281 – 307.

[334] Zygmund A. (1959): *Trigonometric Series*, Cambridge: Cambridge Univ. Press.

GPSR Compliance
The European Union's (EU) General Product Safety Regulation (GPSR) is a set of rules that requires consumer products to be safe and our obligations to ensure this.

If you have any concerns about our products, you can contact us on

ProductSafety@springernature.com

In case Publisher is established outside the EU, the EU authorized representative is:

Springer Nature Customer Service Center GmbH
Europaplatz 3
69115 Heidelberg, Germany

www.ingramcontent.com/pod-product-compliance
Ingram Content Group UK Ltd.
Pitfield, Milton Keynes, MK11 3LW, UK
UKHW021857190726
13853UKWH00003B/1314

* 9 7 8 3 6 4 2 5 3 5 7 1 0 *